Lecture Notes in Computer Science   9724

Commenced Publication in 1973
Founding and Former Series Editors:
Gerhard Goos, Juris Hartmanis, and Jan van Leeuwen

More information about this series at http://www.springer.com/series/7411

Nathalie Mitton · Valeria Loscri
Alexandre Mouradian (Eds.)

Ad-hoc, Mobile, and Wireless Networks

15th International Conference, ADHOC-NOW 2016
Lille, France, July 4–6, 2016
Proceedings

 Springer

Editors
Nathalie Mitton
Inria Lille-Nord Europe
Villeneuve d'Ascq
France

Valeria Loscri
Inria Lille-Nord Europe
Villeneuve d'Ascq
France

Alexandre Mouradian
Université Paris-Sud
Gif sur Yvette
France

ISSN 0302-9743   ISSN 1611-3349   (electronic)
Lecture Notes in Computer Science
ISBN 978-3-319-40508-7   ISBN 978-3-319-40509-4   (eBook)
DOI 10.1007/978-3-319-40509-4

Library of Congress Control Number: 2016941067

LNCS Sublibrary: SL5 – Computer Communication Networks and Telecommunications

Printed on acid-free paper

This Springer imprint is published by Springer Nature
The registered company is Springer International Publishing AG Switzerland

Preface

The International Conference on Ad-Hoc Networks and Wireless (ADHOC-NOW) is one of the mot well-known series of events dedicated to research on wireless sensor networks and mobile computing. Since its creation in 2002, the conference has been held 14 times in six different countries. Its 15th edition in 2016 was held in Lille, France, during July 4–6.

The 15th ADHOC-NOW attracted 64 submissions of which 24 were accepted for presentation after rigorous reviews by external reviewers, Technical Program Committee members, and discussions among technical program chairs. All submissions received at least three reviews. In addition, two excellent keynotes talks completed the scientific program.

This program and organization have been made possible through the strong support of our sponsors: Inria, Euratechnologies, and MEL (Lille's European Metropole). A special thanks to them.

ADHOC-NOW traditionally covers a wide spectrum of topics across all networking layers for sensor networks, localization in various networking environments, and with applications in several domains. The 15th ADHOC-NOW featured eight sessions, structuring the contributions into the following topics: resource allocation, communication, low communication layers in sensor networks and IoT, opportunistic networks, security, VANETS and ITS, robots, and MANETS.

I would like to thank all the people involved in the production of these proceedings. First of all, I am grateful to the program chairs, Valeria Loscri and Alexandre Mouradian, for managing the review process, the Technical Program Committee and the external reviewers for their help in providing detailed reviews of the submissions, Cristina Cano, our proceedings chair, Riccardo Petrolo, our web chair, and Abdoul Aziz Mbacke, our publicity chair. I would like to dedicate a special thanks to Marie Bénédicte Dernoncourt and Anne Rejl for their valuable support in the local organization and arrangements of the event. I also thank Springer's team for their great assistance from the start of the submission process until the production of the proceedings.

Finally, I would like to thank all the authors for their contribution and interesting discussions during the event.

July 2016   Nathalie Mitton

Organization

ADHOC-NOW 2016 was organized by Inria Nord-Lille Europe in Villeneuve D'Ascq, France.

Executive Committee

General Chair

Nathalie Mitton   Inria, France

TPC Chairs

Nathalie Mitton   Inria, France
Valeria Loscri   Inria, France
Alexandre Mouradian   Paris Sud, France

Publicity Arrangements Chair

Abdoul Aziz Mbacke   Inria, France

Proceedings Chair

Cristina Cano   Inria, France

Submissions and Registrations Chair

Valeria Loscri   Inria, France

Local Arrangements Chairs

Marie Bénédicte   Inria, France
  Dernoncourt
Nathalie Mitton   Inria, France
Anne Rejl   Inria, France

Web Chair

Riccardo Petrolo   Inria, France

TPC Members

Marica Amadeo   University Mediterranea of Reggio Calabria, Italy
Flavio Assis   UFBA-Federal University of Bahia, Brazil
Alessia Autolitano   Istituto Superiore Mario Boella, Italy
Evangelos Bampas   Aix-Marseille University, France
Michel Barbeau   Carleton University, Canada
Jose M. Barcelo-Ordinas UPC, Spain

Shaojie Tang	University of Texas at Dallas, USA
Eirini Eleni Tsiropoulou	NTUA, Greece
Volker Turau	Hamburg University of Technology, Germany
Veronique Veque	L2S Supélec, France
Konrad Woran	NCI Agency, The Netherlands
Weigang Wu	Sun Yat-sen University, China
Christos Xenakis	University of Piraeus, Greece
Lynda Zitoune	ESIEE-Paris, France

Additional Reviewers

Frederik Armknecht
Yuanzhu Chen
Gabriele Di Stefano
Dejan Drajic
Dmitry Dugaev
Jason Ernst
K. Mabell Gomez Chavez
Jorge Herrera-Tapia
Christina Karousatou
Mohammad K. Dermany
Anissa Lamani

Sergio Martínez Tornell
Florian Meier
Cem Mergenci
Philipp Morgner
Simin Nadjm-Tehrani
Dominik Pajak
Jacek Rak
Hawar Ramazanali
Alexey Shapin
Marco Steger
Tzu-Chieh Tsai

Contents

DTN/Opportunistic Networks

Sensors/IoT

Security

VANET and ITS

Robots and MANETs

Resource Allocation

Distributed Scheduling of Enhanced Beacons for IEEE802.15.4-TSCH Body Area Networks

Mengchuan Zou[1], Jia-Liang Lu[1], Fan Yang[1], Mathilde Malaspina[2],
Fabrice Theoleyre[3(✉)], and Min-You Wu[1]

[1] Department of Computer Science and Engineering,
Shanghai JiaoTong University, Shanghai, China
[2] Department Télécommunications, INSA-Lyon, Universite de Lyon, Lyon, France
[3] ICube, CNRS/University of Strasbourg, Strasbourg, France
theoleyre@unistra.fr

Abstract. Body Area Networks (BANs) expect to exploit IEEE802.
15.4-2015-TSCH, proposing an efficient MAC layer for wireless industrial
sensor networks. The standard relies on techniques such as channel hopping and bandwidth reservation to ensure both energy savings and reliable transmissions. With the expected growth of the BAN usage, we must
now consider dense topologies, and interference. In this paper, we propose a rescheduling algorithm to avoid the collisions among the Enhanced
Beacons (EB): each coordinator is able to adapt distributively its transmission to avoid interference. Indeed, EB losses impact negatively the
performance of a BAN. We also optimized conjointly the neighbor discovery mechanism since a multichannel MAC would else increase too
much the discovery delay. Our simulations validate the relevance of our
discovery and scheduling mechanisms to cope with a very dense deployment of interfering BANs.

1 Introduction

BANs consist in small wearable wireless devices expected to fulfill the society
needs in a variety of applications such as ubiquitous monitoring, health-care,
entertainment and multimedia, and training. The PAN coordinator often collects
measures transmitted from the wearable devices in its BAN.

The IEEE802.15 working group is currently finalizing the IEEE802.15.4-2015
standard [1]. In particular, the TSCH mode aims at improving the reliability for
industrial sensor networks in noisy environments. A common schedule aims at
reserving a certain amount of bandwidth for each flow, and channel hopping aims
at defeating narrow band noise. This standard is particularly accurate for Body
Area Networks, where devices aim at transmitting periodically their measures
to the PAN coordinator (i.e. BAN gateway).

However, IEEE802.15.4-TSCH was originally designed for a large multihop
wireless sensor network. In Body Area Networks, we rather face to a user cen-
tric topology: one PAN coordinator attached to a collection of devices. However,
because we expect to have a large collection of co-located BANs, we will face to
an explosion of collisions: each BAN computes independently its own schedule.

© Springer International Publishing Switzerland 2016
N. Mitton et al. (Eds.): ADHOC-NOW 2016, LNCS 9724, pp. 3–16, 2016.
DOI: 10.1007/978-3-319-40509-4_1

new device

Fig. 1. Multi-BAN topology

We must propose a certain cooperation among different BANs to detect collisions and to adapt locally their schedule. Besides, BANs require a very short network attachment delay: a person may exchange data sporadically within a group of individuals (aka. opportunistic mobile social networks [2]). Thus, we must propose a mechanism to detect quickly neighboring BANs.

Due to the dynamic environment for BAN communication, one important assumption is that a newly arrived node has no initial information about the neighborhood. Therefore it needs to set up a listening schedule determining when, for how long, and on which channel it should listen to the beacons. When it receives an Enhanced Beacon (EB), it is then able to join the corresponding BAN. We chose here to optimize this discovery delay, i.e. waiting time before receiving the first EB.

In our scenario, we consider co-located BANs, possibly mutually interfering (Fig. 1). However, we consider here only single hop traffic: only neighboring PAN coordinators /devices may exchange data packets. We have consequently a collection of stars, independent concerning the traffic, interdependent concerning interference.

The contributions of this paper are fourfold:

1. we introduce a cooperation among independent BAN: while they don't exchange traffic, we propose mechanisms to locally re-schedule the transmissions of beacons to limit the number of collisions;
2. we reduce the schedule problem (and the detection of collisions) to a classical maximum-weight independent set problem in undirected graphs;
3. we also propose a mechanism so that a new node is able to discover quickly an existing neighboring BAN, even when its EBs use channel hopping;
4. we highlight the relevance of our approach with omnet++ simulations.

2 Related Work

2.1 IEEE802.15.4-2015

IEEE802.15.4-2015 [1] has proposed the TSCH mode for industrial wireless sensor networks. To improve the reliability while maximizing energy savings, a schedule is computed by the Path Computation Engine (PCE) and distributed to all the nodes. The schedule is then repeated periodically, and the ASN (Absolute Sequence Number) counts the number of slots since the beginning.

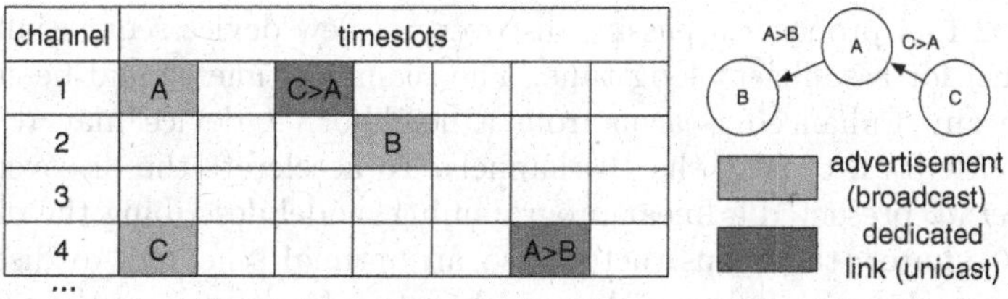

Fig. 2. Schedule in a IEEE802.15.4-TSCH network – illustration of a slotframe (Color figure online)

When a timeslot starts, a device knows if it has to wake-up to transmit or receive a packet. A slot can be either dedicated (without contention) or shared (a CSMA-CA mechanism handles contentions). Thus, a node turns off its radio for all its unused slots, where it is neither receiver nor transmitter.

To improve the reliability, TSCH proposes to implement a channel hopping scheme. To each transmission opportunity is attached a *channel offset*. Let's consider the schedule illustrated in Fig. 2. Three timeslots/channel offsets are dedicated for the advertisements of the nodes A, B and C (in green). Two other slot/channel offsets are dedicated for some radio links (in pink), for their unicast transmissions.

Palatella *et al.* proposed a centralized scheduling for a multihop IEEE802.15.4-TSCH [3] for unicast transmissions. Accetura *et al.* also proposed a decentralized version of their scheduling algorithm [4]. However, these approaches only work within a single BAN, i.e. they don't address the case of multiple interfering and concurrent BANs.

2.2 Neighbor Discovery

Two major methods exist to discover a neighboring node:

Passive discovery: a new node has to listen for the `beacon` packets from its neighbors. The period of `beacon` transmissions directly impacts the discovery delay.

Active discovery: a new node sends an `hello` packet, and neighbors acknowledge it to notify the transmitter of their presence. In multichannel, it is related to the birthday protocol [5]. This method often performs faster, but consumes more resource, and may create collisions with data packets.

A new device has to discover very fast a BAN to attach to. However, multichannel increases the convergence delay, since a receiver is deaf to the transmissions on different channels. In Bluetooth, a *transmitter* sends avertissements and a *listener* must receive this advertisement before exchanging packets. Thus, a node must change randomly its role between listener and transmitter for Peer-To-Peer topologies [6].

In multichannel, scanning all the channels takes a long time. Dasilva *et al.* proposed to use a complete permutation of all possible channels to scan, which decreases the convergence delay [7]. When a group of nodes can cooperate, they can exchange information to accelerate the discovery [8,9].

IEEE802.15.4 proposes a passive discovery: a new device sequentially scans each channel for a sufficient long time. The scanning time should be sufficient to capture any Enhanced Beacon from a neighbor. A device may require in the worst case 8.7 h to scan the 16 channels. To accelerate the discovery time, Karowski *et al.* presented a linear programming model describing the discovery process [10]. Later, they constructed also an optimal schedule, to discover in priority the nodes with a larger duty cycle ratio [11]. However, they focus on the listening device, so that their optimization could only be applied to a fixed advertising schedule.

3   Collision-Free Schedule of Multiple BANs

We consider a dense collection of Body Area Networks, with one PAN coordinator per BAN. Any pair of nodes may interfere with each other, and create collisions. We adopt in this paper the notation described in Table 1.

3.1   Problem Statement

Each BAN may select a different duty cycle ratio. In particular, they may choose a different slotframe length (the cycle duration of the schedule). Each node sends an Enhanced Beacon (EB) at a fixed interval during an advertisement slot of the schedule. More precisely, a node u sends an EB every:

$$T_{EB}(u) = T_{cst} * 2^{BO} \tag{1}$$

where T_{cst} is a duration defined by IEEE802.15.4-2015, and BO is the beacon order (constant) selected by the node u (the rest of the notation is described in Table 1).

IEEE802.15.4-TSCH adopts a FTDMA approach. We consider the PAN coordinator is co-located with the Path Computation Engine (PCE). It computes

Table 1. Notation used in the article

Notation	Meaning
$T_{EB}(u)$	the interval between two Enhanced Beacons (EB) of the node u (denoted further EB interval)
$slot_{EB}(u)$	the cumulative nb. of broadcast slots (EBs) for the node u since the beginning of the scan
nb_{ch}	nb. of channels
nb_{tslots}	nb. of timeslots for all the possible channel offsets
ASN	Absolute Sequence Number ($\approx$time)
T_{scan}	Duration of the scan of the discoverer (nb. of slots)
$nb_{rx}(EBs)$	nb. of EB received during the scan by the discoverer
$S = <(rl,t,c)>$	the IEEE802.15.4-TSCH schedule S is a set of triplets <radio_link, timeslot, channel_offset>

which channels and timeslots have to be used for all the pairs of active nodes, and distributes this schedule. In particular, it assigns an advertisement slot for each node to transmit periodically its EBs.

The PAN coordinator changes dynamically the schedule when it detects a collision with a neighboring network.

A new device has to receive the Enhanced Beacons to decide which network it should join. We adopt consequently a similar objective as [11]: we aim at maximizing the number of EB $nb_{rx}(EBs)$ received during a scanning duration $T_{scan}/nb_{rx}(EBs)$. All the PAN coordinators have to compute a schedule for their network, where the number of collisions between EBs is minimized.

We consider the different BANs are able to maintain a synchronization, so that all the schedules are *aligned*. It may be implemented by maintaining a synchronization with neighboring BA, such as [12] does.

3.2   Problem Formulation

We use the conflict graph model to define an accurate schedule. Let $G_c = (V_c, E_c)$ be an undirected graph, where V is the set of radio links, and E is the set of mutually interfering radio links. G_c is named the *conflict graph*. The schedule consists in the set of activated radio links $v = (rl, t, c) \in V_c$, where rl is the radio link, t a timeslot, and c a channel offset (cf. Fig. 3). For the sake of simplicity, a radio link may also be denoted by the transmitter only if it corresponds to a broadcast (i.e. EB).

An edge $(u, v) \in E_c$ indicates transmitters u and v interfere with each other:

1. either both transmitters own to the same BAN. The PAN coordinator can easily reallocate the transmitters to use different timeslots/ channels;
2. or both transmitters own to different BAN: the corresponding PAN coordinators must cooperate to avoid this kind of overlap.

Let's consider the nodes u and v with their associated beacon periods ($T_{EB}(u)$ and $T_{EB}(v)$). The pair of nodes should not share any common timeslot and channel for *any* of their Enhanced Beacons. If they have different slot frame lengths, a collision may occur if they have the same channel offset and a common active slot to advertise their EB during at least *one* slotframe:

$$\exists (i_v, i_u) \in \mathbb{N}^2 /$$

$$slot_{EB}(v) + i_v * T_{EB}(v) = slot_{EB}(u) + i_u * T_{EB}(u) \quad (2)$$

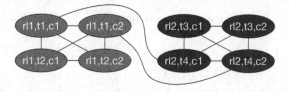

Fig. 3. Conflict graph model for ASP

This equation is an indefinite equation, with a solution (i_u, i_v). We may use the Extended Euclidean algorithm to check the existence of an integer solution. We are then able to construct the conflict graph, with all the edges which respect the Eq. 2.

3.3   Optimization Function

A node may attach to the network as soon as it discovers an accurate neighbor. Let $w(u)$ be a weight associated to a node u, denoting the amount of EBs it transmitted during the scanning time:

$$w(u) = \frac{T_{scan} - slot_{EB}(u)}{T_{EB}(u)} \qquad (3)$$

Our objective consists in maximizing the total amount of EBs received by all the devices during the scanning period:

$$Objective = max \left(\sum_{u \in V} w(u) \right) \qquad (4)$$

3.4   Reduction to the Maximum-Weight Independent Set Problem

When no collision occurs, the number of received EBs equals to the number of sent EBs. More precisely, no edge is present in the conflict graph between any pair of vertices: they form an independent set.

Let us consider the undirected graph $G_c = (V_c, E_c)$. An Independent Set (IS) is a set of vertices such that no edge exists between any pair of vertices. Let $\mathbb{S}$ be the set of all possible independent sets:

$$S \in \mathbb{S} \Leftrightarrow \forall (u, v) \in V_c^2 \cap S^2, \nexists (u, v) \in E_c \qquad (5)$$

If the set of active vertices in G_c (i.e. radio links) forms an independent set, no collision arises: the schedule of EBs is optimal. Let $\mathbb{S}$ be the set of all possible independent sets:

$$if \ S \in \mathbb{S}, \ Objective = \sum_{u \in S} w(u) \qquad (6)$$

Since no conflict exists in the conflict graph, the maximization objective can be removed safely: no EB is dropped because of collisions. In conclusion, the optimization problem turns out to be a maximum-weight independent set problem (MWIS) in a undirected graph. We schedule then a MWIS in a given timeslot/channel offset.

MWIS consists in finding an independent set $S \subseteq V$, which has the maximum weight:

$$max \left(\sum_{v \in S} w(v) \right) \qquad (7)$$

As we aim at maximizing the number of EBs received, our problem may be reduced to the MWIS problem, known as an NP-hard problem [13].

4   Enhanced Beacon Advertising Scheduling (EBAS)

We propose here a localized heuristic. Indeed, a centralized approach presents the following limits:

1. Overhead and Consistency: An exhaustive search requires a global knowledge. All the schedules owning to different BANs must be shared. Practically, a huge volume of information has to be exchanged, and inconsistencies may arise because of packet losses;
2. Complexity: An exhaustive exploration of the solutions is unrealistic for devices with limited capabilities. Besides, we cannot assume a server (connected to the PAN coordinators) is dedicated to this computation.

Our algorithm proceeds in the following way:

Step 1 - Division: we first create groups of devices, sharing the same EB period. We can assign orthogonal resources (i.e. channel offsets) to different groups to avoid collisions. This way, we avoid collisions among devices with different EB periods;

Step 2 - Initialization: a coordinator selects an initial schedule for all the EBs in its BAN. Each device maintains the list of occupied cells in its neighborhood, to detect collisions (i.e. at least *one* interfering device exists);

Step 3 - Collision resolution: we propose heuristics to re-schedule the colliding slots. We use hash tables to detect collisions and we re-allocate the concerned timeslots while limiting the number of changes in the schedule;

Step 4 - Backtracking: if the algorithm does not succeed to allocate a timeslot for each EB, we backtrack to the step 1, by selecting a group with a larger EB period.

4.1   Division: Dealing with Different EB Periods

We consider the same conflict graph as previously: a vertex is the triplet < *device, timeslot, channel_offset* >, and two vertices are linked together if the associated devices interfere.

We can remark the following properties:

1. All vertices representing the same device (for different EBs) form a clique in the conflict graph;
2. when considering a set of vertices with the same EB interval and using the same channel offset, the collision is present for *all* the timeslots, and can be easily detected.

So, we propose a *partition and re-grouping* method:

1. we divide the graph into cliques. We insert an edge between two cliques if they are neighbors in the conflict-graph;
2. we merge the cliques with the same EB interval into a group. We will assign later a given channel offset to a group;

3. then, we assign a given timeslot for each device within a group: they should not interfere since they use the same channel offset. Assigning distinct timeslots is sufficient since they have the same T_{EB}.

By adopting a three-steps approach, we are able to deal with devices with different EB intervals. Indeed, devices with different intervals are partitioned in distinct groups (i.e. different channel offsets). Then, we solve the conflicts among the devices with the same EB interval, using a TDMA repartition. Since times slots are distinct, they cannot collide in *any* slot frame: the period of repetition is the same for all of them – same T_{EB}.

Group Merging. If too many EB intervals exist, partitioning the graph would create too many groups. However, as Eq. (1) states, the EB intervals are multiples of 2. Let's denote by $group_i$ all the devices with an EB period equals to BO^i.

Several devices of the group i may actually be scheduled with the devices in the group $i - 1$. The devices of the group i have either to be scheduled in different timeslots or in different slotframes. Intuitively, a slotframe of the group i contains *two* slotframes of the group $i - 1$.

Obviously, we can merge any group i and j ($i < j$) of devices. Finally, at most $2^{BO^j - BO^i}$ nodes of the group j may be schedule in the final same timeslot. Inversely, a device of the group i uses a given timeslot in *all* the slotframes.

4.2   Initialization: Assigning Timeslots to Each Device

All the devices which share the same EB interval are scheduled in the same channel offset, and the PAN coordinator assigns randomly one free timeslot for each of them. The PAN coordinator has to compute its schedule, and to push it to all the devices. The schedule table is actually the set of timeslots/channel offsets (row), and their corresponding occupation (0 if this timeslot is unoccupied by any transmitter).

Each PAN coordinator broadcasts its schedule periodically in dedicated packets, with an increasing sequence number. A neighbor is thus able to detect a timeslot collision, equivalent to a collision in both hash tables.

A node is able to maintain an up-to-date schedule table of its neighboring PAN coordinators:

- any neighbor for which it stops receiving packets is removed from the neighborhood table after a certain timeout;
- only the last schedule is memorized for each neighbor (i.e. largest sequence number).

Thus, we guarantee a certain consistency in the decisions: each device is able to replace obsolete information from its neighborhood schedule table.

4.3   Collision Resolution in the Same Group - Linear Time Scanning

A PAN coordinator may detect a collision after the reception of the schedule from one neighbor. To solve the collisions, we use hash tables.

More precisely, an hash table makes a correspondence between a key (here, a timeslot) and the value (1 if the timeslot is occupied by at least one node, else 0). A collision of timeslots means also a collision in the hash table. Thus, finding compliant schedules (i.e. interference free) is equivalent to resolving the collisions in the hash table.

Two methods exist to resolve collisions in hash tables:

1. separate chaining: each entry of the has table is a chained list of values;
2. open addressing: the values are stored in contiguous entries in the hash table.

Since the former approach does not maintain the original order in the hash tables, we adopt the open addressing method.

When the network is very dense, many timeslots are occupied. To accelerate the convergence, we propose here to solve several conflicts simultaneously. The following two steps are required:

1. Preprocessing: for any channel, the node combines all the schedules present in the neighborhood table with the OR operator. Then, the node is able to compute the list of slots used by at least one of the BANs, and the list of the idle slots.
2. Linear time scan: we create two pointers (cf. Fig. 4). The first one (in red) points to the first colliding slot in our schedule, and the second one (in green) to the first idle slot in the merged schedules. Then, we modified our schedule: the colliding cell is moved to be placed in the next free cell (in the merged schedule). Then, the pointers keep on iterating, to solve the next conflict.

When resolving the conflicts, both pointers visit each slot exactly once. Thus, our algorithm runs in $O(n)$ time. We have consequently a linear time rescheduling algorithm.

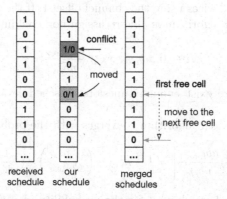

Fig. 4. Linear time scanning (Color figure online)

4.4 Backtracking: Group Change

If the previous search has failed, this means no timeslot is available.

We backtrack the algorithm to the step 3. The device joins a group with a lower EB interval, and re-executes the assignment to find an idle slot in the new group.

This backtracking method is particularly relevant when all the groups do not have the same number of devices. A BAN will be moved to another group to balance the load.

4.5 Analysis

Let's now consider the initialization step. We have nb_{tslots} slots (all the timeslots in a slotframe, across the nb_{ch} possible channel offsets). The number of EBs to be sent is n, and timeslots are assumed to be assigned randomly for initialization.

So, the expected number of collision-free EBs (n_{nocoll}) is the complementary of colliding EBs:

$$n_{coll} = n * \left(1 - \prod_{k=2}^{n} \frac{nb_{tslots} - (k-1)}{nb_{tslots}} \right) \tag{8}$$

We denote by ρ the ratio of the number of devices and the number of timeslots ($\rho = n/nb_{tslots}$). Thus ρ denotes the *pressure* for the scheduling process. The ratio σ of conflicting EBs is consequently:

$$\sigma = \frac{n_{coll}}{n} = 1 - \prod_{k=2}^{n} \frac{n - \rho * (k-1)}{n} \tag{9}$$

Because of the well-known birthday problem [14], the ratio of collisions will be quite large when ρ grows.

Our algorithm solves the conflicts created in the initialization phase. We group the devices by their EB interval, and thus their channel offset. We denote by x_i the number of devices using the channel offset i. If the following condition holds, our scheduling algorithm is able to assign non colliding slots:

$$\forall i \in [0..nb_{ch}], \ x_i \leq \frac{nb_{tslots}}{nb_{ch}}, \tag{10}$$

In other words, there exist enough timeslots to schedule all the devices of a group.

According to [15], our algorithm converges with the probability:

$$P\left[\forall i, x_i \leq \frac{nb_{tslots}}{nb_{ch}} \right] = 1 - nb_{ch} * \left(\frac{n}{\frac{nb_{tslots}}{nb_{ch}}} \right) * \left(\frac{1}{nb_{ch}} \right)^{\frac{nb_{tslots}}{nb_{ch}}} \tag{11}$$

This represents a lower bound for the probability of convergence. For each iteration, we have:

1. After probing and changing a schedule on one channel i, we have $x_i^{t+1} \leq x_i^t$ which means x_i doesn't increase.
2. If the device switches to another channel j, we have

$$x_i^{t+1} \leq x_i^t - 1, x_j^{t+1} \leq \frac{nb_{tslots}}{nb_{ch}} \tag{12}$$

so that there is also $\forall i, x_i \leq \frac{nb_{tslots}}{nb_{ch}}$.

Moreover, if $\exists i, x_i > \frac{nb_{tslots}}{nb_{ch}}$, at least one device is able to switch to another channel offset to find a free timeslot. So Eq. 11 is a lower bound of convergence probability.

5 Performance Evaluation

We used Castalia 3.0 (http://castalia.npc.nicta.com.au), a simulator based on the OMNeT++ framework version 4.1 (http://www.omnetpp.org). Castalia is widely used to simulate a Body Area Network. For the initialization, each coordinator chooses independently and randomly a timeslot to send EBs.

We simulated a multi-BANs scenario where a set of PAN coordinators are directly connected to a random number of devices. The number of coordinators varies between 5 and 20 (with increment of 5). Besides, the ratio of slot occupation varies from 10 % to 90 %, with increment of 20 %. We run 500 simulations for each set of parameters.

We compared the following algorithms:

- PSV: the passive scan mode in IEEE802.15.4-2015 [1]. An arriving device just listens for the maximal possible length of time on each channel;
- SUBOPT: this low-complexity algorithm computes a listening schedule [10]. Intuitively, the device has not to scan all the timeslots when a neighbor sends very fast its EBs. It avoids these *redundant* timeslots;
- EBAS: our rescheduling algorithm detecting collisions and reallocating the colliding devices to free timeslots.

We have the following main parameters/performance criteria:

- ratio of occupation: the ratio of the number of devices (EB to schedule) and the number of timeslots;
- Percentage of EBs received: number of EB received by a discovering device. The higher this value is, the faster a device may discover a network to join;
- Percentage of EB conflicts: ratio of EB which use a timeslot which collides with another EB (i.e. for at least one timeslot when both transmitters have not the same EB interval).

We first measured the number of EBs correctly received (Fig. 5). The efficiency of SUBOPT and PSV decreases when the network comprises more coordinators: SUBOPT has been designed to reduce the discovery delay, but doesn't

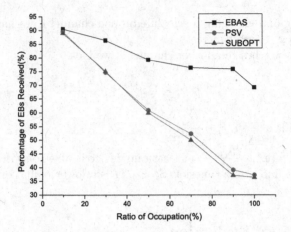

Fig. 5. Ratio of EBs received correctly (i.e. without collision)

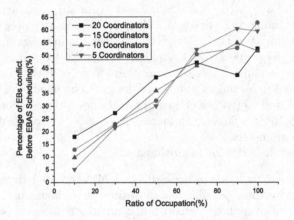

Fig. 6. Ratio of colliding EBs before EBAS algorithm

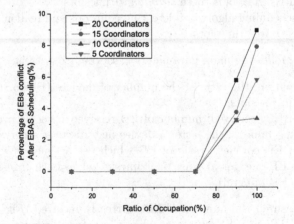

Fig. 7. Ratio of colliding EBs after EBAS algorithm

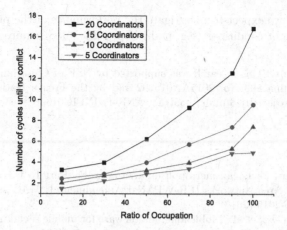

Fig. 8. Number of iterations needed until a conflict-free schedule is found

try to re-schedule the colliding beacons. On the contrary, EBAS is much more scalable and successfully solves most of the conflicts even for a very large occupation ratio. Such feature is vital for very dense BAN deployments.

We also measured the percentage of colliding EBs without (respectively with) EBAS in Fig. 6 (resp. Fig. 7). Comparing both figures, we clearly notice EBAS is very efficient when less than 70% of the slots are occupied: it solves 100% of the collisions, re-allocating the incriminated devices. Even for a very large occupation ratio (e.g. 90%), only 10% of the EBs collide. On the contrary, a distributed random assignment rather leads to a ratio of 60% of collisions.

Finally, we measured the number of iterations with EBAS before obtaining a conflict-free schedule (Fig. 8). For small BANs, only a few iterations are required (at most 4). Without surprise, the number of iterations grows with the occupation ratio. However, EBAS converges in less than 20 iterations, even in very extreme conditions.

6   Conclusion

We proposed here an algorithm to schedule the EBs transmissions when several Body Area Networks co-exist and mutually interfere. We proposed first a dividing strategy, grouping together the devices with the same EB. Then, we propose to detect and solve collisions after a random assignment. Neighboring BANs exchange their schedule to detect conflicts, and reallocate the timeslots while preserving the other collision-free schedules. Simulations prove the relevance of our approach: we reduce the number of collisions even in dense deployments.

In the future, we plan to combine a fast collision detection mechanism with our re-scheduling algorithm. We also aim at validating experimentally our solutions with complex radio propagation (and interference). Besides, we also aim at quantifying the impact of clock drifts on the accuracy of our re-scheduling process. Finally, we expect to investigate the performance of the proposed solution under different conditions (e.g. multi-hop traffic, QoS requirements).

Acknowledgment. This research was supported by NSF of China under grant No. 61100210 and China 863 No. 2015AA015802 and by the French National Research Agency (ANR) project IRIS under contract ANR-11-INFR-016.

References

1. IEEE Standard for Local, metropolitan area networks-Part 15.4: Low-Rate Wireless Personal Area Networks (LR-WPANs) Amendment 1: MAC sublayer. IEEE Std 802.15.4e-2015 (2015)
2. Pietiläinen, A.-K., et al.: Mobiclique: middleware for mobile social networking. In: ACM Workshop on Online Social Networks, pp. 49–54 (2009)
3. Palattella, M.R., et al.: On optimal scheduling in duty-cycled industrial IoT applications using IEEE802.15.4e TSCH. Sens. J. IEEE **13**(10), 3655–3666 (2013)
4. Accettura, N., et al.: Decentralized traffic aware scheduling for multi-hop low power lossy networks in the internet of things. In: IEEE WoWMoM (2013)
5. McGlynn, M.J., Borbash, S.A.: Birthday protocols for low energy deployment and flexible neighbor discovery in ad hoc wireless networks. In: Proceedings of the 2nd ACM International Symposium on Mobile Ad Hoc Networking & Computing, pp. 137–145. ACM (2001)
6. Drula, C., Amza, C., Rousseau, F., Duda, A.: Adaptive energy conserving algorithms for neighbor discovery in opportunistic bluetooth networks. IEEE J. Sel. Areas Commun. **25**(1), 96–107 (2007)
7. DaSilva, L.A., et al.: Sequence-based rendezvous for dynamic spectrum access. In: New Frontiers in Dynamic Spectrum Access Networks (DySPAN), pp. 1–7. IEEE (2008)
8. Chen, L., et al.: Group-based discovery in low-duty-cycle mobile sensor networks. In: IEEE SECON, pp. 542–550. IEEE (2012)
9. De Nardis, L., et al.: Role of neighbour discovery in distributed learning and knowledge sharing algorithms for cognitive wireless networks. In: ISWCS, pp. 421–425. IEEE (2012)
10. Karowski, N., et al.: Optimized asynchronous multi-channel neighbor discovery. In: IEEE INFOCOM, April 2011
11. Karowski, N., et al.: Optimized asynchronous multichannel discovery of IEEE 802.15. 4-based wireless personal area networks. IEEE Trans. Mob. Comput. **12**(10), 1972–1985 (2013)
12. Lipa, N., Mannes, E., Santos, A., Nogueira, M.: Firefly-inspired and robust time synchronization for cognitive radio ad hoc networks. Comput. Commun. **66**, 36–44 (2015)
13. Trevisan, L.: Inapproximability of combinatorial optimization problems. The Computing Research Repository (2004)
14. Mathis, F.H.: A generalized birthday problem. SIAM Rev. **33**(2), 265–270 (1991)
15. Borodin, A., et al.: A time-space tradeoff for sorting on a general sequential model of computation. In: ACM Symposium on Theory of Computing. ACM (1980)

The Cost of Installing a 6TiSCH Schedule

Erwan Livolant[✉], Pascale Minet, and Thomas Watteyne

Inria-Paris, EVA Team, Paris, France
{erwan.livolant,pascale.minet,thomas.watteyne}@inria.fr

Abstract. Scheduling in an IEEE802.15.4e TSCH (6TiSCH) low-power wireless mesh network can be done in a centralized or distributed way. When using centralized scheduling, a scheduler computes a communication schedule, which then needs to be installed into the network. This can be done using standards such CoAP and CoMI, or using a custom protocol such as OCARI. In this paper, we compute the number of messages installing and updating the schedule takes, using both approaches, on a realistic example scenario. The cost of using today's standards is high. In some cases, a standards-based solution requires approximately 4 times more messages to be transmitted in the network, than when using a custom protocol. This paper makes three simple recommended changes to the standards which, when integrated, reduce the cost of a standards-based solution by 18 % to 74 %. Since they are still being developed, these recommendations can easily be integrated into the standards.

Keywords: Low-power wireless mesh networks · IEEE802.15.4e TSCH · 6TiSCH · CoAP · CoMI · OCARI

1 Introduction

Industrial low-power wireless mesh network applications have strong requirements in terms of latency, energy efficiency and reliability. To cope with these requirements, the IEEE802.15.4e amendment [7] introduces the Time Slotted Channel Hopping (TSCH) mode. In a TSCH network, nodes are synchronized, and time is cut into timeslots, each typically 10 ms long. All communication is orchestrated by a communication schedule, which indicates to each node what to do in each slot: transmit, listen or sleep. This schedule can be built to enable collision-free communication, yielding predictable behavior, ultra-high reliability and years of battery lifetime.

A schedule consists of a number of timeslots which continuously repeat over time. An example schedule is depicted in Fig. 2 with 9 timeslots and 3 logical channels. The index of a timeslot (the x-axis of the matrix in Fig. 2) is called `slotOffset`. The `channelOffset` represents the communication channel (the y-axis of the matrix in Fig. 2).

Building the schedule consists in assigning a source node, a destination node and a channel to cells in the schedule. Installing the schedule into the network means indicating to each node the list of cells it is involved in, either

© Springer International Publishing Switzerland 2016
N. Mitton et al. (Eds.): ADHOC-NOW 2016, LNCS 9724, pp. 17–31, 2016.
DOI: 10.1007/978-3-319-40509-4_2

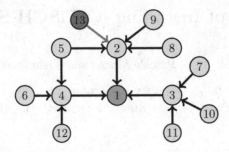

Fig. 1. Logical network topology.

chan. \ slot	0	1	2	3	4	5	6	7	8
0	4→1	2→1	4→1	2→1	4→1	2→1	4→1	2→1	3→1
1	3→1	10→3	3→1	11→3	3→1	10→3	3→1	7→3	
2	5→2	6→4	2→1	12→4	9→2	5→4	8→2		

Fig. 2. Schedule computed by MODESA for 12 nodes.

chan. \ slot	0	1	2	3	4	5	6	7	8
0	4→1	2→1	4→1	2→1	4→1	2→1	4→1	2→1	4→1
1	3→1	10→3	3→1	11→3	3→1	12→4	3→1	5→4	3→1
2	2→1	5→4	9→2	6→4	8→2	10→3	13→2	7→3	

Fig. 3. Updated schedule after node 13 is added. The black cells are the ones that differ from the schedule for 12 nodes.

as transmitter or as receiver. Each cell is represented by a tuple [slotOffset, channelOffset, nodeAddress, linkType] (linkType indicates whether it is a transmit – TX – and/or receive – RX – cell).

We want to compare the number of packets it takes a central scheduler to install and update a schedule, using the different approaches listed in Sect. 2. We are particularly interested in comparing standards-based and custom-built protocols. The goal is *not* to explore edge cases, but rather to take an example representative enough that we can learn lessons and made recommendations.

We consider the topology depicted in Fig. 1. The network first consists of 12 nodes where Node 1 is the root of the network. Arrows represent the links that are used for communication and that therefore need to appear in the schedule. The goal is to *install* the schedule from Fig. 2 into the network. When node 13 is added, the scheduler computes the schedule depicted in Fig. 3, in which 12 cells differ. The goal is then to *update* the schedule in the network.

Several centralized scheduling algorithms exist. This paper does *not* recommend one or the other, nor does it attempt to survey them. Rather, we use a particular scheduling algorithm, MODESA [10], and measure the number of packets to install the schedule *once it is computed*. The choice of MODESA is, as far as this paper is concerned, arbitrary.

We focus on two approaches to install the schedule: using standards (detailed in Sect. 2.2), and using a custom protocol (detailed in Sect. 2.3). By executing both approaches on a representative example scenario, we show that using a standards-based approach can cost 4 times more frames than using a custom-built protocol. In Sect. 3, we make three simple recommended changes to the standards which reduces the overhead of the standards-based approach by 18 % to 74 % percent.

2   Approaches to Install the Schedule

We consider different approaches to install a schedule: either using the CoAP and CoMI standards (Sect. 2.2), or using a custom protocol called OCARI (Sect. 2.3). Both use IEEE802.15.4 [6] as the underlying physical layer. Section 2.1 first details the notation used.

2.1   Notation

We define the following notations:

- $Depth(u)$, the depth of node u in the topology.
- $Child(u)$, the set of children of node u.
- $Trans(u)$, the number of user data frames transmitted by node u, including both the ones generated by node u itself and the ones received from its children and forwarded.
- P, the number of nodes which are not leaf nodes.
- $B(u)$, number of CoAP blocks sent to node u.
- B_{sched}, number of CoAP blocks needed to broadcast the complete schedule.
- Let N_{field} number of fields updated per cell.
- $ScheduleNumber$, version of the schedule, incremented each time the schedule is computed.

2.2   CoAP and CoMI: A Standards-Based Approach

6TiSCH is an active IETF working group which standardizes how to build and maintain a TSCH communication schedule [4]. While 6TiSCH supports both centralized and distributed scheduling, this paper focuses on the former. The protocol stack considered by 6TiSCH consists of IEEE802.15.4e [7] (*physical and link layers*), 6LoWPAN [5], RPL [14] (*network layer*), CoAP [3,9,12] and CoMI [11] (*application layer*). The resulting protocol stack is depicted in Fig. 4.

Table 1 depicts the format of an IEEE802.15.4 frame which encapsulates those protocols. The Maximum Transmission Unit (MTU) at the IEEE802.15.4 [6] PHY layer is 127 bytes. MAC header and footer require a total of 29 bytes in the context of the IEEE802.15.4e TSCH [7] mode[1].

---

[1] [13] indicates that data messages must provide in their MAC header 64-bit addresses for the destination and the source.

```
+-------------------------------------------+
|                    CoMI                   |
+-------------------------------------------+
|                    CoAP                   |
+-------------------------------------------+
|                    UDP                    |
+-------------------------------------------+
|                    RPL                    |
+-------------------------------------------+
|                    IPv6                   |
+-------------------------------------------+
|                 6LoWPAN HC                |
+-------------------------------------------+
|                    6top                   |
+-------------------------------------------+
|             IEEE 802.15.4e TSCH           |
+-------------------------------------------+
|             IEEE 802.15.4 PHY             |
+-------------------------------------------+
```

Fig. 4. The 6TiSCH protocol stack.

Table 1. Packet format when using the standards-based approach. Numbers indicate the number of bytes in the fields.

	Frame Control	Seq Num	Dest PAN	Dest Address	Src Address	Aux Sec Header	MIC	Payload	FCS
IEEE802.15.4e TSCH	2	1	2	8	8	2	4	98	2

	LOWPAN_IPHC	Hop limit	Src Address	Dest Address	Payload			
6loWPAN - IP	2	1	8	8	79			

	LOWPAN_NHC	Port src + dest	Payload	
6loWPAN - UDP	1	1	77	

	Header	URI	Payload marker	Payload without block
COAP without frag	4	11	1	61

	Header	URI	Option delta + Length	Option Delta extended	Option value	Payload marker	Payload with block
COAP with ≤ 16 frag	4	11	1	1	1	1	32
COAP with ≤ 4096 frag	4	11	1	1	2	1	32

The Auxiliary Security header is encoded with 2 bytes and the Message Integrity Code (MIC) is coded with 4 bytes. The MAC payload can contain up to 98 bytes. At the Network layer, the protocol requires 19 bytes for the IP compressed header (2 bytes), the Hop limit (1 byte) and the Source and Destination addresses coded on 8 bytes each. At the Transport layer, the payload is reduced by 2 bytes corresponding to the UDP compressed header and the destination and source port coded together in 1 byte. At the Application layer, the size of the applicative payload depends on the CoAP options used. 4 bytes of header and 1 byte of payload marker are required. The URI targeting the resource is defined as follows: /mg/6t/hash where /mg/6t/ is the main path to the 6top management resources and hash is a 30-bit hash (encoded on 4 bytes) defining the rest of the path towards the target resource. We look at three options for using CoAP:

without fragmentation, less than 16 fragments and with than 4096 fragments. In any case, we get a useful payload whose size is less than 61 bytes. For the *Block* option, it is specified in CoAP that the block size should be a power of two. As a consequence, the only possibility with this payload size is a block size of 32 bytes.

As a conclusion, in the 127-byte IEEE802.15.4 frame, only 32 bytes are available for the actual encoding of the schedule, as the remainder of the frame is occupied by the headers of the different standards.

The standards do not indicate the example approach to install the schedule. We consider three approaches, which we call "Single", "PATCH" and "Broadcast". Each is detailed below.

Updating a Single Field at a Time ("Single"). The central scheduler issues a separate confirmable CoAP POST message to write each field of each cell it installs into a node. This solution assumes that for any network node a route to reach it from the scheduler node is known by all nodes within the route. Figure 5 shows an example CoAP POST to set nodeAddress = 3 for the cell at slotOffset = 3 and channelOffset = 1 in the schedule.

The central scheduler needs to install a cell on both communicating neighbors. For each field, the scheduler issues a CoAP CON and receives CoAP ACK. Since each cell contains N_{field} fields, the scheduler sends N_{field} CoAP CON messages and receives N_{field} CoAP ACK acknowledgments. When sent to node u, each of these messages travels over $Depth(u)$ hops. Moreover, node u is involved in $Trans(u)$ transmissions and $\sum_{v \in Child(u)} Trans(v)$ receptions. This results in the total number of messages in (1).

$$numFrames_{Single} = 2 \cdot Nfield \cdot$$
$$\sum_{u \neq sink} Depth(u) \left(Trans(u) + \sum_{v \in Child(u)} Trans(v) \right) \tag{1}$$

Table 2 presents the number of messages required for first installing the schedule (see Fig. 2) computed for a network of 12 nodes, then updating this schedule taking into account the new node 13 (see Fig. 3). When the schedule is updated, node 2 has two cells where two fields are modified and two other cells with only one field changed. Hence in Table 2, the element at line 2 and column "Cells * $Nfield$" contains 2*2 + 2*1.

```
REQ:
    POST
    url:  coap://<ip>:5683/mg/6t/cellList/nodeAddress/?slotOffset=3&channelOffset=1
    body: 3
```

Fig. 5. A CoAP POST addressing the nodeAddress value of a cell.

Table 2. Total number of messages required for installing and updating the schedule with the Single method.

| Node | Install | | | Update | |
	Cells	$Nfield$	Messages	Cells * $Nfield$	Messages
2	8	4	64	2*2 + 2*1	8 + 4
3	9	4	72	3*1	6
4	7	4	56	3*1 + 2*4	6 + 16
5	2	4	32	2*2	16
6	1	4	16	1*1	4
7	1	4	16	1*1	4
8	1	4	16	1*1	4
9	1	4	16	1*1	4
10	2	4	32	1*1	4
11	1	4	16	0	0
12	1	4	16	1*2	8
13	-	-	-	1*4	16
Total	-	-	**352**	-	**100**

Sending a PATCH to Each Node ("PATCH"). The scheduler contacts each node once and transfers a list of cells that must be updated, encoded as a CoAP PATCH. This allows it to send only the differences between the current schedule and the new one. An example CoAP PATCH which modifies the nodeAddress and linkType fields of the cell with slotOffset = 1 and channelOffset = 2 is presented in Fig. 6.

When the payload of the PATCH is too long for a single frame, CoAP Block is used for application-layer fragmenting. $B(u)$ blocks are transmitted to node u, the scheduler receives an acknowledgment per block, as recommended in [11]. The resulting total number of messages is given in (2).

```
REQ:
    PATCH
    url:  coap://<ip>:5683/mg/6t/cellList
    body: [
            {
                "op":    "replace",
                "path":  "/nodeAddress?slotOffset=1&channelOffset=2",
                "value": 4
            },
            {
                "op":    "replace",
                "path":  "/linkType?slotOffset=1&channelOffset=2",
                "value": 1
            }
        ]
```

Fig. 6. A CoAP PATCH modifying the nodeAddress and linkType fields of the cell.

Table 3. Total number of messages required for installing and updating the schedules with the `PATCH` method.

Node	Install			Update		
	CBOR bytes	Blocks	Messages	CBOR bytes	Blocks	Messages
2	1049	33	66	397	13	26
3	918	29	58	208	7	14
4	918	29	58	533	17	34
5	263	9	36	269	9	36
6	132	5	20	67	3	12
7	132	5	20	70	3	12
8	132	5	20	67	3	12
9	132	5	20	67	3	12
10	263	9	36	70	3	12
11	132	5	20	0	0	0
12	132	5	20	136	5	20
13	-	—	—	132	5	20
Total	-	—	**374**	—	—	**210**

$$numFrames_{PATCH} = 2 \cdot \sum_{u \neq sink} B(u) \cdot Depth(u) \qquad (2)$$

Table 3 presents the number of messages required for installing and updating the schedule computed for our illustrative network (see Fig. 1). Detailed CoAP `PATCH` messages for installing the schedule can be found in [8].

Broadcasting the Complete Schedule ("Broadcast"). The scheduler broadcasts the complete schedule over CoAP. The schedule is represented as a CBOR-encoded [2] JSON document. Each node u filters the cells it is involved in as transmitter or receiver. An example of broadcast with the complete schedule is proposed in Fig. 7 where each tuple is [`slotOffset`, `channelOffset`, source node, destination node].

To ensure that each node correctly receives the schedule, the scheduler uses CoAP Observe to monitor the value of the schedule number on each node. Each time a new schedule is pushed to a node, the scheduler expects to receive a CoAP Observe notification, confirming the successful reception of the schedule update by the node.

At network initialization, the scheduler issues an CoAP Observe request on each of the nodes. Broadcasting the schedule requires B_{sched} blocks of 32 bytes to be broadcast into the network, resulting in $P \cdot B_{sched}$ frames, assuming ideal flooding. Each time a schedule is installed, the waves of CoAP Observe notifications account for $\sum_{u \neq sink} Depth(u)$ messages. The total number of frames to

```
REQ:
    POST
    url:  coap://<ip>:5683/mg/6t/schedule
    body: {
            "ScheduleNumber": "1",
            "Schedule":
                [
                    [0, 0,  4, 1],
                    [0, 1,  3, 1],
                    [0, 2,  5, 2],
                    [1, 0,  2, 1],
                    [1, 1, 10, 3],
                    [1, 2,  6, 4],
                    [2, 0,  4, 1],
                    [2, 1,  3, 1],
                    [2, 2,  2, 1],
                    [3, 0,  2, 1],
                    [3, 1, 11, 3],
                    [3, 2, 12, 4],
                    [4, 0,  4, 1],
                    [4, 1,  3, 1],
                    [4, 2,  9, 2],
                    [5, 0,  2, 1],
                    [5, 1, 10, 3],
                    [5, 2,  5, 4],
                    [6, 0,  4, 1],
                    [6, 1,  3, 1],
                    [6, 2,  8, 2],
                    [7, 0,  2, 1],
                    [7, 1,  7, 3],
                    [8, 0,  3, 1]
                ]
    }
```

Fig. 7. A CoaP POST broadcasting the complete schedule.

install a schedule is given in (3); δ_P is equal to 1 when at least one new node has joined the network since the last transmission CoAP Observe notification, 0 otherwise.

$$numFrames_{Broadcast} = P \cdot Bsched + \sum_{u \neq sink} Depth(u) + \delta_P \cdot P \qquad (3)$$

The length of the CBOR transcription of the JSON document describing our schedule (see Fig. 7) is 149 bytes. This CBOR transcription is divided into 5 fragments of 32 bytes, hence $Bsched = 5$. In our example topology, the number of parents broadcasting the schedule is $P = 4$ (node 1, 2, 3, 4) and the sum of depths of nodes is $\sum_{u \neq sink} Depth(u) = 19$. For this first schedule, as all nodes just joined the network $\delta_P = 1$. Finally, $msg(Broadcast) = 4 \times 5 + 19 + 1 \times 4 = 43$.

For the second schedule depicted in Fig. 3, the length of the CBOR transcription of the JSON document describing our schedule is 159 bytes. The detailed CoAP POST for the broadcast of this second schedule can be found in [8]. As in the previous case, this CBOR transcription is divided into 5 fragments of 32 bytes, hence $Bsched = 5$. The number of parents broadcasting the schedule is the same $P = 4$ but now the sum of depths of nodes is

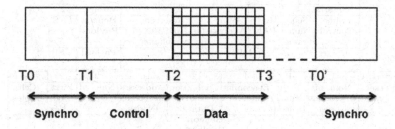

Fig. 8. Cycle provided by OCARI

$\sum_{u \neq sink} Depth(u) = 21$. Node 13 just joined the network, hence $\delta_P = 1$. Finally, $msg(Broadcast) = 4 \times 5 + 21 + 1 \times 4 = 45$.

2.3 OCARI, a Custom Protocol

An alternative is to use OCARI [1], a custom-built non-standards-based protocol. We compare a standards-based approach to it to provide a lower bound on the number of packets needed to install a schedule.

OCARI is compliant with IEEE802.15.4, but not with upper-layer protocols such as 6LoWPAN or CoAP. Each OCARI node is assigned a 2-byte identifier, unique in the network and given during its association.

OCARI schedules activities in the network in a cycle organized in four periods, as depicted in Fig. 8. First, OCARI synchronizes the network in a collision-free multi-hop way during the [T0,T1] period. All nodes periodically send beacons to maintain this synchronization. The period [T1,T2] is dedicated to control traffic used to collect network characteristics in order to compute a schedule for user data. Period [T2,T3] allows user data gathering. Finally, period [T3,T0'] is a sleep period, all nodes sleep to save energy.

A new schedule computation is kicked off when the topology or the application-needs change. When a new schedule is ready, it is broadcast into the network by piggy-backing the entire schedule into the beacons sent by all parent nodes. The schedule is a sequence of 7-byte cell tuples [slotOffset, channelOffset, SourceNode, DestinationNode], each describing a single cell in the schedule.

Figure 9 summarizes the format of an OCARI beacon. For the topology depicted in Fig. 1, the first schedule (see Fig. 2) to install is 24 tuples long, the second one (see Fig. 3) is 26 tuples long. The scheduler hence needs to transmit $24 \times 7 = 168$ bytes for the first schedule, $26 \times 7 = 182$ bytes for the second. Since in this typical case the maximum payload available for cell tuples in a beacon is 80 bytes, up to 11 cell tuples can be transported in a single beacon. If the schedule contains more than 11 cells, the full schedule is fragmented across different beacons. The entire schedule hence need 3 beacons to be transferred, or a total of $3 \times 4 = 12$ messages, since in our example $P = 4$ parent nodes.

Details about OCARI can be found in [1].

Octets: 2	1	2	2	4	variable	2
Frame Control	Seq. Num. Frame	Source PAN Id.	Source Address	Reserved *for IEEE802.15.4 compliance*	**Beacon Payload**	FCS
MHR					Payload	MFR

Octets: 1	1	1	4	2	variable	2	2	variable
Packet Type	Num. of Beaconing Nodes	Seq. Num. Beacon	Beaconing Interval	Contention Slot Duration	Addresses Beaconing Nodes	Num. of Cells	First Cell Index	Cell Tuples
Beacon Payload								

Fig. 9. Packet format when using the OCARI custom protocol.

3 Recommended Optimizations

Table 4 summarizes the total number of messages to install and update the schedule. The detailed computation of the number of messages can be found in [8]. It shows that a standard-based approach is less efficient than a custom-built protocol by a factor of 4 or more.

Simple changes to the standards allow them to be much more efficient. This section lists three simple optimizations which, when applied, yield a reduction in number of messages between 18 % and 74 % as depicted in Table 4.

3.1 Use Short MAC Addresses

In today's 6TiSCH standards, only 64-bit long MAC addresses are used. Implementing an association mechanism would enable a coordinator (typically the root of the network) to assign a 16-bit address unique in the network to each device. This association mechanism would save 12 bytes in each frame. The payload at the MAC layer can then be extended to 110 bytes and the maximum applicative payload could be equal to 73 bytes, allowing a 64-byte CoAP Block size.

Table 4. Total number of messages traveling in the network for installing and updating the schedule, with none, some or all optimizations.

Approach	Optimizations	Install schedule	Relative gain	Absolute gain	Update schedule	Relative gain	Absolute gain
Single	*None*	352	N.A	N.A	100	N.A	N.A.
PATCH	*None*	374	N.A	N.A	206	N.A	N.A.
	+ Short addresses (3.1)	206	45 %	45 %	104	50 %	50 %
	+ CellId (3.2)	136	34 %	64 %	72	31 %	65 %
	+ PATCH syntax (3.3)	98	28 %	74 %	58	19 %	72 %
Broadcast	*None*	43	N.A	N.A	45	N.A	N.A.
	+ Short addresses (3.1)	35	19 %	19 %	37	18 %	18 %
	+ CellId (3.2)	35	0 %	19 %	37	0 %	18 %
Custom	*None*	12	N.A	N.A	12	N.A	N.A

Table 5. Total number of messages required for installing and updating the schedules with the PATCH method.

	Without node 13			With node 13		
Node	CBOR bytes	Blocks	Messages	CBOR bytes	Blocks	Messages
2	1049	17	34	397	7	14
3	918	15	30	208	4	8
4	918	15	30	533	9	18
5	263	5	20	269	5	20
6	132	3	12	67	1	4
7	132	3	12	70	1	4
8	132	3	12	67	1	4
9	132	3	12	67	1	4
10	263	5	20	70	1	4
11	132	3	12	0	0	0
12	132	3	12	136	3	12
13	–	–	–	132	3	12
Total	–	–	**206**	–	-	**104**

Table 5 presents the number of messages required for installing and updating the schedule computed for our illustrative network (Fig. 1). We see a dramatic saving for this PATCH method compared to the previous results in Sect. 2.2.

With the method of broadcasting a schedule to all nodes in the network, the savings in term of the number of messages is also significant. In our example topology, we obtain $msg(Broadcast) = 4 \times Bsched + 19 + 1 \times 4 = 35$ messages with $Bsched = 3$ for the installing the schedule. For updating the schedule, we obtain $msg(Broadcast) = 4 \times Bsched + 21 + 1 \times 4 = 37$ messages with $Bsched = 3$.

Per Table 4, using short MAC addresses reduces the number of frames by 45 % (respectively 50 %) to Install (respectively Update) the schedule when using the PATCH approach, and 19 % (respectively 18 %) when using the Broadcast approach.

3.2 More Efficient CellId Representation

According to the IEEE802.15.4e standard [7], slotOffset and channelOffset are each encoded on 2 bytes. Since there are only 16 channels available for a IEEE802.15.4 radio operating at 2.4 GHz, only 4 bits are needed to encode the channelOffset. Moreover, since the slotframe length in an IEEE802.15.4e TSCH network is rarely longer than 1000 timeslots (10 s when using 10 ms timeslots), channelOffset can be encoded using 12 bits. A single 16-bit number called "CellId" can encode slotOffset and channelOffset, in which the 4 most significant bits represent the channelOffset and the remaining 12 bits the slotOffset.

Table 6. Total number of messages required for installing and updating the schedules with the PATCH method using a more efficient CellId Representation.

	Without node 13			With node 13		
Node	CBOR bytes	Blocks	Messages	CBOR bytes	Blocks	Messages
2	729	12	24	275	5	10
3	636	10	20	132	3	6
4	636	10	20	371	6	12
5	181	3	12	132	3	12
6	92	2	8	44	1	4
7	92	2	8	45	1	4
8	92	2	8	44	1	4
9	92	2	8	44	1	4
10	183	3	12	44	1	4
11	92	2	8	0	0	0
12	92	2	8	44	1	4
13	–	–	–	92	2	8
Total	–	–	**136**	–	-	**72**

```
REQ:
   POST
   url:  coap://<ip>:5683/mg/6t/schedule
   body: {
            "ScheduleNumber": "1",
            "Schedule":
            [
               [0, 4, 1],
               [1, 3, 1],
               [2, 5, 2],
               [16, 2, 1],
               [17, 10, 3],
               [18, 6, 4],
               [32, 4, 1],
               [33, 3, 1],
               [34, 2, 1],
               [48, 2, 1],
               [49, 11, 3],
               [50, 12, 4],
               [64, 4, 1],
               [65, 3, 1],
               [66, 9, 2],
               [80, 2, 1],
               [81, 10, 3],
               [82, 5, 4],
               [96, 4, 1],
               [97, 3, 1],
               [98, 8, 2],
               [112, 2, 1],
               [113, 7, 3],
               [128, 3, 1]
            ]
         }
```

Fig. 10. A CoAP POST broadcasting the first schedule using CellId.

```
REQ:
      PATCH
      url:  coap://<ip>:5683/mg/6t/cellList
      body: [
              {
                "o": "rpl",
                "p": "/nodeAddress?CellId=18",
                "v": 4
              },
              {
                "o": "rpl",
                "p": "/linkType?CellId=18",
                "v": 1
              }
           ]
```

Fig. 11. A CoAP PATCH with shorter operators.

Table 7. Total number of messages required for installing and updating the schedules with the PATCH method with shorter operators.

Node	Without node 13			With node 13		
	CBOR bytes	Blocks	Messages	CBOR bytes	Blocks	Messages
2	537	9	18	203	4	8
3	468	8	16	96	2	4
4	468	8	16	275	5	10
5	133	3	12	96	2	8
6	68	1	4	32	1	4
7	70	1	4	33	1	4
8	68	1	4	32	1	4
9	68	1	4	32	1	4
10	135	3	12	32	1	4
11	68	1	4	0	0	0
12	68	1	4	32	1	4
13	–	-	–	68	1	4
Total	–	-	**98**	–	-	**58**

Table 6 presents the number of messages required for installing and updating the schedule computed for our illustrative network (see Fig. 1). We notice that this simple coding mechanism brings a greater gain compared to the previous results in Sect. 2.2.

Figure 10 depicts the schedule description using the CellId coding in the context of the broadcasting method.

Per Table 4, using a CellId rather than a [slotOffset, channelOffset] tuple saves an additional 34 % (respectively 31 %) when installing (respectively updating) the schedule using the PATCH approach. Its impact is negligible when using the Broadcast approach.

3.3 Shorter PATCH Operators

CoAP PATCH [12], even if if were designed for constrained devices, still uses the multi-character operators "op", "path", "value", "replace". Reducing those to the shorter operators "o", "p", "v", "rpl" would be syntactically equivalent, but more efficient. An example resulting CoAP PATCH is described in Fig. 11.

Table 7 presents the number of messages required for installing and updating the schedule computed for our illustrative network (Fig. 1) with the recommended shorter operators.

The CBOR representation of the JSON document above is 26 % shorter than the CBOR representation of the same JSON document with longer operator string. According to Table 4, using shorter PATCH operators saves an additional 28 % (respectively 19 %) when installing (respectively updating) the schedule using the PATCH approach. Its impact is negligible when using the Broadcast approach.

4   Conclusion

This paper discusses the efficiency of installing and updating a communication schedule in a 6TiSCH network with a central scheduler. Not surprisingly, it shows that a set of (generic) standards-based protocols is less efficient than a custom-built protocol. It shows how simple optimizations to the standards can reduce the number of frames by up to 74 %.

It is, to the best of our knowledge, the first work to provide a system-wide analysis of the different protocols involved in a 6TiSCH network, and highlight the inefficiencies when "putting them all together". We believe this paper makes a strong contribution to the standardization activities at the IETF, in particular in the 6TiSCH, RPL and 6lo working group. It makes a strong (and quantified) case for adding an association step to the 6TiSCH protocol suite (which requires an additional protocol but reduces the overhead by up to 50 %), and reducing the length to the CoAP PATCH operators (a trivial change which saves up to 28 % overhead).

Our future work will focus on distributed scheduling in 6TiSCH networks. This paper is an example of the collaboration between the IETF and the academic world which the newly created Thing-to-Thing Research Group is fostering.

References

1. Al Agha, K., Chalhoub, G., Guitton, A., Livolant, E., Mahfoudh, S., Minet, P., Misson, M., Rahmé, J., Val, T., van den Bossche, A.: Cross-layering in an industrial wireless sensor network: case study of OCARI. J. Netw. 4(6), 411–420 (2009)
2. Bormann, C., Hoffman, P.: Concise Binary Object Representation (CBOR). IETF RFC7049, October 2013
3. Bormann, C., Shelby, Z.: Block-wise Transfers in CoAP. draft-ietf-core-block-18 [work-in-progress], September 2015

4. Dujovne, D., Watteyne, T., Vilajosana, X., Thubert, P.: 6TiSCH: deterministic IP-enabled industrial internet (of Things). IEEE Commun. Mag. **52**(12), December 2014
5. Hui, J., Thubert, P.: Compression Format for IPv6 Datagrams over IEEE 802.15.4-Based Networks. IETF RFC6282, September 2011
6. IEEE: IEEE Standard for Local and metropolitan area networks-Part 15.4: Low-Rate Wireless Personal Area Networks (LR-WPANs). IEEE Std 802.15.4-2011 (Revision of IEEE Std 802.15.4-2006), September 2011
7. IEEE: IEEE Standard for Local and metropolitan area networks-Part 15.4: Low-Rate Wireless Personal Area Networks (LR-WPANs) Amendment 1: MAC Sublayer. IEEE Std 802.15.4e-2012 (Amendment to IEEE Std 802.15.4-2011), April 2012
8. Livolant, E., Minet, P., Watteyne, T.: The Cost of Installing a New Communication Schedule in a 6TiSCH Low-Power Wireless Network using CoAP. Technical report RR-8817, Inria, 10 December 2015
9. Shelby, Z., Hartke, K., Bormann, C.: The Constrained Application Protocol (CoAP). IETF RFC7252, June 2014
10. Soua, R., Minet, P., Livolant, E.: MODESA: an optimized multichannel slot assignment for raw data convergecast in wireless sensor networks. In: International Performance Computing and Communications Conference (IPCCC), pp. 91–100. IEEE, Austin (2012)
11. Van der Stok, P., Andy, B., Schönwälder, J., Sehgal, A.: CoAP Management Interface. draft-vanderstok-core-comi-07 [work-in-progress], July 2015
12. Van der Stok, P., Sehgal, A.: Patch Method for Constrained Application Protocol (CoAP). draft-vanderstok-core-patch-02 [work-in-progress], October 2015
13. Vilajosana, X., Pister, K.: Minimal 6TiSCH configuration. draft-ietf-6tisch-minimal-12, September 2015
14. Winter, T., Thubert, P., Brandt, A., Hui, J., Kelsey, R., Levis, P., Pister, K., Struik, R., Vasseur, J., Alexander, R.: RPL: IPv6 Routing Protocol for Low-Power and Lossy Networks. IETF RFC6550, March 2012

Resource Allocation in Visible Light Communication Networks: NOMA vs OFDMA Transmission Techniques

Eirini Eleni Tsiropoulou[1]([✉]), Iakovos Gialagkolidis[2],
Panagiotis Vamvakas[2], and Symeon Papavassiliou[2]

[1] Erik Jonsson School of Engineering and Computer Science,
University of Texas at Dallas, 800 W Campbell Rd,
Richardson, TX 75080, USA
eetsirop@utdallas.edu
[2] School of Electrical and Computer Engineering Institute
of Communication and Computer Systems (ICCS), National Technical
University of Athens (NTUA), 9 Iroon Polytechniou Street Zografou,
15773 Athens, Greece
{ell0064, pvamvaka}@central.ntua.gr,
papavass@mail.ntua.gr

Abstract. Energy-efficiency, high transmission data rates and Quality of Service (QoS) awareness are the major challenges for resource management in the uplink of Visible Light Communication Personal Area Networks (VPANs). This paper investigates the problem of Optical Access Point (OAP) selection and resource allocation in the uplink of VPANs under two different transmission techniques, namely Orthogonal Frequency Division Multiple Access (OFDMA) and Non-Orthogonal Multiple Access (NOMA). Each user is associated with a generic utility function, which represents his perceived satisfaction with respect to the overall resource allocation problem. OAP selection adopts Maximum Gain Selection (MGS) policy, i.e. users select an OAP to connect to based on the highest path gain. A distributed resource allocation problem in VPANs is formulated and solved as an optimization problem. Following this analysis, a decentralized iterative and low-complexity algorithm for determining OAP selection and resource allocation is proposed, while the overall approach's efficiency is illustrated via modeling and simulation, highlighting and assessing the advantages and drawbacks of each adopted transmission technique, i.e. OFDMA and NOMA.

Keywords: Users association · Resource allocation · Visible light communication · OFDMA · NOMA

1 Introduction

With the growing demand for high data rate, support of multiple services with various Quality of Service (QoS) requirements, as well as the advent of numerous wireless devices, the spectrum demand is increasing rapidly. Moreover, energy-efficient

© Springer International Publishing Switzerland 2016
N. Mitton et al. (Eds.): ADHOC-NOW 2016, LNCS 9724, pp. 32–46, 2016.
DOI: 10.1007/978-3-319-40509-4_3

solutions are required to extend mobile users' battery life and support the vision of green wireless networking. Towards this direction Visible Light Communication (VLC) is acknowledged as a promising wireless technology that solves the bandwidth scarcity problem and enables decreased mobile users' transmission power [1]. VLC can ensure ubiquitous ultrahigh speed wireless communication in any indoor environment (e.g. aircraft, hospital, museum etc.) and is characterized by numerous advantages, such as high data rate, mitigated interference, harmlessness to human health, strong security and unlicensed frequency band usage.

Prior studies have focused mainly on physical (PHY) and medium access control (MAC) optimal design, without considering mobile users' QoS provisioning [2, 3]. Therefore, in the last years the interest of research community has turned towards optimal resource allocation in order to fulfill users' QoS prerequisites. Two main transmission technologies are adopted in VPANs, especially within the emerging era of 5G networks: Orthogonal Frequency Division Multiple Access (OFDMA) and Non-Orthogonal Multiple Access (NOMA).

In the recent literature, considerable research efforts have been devoted to the problem of optimal resource allocation in the downlink of VPANs, where OFDMA technology is adopted. In [4] the authors propose an interference bounded opportunistic channel allocation for VPANs towards maximizing the sum rate capacity in the downlink and in parallel allowing maximum number of users with a minimum QoS prerequisite. In [5] an intercell interference coordination heuristic approach is proposed towards realizing subcarrier reuse between different transmitters, as well as power redistribution between different subcarriers. In [6], a centralized joint power and bandwidth allocation for downlink transmission is introduced to maximize system's overall capacity. The authors formulate a user's weighted sum rate maximization problem, which is solved in a centralized manner and concludes to a power and bandwidth allocation. In [7], the authors propose a joint scheduling and optimal resource allocation, via formulating a cross-layer optimization problem towards maximizing the system average throughput under the SINR constraints and solving it in a centralized manner. An interference aware time-frequency slots allocation using busy burst signaling for transmitting data is proposed in [8], where the assignment of time-frequency slots is adjusted dynamically depending on the location of an active user in the cell. In [9], the authors introduce a novel framework towards supporting rapid link recovery and visibility that minimize system's performance degradation. A resource allocation scheme that selects non-overlapping channels having best signal-to-interference plus noise ratio and acting indirectly on the interference levels has been proposed in [10], where the visible light multi-color logical channels are allocated in order to minimize the co-channel interference.

Considering the proposed resource allocation problems in VPANs adopting NOMA technology, the performed research is still primitive, due to the fact that NOMA technology has been recently proposed in the literature as a promising candidate for 5G wireless networks. In [11], a channel-dependent power allocation (called gain ratio power allocation) is proposed, considering users' channel conditions to ensure efficient and fair power allocation and targeting at enhancing the achievable throughput in high-rate VLC downlink networks. In [12], the authors propose theoretical expressions on the system performance, considering two different scenarios: (i) ensuring guaranteed quality of service (QoS) and providing system coverage probability and (ii) multiple users request opportunistic best-effort service and calculating system ergodic sum rate in a closed form. However, all the above approaches are centralized and mainly network-centric and their goal is to guarantee system's welfare and efficiency, e.g. maximization of system's sum rate, system's overall interference mitigation, service provider's profit etc.

In this paper, we address the problem of OAP selection and resource allocation, with respect to bandwidth and power, in the uplink of VPANs under two different enabling transmission techniques, i.e. OFDMA and NOMA. Our approach leads towards the creation of a framework where the decision making process lie at the mobile user, irrespectively of the adopted transmission technique, i.e. OFDMA or NOMA, and is well aligned with the current efforts for the realization of mobile user self-optimization functionalities. To the best of our knowledge, this is the first time in the recent literature that the problem of optimal resource allocation in the uplink of VPANs is studied. The main recent research efforts have focused on the problem of optimal resource allocation (e.g. power, bandwidth and subcarriers) in the downlink, while the corresponding resource allocation problem in the uplink has not yet been studied. The other fundamental contribution of this work is the detailed presented comparison among OFDMA and NOMA transmission techniques, illustrating the advantages and drawbacks of adopting OFDMA or NOMA as the transmission technique in the uplink of VPANs. Detailed numerical results are provided towards better identifying the advantages of each transmission technique with respect to the resource allocation problem.

The rest of this paper is organized as follows. In Sect. 2, VPAN's system model is introduced, while the OFDMA and NOMA transmission techniques are briefly presented. In Sect. 3, users' various QoS requirements are mapped to a generic utility function, and in Sect. 4 the OAP selection method is presented. In Sect. 5, the resource allocation problem is formulated and solved, both in NOMA and OFDMA VPANs. In Sect. 6, a low-complexity OAP selection and resource allocation algorithm is presented. Finally, in Sect. 7, detailed numerical results that illustrate the operation and features of the proposed distributed framework are presented, while Sect. 8 concludes the paper.

2 System Model and Transmission Techniques

2.1 VPAN System Model

We consider the uplink of a multi-cell VPAN, consisting of T OAPs and U mobile users, where their corresponding sets are denoted as $\mathrm{T} = \{t = 1, 2, \ldots, T\}$ and $\mathrm{U} = \{u = 1, 2, \ldots, U\}$, respectively. Each OAP serves a total number of mobile users N_t. A spectrum of total bandwidth W is devoted per each OAP (and its corresponding cell) and is available for transmissions from the OAP to the mobile users and vice versa. Each mobile user $u \in \mathrm{U}$ communicates directly with an OAP $t \in \mathrm{T}$ via a communication link $l = \{u, t\}$. User's uplink transmission power $P_{u,t}$ is upper and lower bounded, i.e. $0 \prec P_{u,t} \leq P_{u,t}^{Max}$, due to user's physical and technical limitations.

Let $H_{u,t}$ be the gain of optical communication link l between user $u \in U$ to OAP $t \in T$. The line-of-sight (LOS) DC gain between mobile user $u \in U$ and OAP $t \in T$ is given by [13]:

$$H_{u,t} = \begin{cases} \dfrac{(m+1)A}{2\pi d^2} \cos^m(\varphi)T_s(\psi)g(\psi)\cos(\psi), & 0 \leq \psi \leq \psi_c \\ 0, & otherwise \end{cases} \tag{1}$$

where A denotes the photodetector area, φ the angle of irradiance, $T_s(\psi)$ the signal transmission coefficient of an optical filter, ψ the angle of incidence and m is the order of the Lambertian emission, which is given by

$$m = -\frac{\ln 2}{\ln\left(\cos\phi_{1/2}\right)} \tag{2}$$

where $\phi_{1/2}$ is the transmitter semi-angle at half power. Moreover, $g(\psi)$ denotes the channel gain of an optical concentrator and is given by

$$g(\psi) = \begin{cases} \dfrac{\alpha^2}{\sin^2(\psi_c)}, & 0 \leq \psi \leq \psi_c \\ 0, & otherwise \end{cases} \tag{3}$$

where α denotes the refractive index of the optical concentrator, ψ_c user's field of view (FOV) and d the distance between the OAP and the user. Throughout the analysis in this paper we make the assumption that the proposed OAP selection and resource allocation is based on perfect knowledge of path gain information both at the user and at the OAP. The corresponding optical communication link geometry, as well as the overall topology of the multi-cell VPAN are presented in Fig. 1.

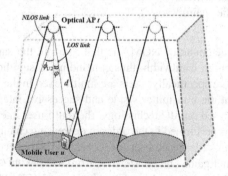

Fig. 1. Visible light Communication Personal Area Network (VPAN) topology.

2.2   Transmission Techniques in VPAN

In VPAN, two different transmission technologies have been considered, i.e. OFDMA and NOMA. In the following, we provide some basic background information considering each transmission technology.

A. OFDMA

In OFDMA, the bandwidth of each OAP is divided into subcarriers, which are organized into resource blocks (RBs). The RBs are allocated to the users. Within the same OAP, each RB is uniquely reserved by one user of the same OAP. The same RBs can be reused among different OAPs, which in turn can cause co-channel interference. Let $R = \{r = 1, 2, \ldots, R\}$ be the set of available RBs. In the proposed system model, we assume the flat-fading model, i.e. mobile user's channel gain is not differentiated per RB.

The signal-to-interference-plus-noise ratio (SINR), $\gamma_{u,t}^{(r)}$ of user $u \in U$ served by the OAP $t \in T$ at RB r can be expressed as follows:

$$\gamma_{u,t}^{(r)} = \frac{R_{PD} H_{u,t} P_{u,t}^{(r)}}{\sum\limits_{u'=1, u' \neq u, t \in \mathrm{T}}^{U} H_{u',t} P_{u',t}^{(r)} + \xi} \tag{4}$$

where R_{PD} denotes the responsitivity of the photodiode, $H_{u,t}$ the line-of-sight (LOS) path gain between user $u \in \mathrm{U}$ and OAP $t \in T$, $P_{u,t}^{(r)}$ user's $u \in U$ uplink transmission power to the OAP $t \in T$ for RB r and ξ is the cumulative noise power, which is given by

$$\xi = 2q R_{PD} I_{amb} B_{noise} + \frac{4 K_B T B}{R_F} \tag{5}$$

where $q = 1.6 \cdot 10^{-19} C$, I_{amb} denotes the ambient light intensity, B_{noise} the equivalent noise bandwidth, K_B Boltzmann's constant, T the absolute temperature and R_F the transimpedance amplifier gain.

B. NOMA

The fundamental characteristic of NOMA is that users in the same OAP can simultaneously exploit the entire bandwidth, thus concluding to significant enhancement in the achievable rate. More specifically, users are multiplexed in the power domain using superposition coding at the transmitter's side and successive interference cancellation (SIC) at the receiver. Based on SIC technology, the multi-user interference is mitigated and signals' decoding is performed in the order of decreasing channel gain in the uplink of VPAN. Based on this order, each user $u \in \mathrm{U}$ can correctly decode the signals of all users with better channel gain. The interference from users with worse channel gain is not mitigated, thus it is treated as noise. Therefore, user's SINR in the NOMA technology can be calculated as follows.

$$\gamma_{u,t} = \frac{R_{PD}H_{u,t}P_{u,t}}{\sum\limits_{u' > u, t \in T}^{U} H_{u',t}P_{u',t} + \xi} \tag{6}$$

where u' is the first user in the OAP having worse channel gain compared to u.

3   User's QoS Requirements and Utility Function

The concept of utility function is adopted from the field of microeconomics and reflects user's degree of satisfaction as a result of his actions. The utility function that describes the set of users' preferences rules is not unique in principle, thus we first verify users' preference relations that are specific to our problem and then propose a utility function that satisfies this structure. A mobile user should transmit with low uplink transmission power in order to extend his battery lifetime and at the same time cause less inter-ference in the multi-cell VPAN environment. Furthermore, mobile users' satisfaction increases by achieving high uplink transmission data rate, which is appropriately reflected by the number of reliably transmitted bits to the OAP. Based on these observations, each user adopts a utility function, which represents his degree of sat-isfaction in relation to the expected tradeoff between the number of information bits that are successfully transmitted to the OAP and user's corresponding power con-sumption. Therefore in alignment with the above claims the utility functions assumed here for the OFDMA and NOMA cases are as follows:

$$OFDMA : U_{u,t}^{(r)} = \frac{W \cdot f_u\left(\gamma_{u,t}^{(r)}\right)}{N_t \cdot P_{u,t}^{(r)}}$$

$$\tag{7}$$

$$NOMA : U_{u,t} = \frac{W \cdot f_u\left(\gamma_{u,t}\right)}{P_{u,t}}$$

where N_t denotes the number of users served by the OAP t, W is OAP's $t \in T$ band-width and $f_u(\cdot)$ is the efficiency function. The efficiency function represents the probability of a successful packet transmission for user $u \in U$ and is an increasing, continuous, twice differentiable sigmoidal function of his SINR $\gamma_{u,t}^{(r)}$ (resp. $\gamma_{u,t}$). A user's function for the probability of a successful packet transmission depends on the transmission schemes used, i.e. modulation and coding schemes [14].

4   Optical Access Point Selection

The topology that is considered in this paper is a multicell VPAN. As presented in Fig. 1, a mobile user can reside within the coverage area of multiple OAPs. Therefore, a sophisticated OAP selection mechanism should be proposed towards mitigating the overall interference in the multicell VPAN. The proposed OAP selection methodology in this paper is based on Maximum Gain Selection (MGS) policy. According to the MGS policy, each user residing in the multi-cell VPAN selects the OAP to which he will be associated based on the highest path gain $H_{u,t}$ between himself, i.e. $u \in U$, and all available OAPs $t \in T$ that he is able to connect to. This method provides a near-optimum solution achieving multi-user, as well as multiple OAPs diversity and channel gain diversity. However, it should be noted that MGS policy is neither necessary nor sufficient for optimality. More specifically, we create the matrix $\boldsymbol{H}$, which is structured based on the line-of-sight (LOS) channel gain $H_{u,t}$ between each user $u \in U$ and OAP $t \in T$.

$$
\mathrm{H}(u,t) = \begin{bmatrix} H_{1,1} & H_{1,2} & \cdots & H_{1,T} \\ H_{2,1} & H_{2,2} & \cdots & \cdot \\ \cdot & \cdot & \cdots & \cdot \\ \cdot & \cdot & \cdots & \cdot \\ H_{U,1} & H_{U,2} & \cdots & H_{U,T} \end{bmatrix} \tag{8}
$$

According to users' path gain matrix $\boldsymbol{H}$, each user selects the OAP to connect to and communicate with based on MGS policy as follows:

$$
l^* = \{u^*, t^*\} = \arg \max_{t \in T} \boldsymbol{H}(u,t) \tag{9}
$$

It should be noted that after the OAP selection process, each OAP $t \in T$ knows the number of users N_t that are connected to its corresponding cell. It is obvious that the number of users N_t residing in each OAP's cell is a time-varying parameter due to users' mobility. Moreover, users' path gains $H_{u,t}$ are also time-varying variables, thus the problem of OAP selection should be solved periodically per time-slot. In this paper, we assume static users' positions within the multi-cell indoor environment.

5 Resource Allocation in the Uplink of VPAN

5.1 Resource Blocks Association in OFDMA-Based VPANs

In an OFDMA VLC wireless network, two main resources should be allocated to the users considering the uplink scenario: (i) resource blocks and (ii) user's uplink transmission power. Based on the number of RBs, i.e. R, and considering non-reusability of the RBs within the same OAP, the maximum capacity of the OAP in terms of number of users is fixed, i.e. R. In this paper, given that the flat-fading model is adopted, the RBs are not differentiated considering user's channel gain per RB, thus a simplified RBs allocation is adopted.

In the case that $U < R$, we ensure that each user $u \in U$ occupies at least one RB in the OAP $t \in T$ and the remaining RBs are allocated to the users based on their order with respect to the channel gain $H_{u,t}$. In the case of $U > R$, users are ordered based on their channel gain and the first R users occupy the R RBs, while the rest of the users, i.e. $U - R$, cannot be served by the OAP and they are rejected. Finally, it should be noted that in the case of selective-fading channel model, an RBs allocation based on MGS policy could be also adopted. Therefore, the overall framework proposed in this paper is not applicable only in the flat-fading channel model.

5.2 Power Allocation

In this section, we aim to formulate users' transmission power allocation in the uplink of VPANs either adopting OFDMA or NOMA transmission technique. More specifically, as the system evolves, at the beginning of each time slot a user's Non-cooperative OAP selection and Resource Allocation algorithm (NOAPRA algorithm) is responsible for determining an optimal uplink transmission power $P_{u,t}^{(r)}$ for the case of OFDMA (resp. $P_{u,t}$ for the case of NOMA). User's optimal uplink transmission power maximizes his overall perceived satisfaction, which is appropriately represented via the corresponding values of the utility function, as it is given by Eq. (7) considering either OFDMA or NOMA transmission technique. Towards supporting mobile user's autonomicity, as well as to consider user's selfish behavior, we propose a distributed approach.

A. Problem Formulation

As mentioned before, user's uplink transmission power is upper bounded, due to some physical and technical limitations. Therefore, the feasible set of user's uplink transmission power is denoted as $A_u = \left(0, P_{u,t}^{Max}\right]$. Let $R_{u,t}$ denote the number of RBs allocated to user $u \in U$ residing in the OAP $t \in T$. In the power allocation problem, users have a selfish behavior and their goal is to maximize their utility considering the occupied resources.

Therefore, the power allocation problem is formulated as a distributed maximization problem as follows:

$$OFDMA: \quad \max_{P_{u,t}^{(r)}} U_{u,t}^{(r)}\left(P_{u,t}^{(r)}, P_{-u,t}^{(r)}\right) \qquad NOMA: \quad \max_{P_{u,t}} U_{u,t}\left(P_{u,t}, P_{-u,t}\right)$$

$$s.t. \quad 0 \quad P_{u,t}^{(r)} \leq \frac{P_{u,t}^{Max}}{R_{u,t}} \qquad\qquad s.t. \quad 0 \prec P_{u,t} \leq P_{u,t}^{Max} \tag{10}$$

The distinctive novelty of the above proposed power control problem, which was formulated as a distributed optimization problem, lies in the fact that this is the first approach in the recent literature, where the problem of users' uplink transmission power allocation in VPANs towards supporting QoS provisioning is addressed.

B. Solution

Towards solving the proposed optimization problem in Eq. (10), an optimization method is followed. Let $P^{(r)*} = \left[P_{1,t}^{(r)*}, P_{2,t}^{(r)*}, \ldots, P_{u,t}^{(r)*}, \ldots, P_{U,t}^{(r)*}\right]$ for the case of OFDMA (resp. $P^* = \left[P_{1,t}^*, P_{2,t}^*, \ldots, P_{u,t}^*, \ldots, P_{U,t}^*\right]$ for the case of NOMA) denote the optimal solution of the optimization problem in Eq. (10). Considering the first order derivative of $U_{u,t}^{(r)}$ (resp. $U_{u,t}$) with respect to $P_{u,t}^{(r)}$ (resp. $P_{u,t}$) we have the following expression:

$$\frac{\partial U_{u,t}^{(r)}}{\partial P_{u,t}^{(r)}} = 0 \Leftrightarrow \frac{\partial f_u\left(\gamma_{u,t}^r\right)}{\partial \gamma_{u,t}^{(r)}} \cdot \gamma_{u,t}^r - f_u\left(\gamma_{u,t}^r\right) = 0$$

$$\left(resp. \ \frac{\partial U_{u,t}}{\partial P_{u,t}} = 0 \Leftrightarrow \frac{\partial f_u\left(\gamma_{u,t}\right)}{\partial \gamma_{u,t}} \cdot \gamma_{u,t} - f_u\left(\gamma_{u,t}\right) = 0 \right)$$

(11)

where the equation $\frac{\partial f_u\left(\gamma_{u,t}^{(r)}\right)}{\partial \gamma_{u,t}^{(r)}} \cdot \gamma_{u,t}^{(r)} - f_u\left(\gamma_{u,t}^{(r)}\right) = 0$ (resp

$\frac{\partial U_{u,t}}{\partial P_{u,t}} = 0 \Leftrightarrow \frac{\partial f_u\left(\gamma_{u,t}\right)}{\partial \gamma_{u,t}} \cdot \gamma_{u,t} - f_u\left(\gamma_{u,t}\right) = 0$) has unique positive solution $\gamma_{u,t}^{(r)*}$ (resp. $\gamma_{u,t}^*$),

due to the sigmoidal form of $f_u(\cdot)$ with respect to $\gamma_{u,t}^{(r)}$ (resp. $\gamma_{u,t}$). Furthermore, the SINR is an one-to-one function with respect to user's uplink transmission power, thus resulting to a unique solution, as follows:

$$OFDMA : P_{u,t}^{(r)*} = min \left\{ \frac{\gamma_{u,t}^{(r)*} \cdot \left(\sum_{\substack{u'=1 \\ u' \neq u \\ t \in T}}^{U} H_{u',t} P_{u',t}^{(r)} + \xi \right)}{R_{PD} H_{u,t}}, \frac{P_{u,t}^{Max}}{R_u} \right\}$$

$$NOMA : P_{u,t}^* = min \left\{ \frac{\gamma_{u,t}^* \cdot \left(\sum_{\substack{u' > u \\ t \in T}}^{U} H_{u',t} P_{u',t} + \xi \right)}{R_{PD} H_{u,t}}, P_{u,t}^{Max} \right\}$$

(12)

6 Non-Cooperative OAP Selection and Resource Allocation Algorithm (NOAPRA Algorithm)

In this section, we propose a decentralized, iterative and low-complexity algorithm, which determines users' OAP selection and resource allocation in a distributed manner. The proposed algorithm is called NOAPRA algorithm and is divided into two basic parts. In the Optical Access Point Selection Part, each user selects the OAP to connect to based on the Maximum Gain Selection (MGS) policy, as discussed in Sect. 4. Additionally, in the Resource Allocation Part the RBs are allocated to the users in the OFDMA module and user's uplink transmission power is determined in accordance to (12) in a distributed manner. The proposed procedure in the second part of the algorithm is repeated iteratively till the algorithm converges to its unique solution. The above described parts of the overall NOAPRA algorithm are as follows.

NOAPRA Algorithm
Optical Access Point Selection Part
Step 1: At the beginning of each time slot, given that we assume that each user u, $u \in U$, residing in the multicell environment has perfect knowledge of path gain information, create the matrix $H(u,t)$, as presented in Eq. (8). Set $k = 1$ and $U^{(0)} = \{1, 2, \ldots, U\}$.

Step 2: Each user u^* selects to connect to the OAP t^* via creating the communication link $l^* = \{u^*, t^*\}$ based on the highest path gain $H_{u,t}$ (MGS policy), as follows:

$$l^* = \{u^*, t^*\} = \arg\max_{u^* \in U, t^* \in T} H(u,t)$$

Step 3: Set $k := k + 1$, delete user u^* in the set of users, i.e. $U^{(k+1)} = U^{(k)} - \{u^*\}$

Step 4: If $U^{(k+1)} = \emptyset$ then stop. Otherwise go to step 2.

Resource Allocation Part
RBs Allocation (only for OFDMA transmission technique)
Step 1: Based on the previous part of the algorithm, each OAP $t \in T$ is aware of the number of users residing in it, i.e. U.

Step 2: If $R = U$ allocate one RB $r \in R$ to each user $u \in U$, else if $R < U$, allocate at least one RB $r \in R$ to each user $u \in U$, sort the users based on their channel gain $H_{u,t}$ and allocate the remaining RBs, i.e. R-U, to the users based on the order of their channel gain, else if $R > U$, sort the users based on their channel gain $H_{u,t}$ and allocate the RBs R to the first R users with the best channel gain conditions, while the rest of the users are rejected from the OAP $t \in T$.

Users' Uplink Transmission Power Allocation
Step 1: Each user, u, $u \in U$ has already decided the OAP that he is connected to and initially he transmits with a randomly selected feasible uplink transmission power, i.e.
$0 \prec P_{u,t}^{*(r)(0)} \prec \frac{P_{u,t}^{Max}}{R_{u,t}}$ (resp. $0 \prec P_{u,t}^{*(0)} \prec P_{u,t}^{Max}$). Set $k:=0$ and hence $P_{u,t}^{*(r)(0)}$ (resp. $P_{u,t}^{*(0)}$), $u \in U$ and $t \in T$.

Step 2: Given that the single central controller of the OAPs collects the information of the overall interference within the multi-cell environment, each OAP announces this

information, i.e. $\sum\limits_{\substack{u=1 \\ t \in T}}^{U} H_{u,t}P_{u,t}^{(r)}$ (resp. $\sum\limits_{\substack{u=1 \\ t \in T}}^{U} H_{u,t}P_{u,t}$) to all users residing within its

coverage area, via broadcasting. Each user computes his sensed interference

$\sum\limits_{\substack{u'=1 \\ u' \neq u \\ t \in T}}^{U} H_{u',t}P_{u',t}^{(r)}$ (resp. $\sum\limits_{\substack{u'=1 \\ u' > u \\ t \in T}}^{U} H_{u',t}P_{u',t}$).

Step 3: Set $k := k + 1$. Each user updates his uplink transmission power, i.e. $P_{u,t}^{*(r)(k)}$ (resp. $P_{u,t}^{*(k)}$), $u \in U$ and $t \in T$, based on Eq. (12).

Step 4: If $\left| P_{u,t}^{*(r)(k+1)} - P_{u,t}^{*(r)(k)} \right| \leq \varepsilon$ (resp. $\left| P_{u,t}^{*(k+1)} - P_{u,t}^{*(k)} \right| \leq \varepsilon$), e.g. $\varepsilon = 10^{-5}$ then stop. Otherwise go to step 2

7 Numerical Results

In this section we provide some indicative numerical results illustrating the operation and features of the proposed framework and the NOAPRA algorithm. Throughout our study, we consider a VPAN, where $T = 8$ OAPs are established within a room of size $10\ m \times 6\ m \times 3\ m$ ($W \times L \times H$) and U users are distributed within the room. In the OFDMA transmission technique, the number of RBs is set to $R = 7$. The responsitivity of the photodiode is set to $R_{PD} = 0.63\ A/W$, users' maximum feasible uplink transmission power is $P_{u,t}^{Max} = 1\ W$ and the OAP's bandwidth is $W = 20\ MHz$. The cumulative noise power is $\xi = 0.03 \cdot 10^{-18}\ W$, the photodetector area is $A = 3 \cdot 10^{-6} m^2$, the transmitter semi-angle at half power $\phi_{1/2} = \Pi/6\ rad$, user's field of view (FOV) $\psi_c = \pi/3$ rad, the signal transmission coefficient of an optical filter $T_s(\psi) = 1$, the order of the Lambertian emission $m = 1$ and the refractive index of the optical concentrator is $\alpha = 1.5$. The adopted efficiency function is $f_u\left(\gamma_{u,t}^{(r)}\right) = \left(1 - e^{-A \cdot \gamma_{u,t}^{(r)}}\right)^M$ (resp. $f_u\left(\gamma_{u,t}\right) = \left(1 - e^{-A \cdot \gamma_{u,t}}\right)^M$), where $A = 0.2$ and $M = 1.61$.

In the following, we consider that $U = 56$ users reside within the room under two different distributions: (i) uniform distribution, i.e. 7 users reside per OAP (which is the OAP's maximum capacity in terms of users for the OFDMA transmission technique due to the fact that the number of RBs is $R = 7$) and (ii) non-uniform (unbalanced) distribution, i.e. some OAPs are overloaded in terms of users while some others are less congested, as presented in Fig. 2(a) and (b), respectively.

In Fig. 3(a) and (b), users' average uplink transmission power per OAP is presented for both users' distributions. Similarly corresponding users' average uplink transmission rate per OAP is presented in Fig. 4(a) and (b), respectively. Considering users' uniform distribution, it is observed that for the OFDMA transmission technique, users' average uplink transmission power is lower compared to the corresponding values in the unbalanced users' distribution. This observation stems from the fact that in the latter case more users are concentrated in a specific physical area of the room, thus users from different OAPs who share the same RB will cause extremely high interference to their neighbors, resulting in increasing uplink transmission power. On the other hand, considering the NOMA transmission technique, users' average transmission power also increases in the unbalanced users' distribution compared to the uniform one, however users' average transmission power's increase is smoother compared to the aforementioned increase in the OFDMA transmission technique. This observation holds true due to the successive interference cancellation (SIC) technique, which is an embedded module at the receivers in NOMA. Based on SIC, even if users' concentration in a specific physical area is increased, i.e. unbalanced users' distribution, the receiver is

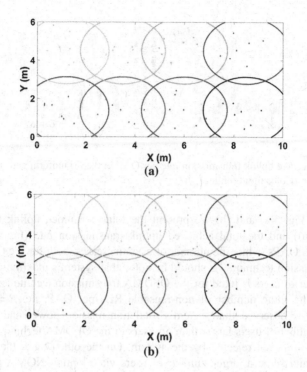

Fig. 2. (a) Uniform and (b) unbalanced users' distribution

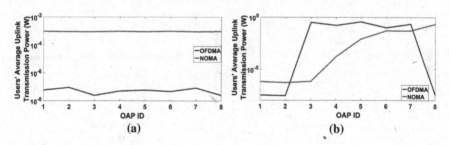

Fig. 3. Users' average uplink transmission power per OAP in the (a) uniform and (b) unbalanced users' distribution (Color figure online)

able to cancel part of the increased caused interference, fact that does not hold true in the OFDMA transmission technique.

On the other hand, considering users' average uplink transmission rate per OAP, it is observed that both for the uniform (Fig. 4(a)) and the unbalanced (Fig. 4(b)) users' distribution, NOMA transmission technique results in higher users' average uplink transmission rate compared to OFDMA transmission technique, due to the fact that in the former transmission technique, i.e. NOMA, users exploit the whole bandwidth towards transmitting.

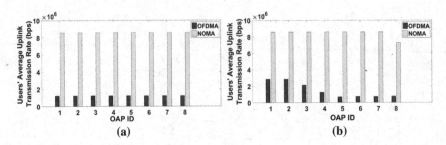

Fig. 4. Users' average uplink transmission rate per OAP in the (a) uniform and (b) unbalanced users' distribution (Color figure online)

Finally, in Fig. 5(a) and (b) we present the total consumed uplink transmission power (Fig. 5(a)) and the total achieved uplink transmission rate (Fig. 5(b)) in the system of $T = 8$ OAPs, while the number of users increases, both for the OFDMA and NOMA transmission technique. It should be noted that system's maximum capacity in terms of number of users is fixed for the OFDMA transmission technique, i.e. $U = 56$ users, due to the finite number of non-reusable RBs per OAP, i.e. $R = 7$ RBs per OAP. Thus, it is observed that system's total transmission power and rate remain constant for number of users larger than 56 users in the OFDMA technique, while the supplementary users are rejected by the system. On the other hand, the system can accommodate and serve a larger number of users via adopting NOMA transmission technique.

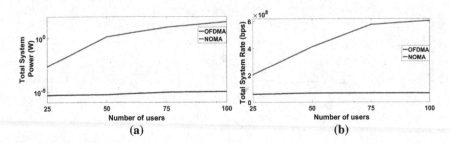

Fig. 5. System's total uplink transmission (a) power and (b) rate versus the increasing number of users (Color figure online)

8   Conclusions and Future Work

In this paper, the problem of users' association to OAPs and resource allocation in the uplink of VPANs was examined under two different transmission techniques, i.e. OFDMA and NOMA. Maximum Gain Selection (MGS) policy is used for treating the users' to OAPs association problem, where users select an OAP to connect to based on the highest path gain. Furthermore, in order to tackle the resource allocation problem in a unified and formalized manner, a generic utility function was adopted by each user, reflecting each user's perceived satisfaction from resources' allocation. The resource allocation problem was formulated and solved as a distributed optimization problem. An iterative and low-complexity algorithm was proposed towards associating users to the OAPs and allocating the resources to them. Detailed numerical results were presented illustrating the benefits of the proposed framework and comparing the impact of OFDMA versus NOMA transmission technique in the resource allocation process.

It should be noted that the overall resource allocation problem was addressed in this problem under the assumption that all users request the same type of service. However, considering various types of requested services, which can be formulated by different utility functions and confronting the corresponding resource allocation problem towards achieving users' Quality of Service (QoS) prerequisites in VLC wireless networks is of high interest and part of our current work.

References

1. IEEE Standard 802.15.7, IEEE Standard for Local and Metropolitan Area Networks 15.7: PHY and MAC Standard for Short Range Wireless Optical Communication Using Visible Light. IEEE Std. IEEE Standard 802.15.7
2. Chen Z., Tsonev, D., Haas, H.: Improving SINR in indoor cellular visible light communication networks. In: IEEE International Conference on Communications (ICC), pp. 3383–3388, June 2014
3. Dimitrov, S., Haas, H.: Information rate of OFDM-based optical wireless communication systems with nonlinear distortion. J. Lightwave Technol. **31**(6), 918–929 (2013)
4. Saha, N., Mondal, R.K., Jang, Y.M.: Opportunistic channel reuse for a self-organized visible light communication personal area network. In: Fifth International Conference on Ubiquitous and Future Networks (ICUFN), pp. 131–134, July 2013
5. Bykhovsky, D., Arnon, S.: Multiple access resource allocation in visible light communication systems. J. Lightwave Technol. **32**(8), 1594–1600 (2014)
6. Saha, N., Mondal, R.K., Ifthekhar, M.S., Jang, Y.M.: Dynamic resource allocation for visible light based wireless sensor network. In: International Conference on Information Networking (ICOIN), pp. 75–78, February 2014
7. Mondal, R.K., Saha, N., Le, N.-T., Jang, Y.M.: SINR-constrained joint scheduling and optimal resource allocation in VLC based WPAN. Wirel. Pers. Commun. **78**(4), 1935–1951 (2014)
8. Ghimire, B., Haas, H.: Self-organising interference coordination in optical wireless networks. EURASIP J. Wirel. Commun. Netw. **1**, 2012 (2012)
9. Kim, W.-C., Bae, C.-S., Jeon, S.-Y., Pyun, S.-Y., Cho, D.-H.: Efficient resource allocation for rapid link recovery and visibility in visible-light local area networks. IEEE Trans. Consum. Electron. **56**(2), 524–531 (2010)
10. Mondal, R.K., Chowdhury, M.Z., Saha, N., Jang, Y.M.: Interference-aware optical resource allocation in visible light communication. In: International Conference on ICT Convergence (ICTC), pp. 155–158, October 2012
11. Marshoud, H., Kapinas, V.M., Karagiannidis, G.K., Muhaidat, S.: Non-Orthogonal multiple access for visible light communications. IEEE Photonics Technol. Lett. **28**(1), 51–54 (2016)

12. Yin, L., Wu, X., Haas, H.: On the performance of non-orthogonal multiple access in visible light communication. In: IEEE 26th Annual International Symposium on Personal, Indoor, and Mobile Radio Communications (PIMRC), pp. 1354–1359 (2015)
13. Kahn, J., Barry, J.: Wireless Infrared Communications. Proc. IEEE **85**(2), 265–298 (1997)
14. Lee, J.-W., Mazumdar, R.R., Shroff, N.B.: Joint resource allocation and base-station assignment for the downlink in CDMA networks. IEEE/ACM Trans. Netw. **14**(1), 1–14 (2006)

Kausa: KPI-aware Scheduling Algorithm for Multi-flow in Multi-hop IoT Networks

Guillaume Gaillard[1]([✉]), Dominique Barthel[2], Fabrice Theoleyre[3], and Fabrice Valois[1]

[1] Univ. Lyon, INSA Lyon, Inria, CITI, 69621 Villeurbanne, France
{guillaume.gaillard1,fabrice.valois}@insa-lyon.fr
[2] Orange Labs R&D, Meylan, France
dominique.barthel@orange.com
[3] ICube, Université de Strasbourg/CNRS, Strasbourg, France
theoleyre@unistra.fr

Abstract. The telecommunication operators focus on the Internet of Things (IoT) and route the traffic of several clients on a multi-hop infrastructure. Operators need to offer Service Level Agreements (SLAs) to each client, guaranteeing a minimum reliability or a maximum delay for each application. The deterministic IETF 6TiSCH protocol stack is particularly appropriate to provide SLA guarantees, because it allocates dedicated time-frequency blocks for a given traffic. We propose *Kausa*, a scheduling algorithm to assign a route and allocate resources to each client flow. We optimize the network lifetime while respecting the flow-level requirements. Kausa efficiently deals with lossy links, by scheduling ad-hoc retransmission opportunities. It limits both the buffer occupation and the end-to-end delay. Our simulations mimic multiple scenarios on multi-hop topologies, highlighting the relevance of our approach.

Keywords: SLA · IoT · Multi-hop · FTDMA · Reliability · Scheduling · Resource allocation · Delivery · Delay · Backtracking

1 Introduction

With the Internet of Things (IoT), the density of wireless devices and data traffic grows in the cities, increasing the radio channel occupation and the interference. The new digital services such as telemetering or smart parking require reliability and timeliness. In this context, the delivery of the data in short time is harder.

Instead of having IoT applications compete against one another to access the radio channel, we adopt the point of view of an operator dedicated to the IoT. In order to scale with the increasing traffic demand, the operator shares a relay infrastructure for its clients over the city. The characteristics of each application are specified through specific Service Level Agreements (SLA) [2]. We assume that the operator globally allocates the resource for each client. Centralized scheduling allows a better spectrum usage and a larger network capacity.

© Springer International Publishing Switzerland 2016
N. Mitton et al. (Eds.): ADHOC-NOW 2016, LNCS 9724, pp. 47–61, 2016.
DOI: 10.1007/978-3-319-40509-4_4

The operator must satisfy a specific level of Quality of Service (QoS) for each client flow, as defined in the SLAs [2]. SLAs specify both the maximum intensity of the offered traffic and the expected Key Performance Indicators (KPIs): the end-to-end delay and the Packet Delivery Ratio (PDR). For instance, data collection applications such as gas metering, require all the generated information: the effective delivery of each and every meter value is the main KPI. Applications such as pollution or light monitoring require a quick detection of a change: the main KPI is the delay.

The Packet Error Rate (PER) of the links depends on the radio environment. It has variable impact on the end-to-end delivery. Retransmissions allow to satisfy the PDR constraint, but they increase the delay and the traffic load. Indeed, the operator needs to over-provision resource to satisfy the QoS.

The IETF 6TiSCH Working Group brings IPv6 over the Time Slotted Channel Hopping mode of IEEE 802.15.4e. 6TiSCH is a good candidate technology because it benefits from the Frequency and Time Division Multiple Access (FTDMA) technology [11]. The operator is able to allocate resources to each flow, to quantify the remaining network capacity, and to adapt to interference.

One can build the FTDMA schedule with the Traffic-Aware Scheduling Algorithm (TASA) [6] and the routing graph provided by RPL [6]. Using the Expected Transmission Count (ETX) metric, the scheduler over-provisions enough resource to enable retransmissions and hence satisfies the reliability constraint. However, this approach reduces the set of used routes to the ones provided by RPL: the load is not correctly balanced over the nodes. This also increases the buffer occupation and reduces the network efficiency.

Furthermore, TASA does not consider long messages that are fragmented in order to be transmitted on various time-frequency blocks. In this case the end-to-end delay is important because the fragments are independently scheduled.

As far as we know, no resource allocation algorithm exists that both considers different flow-level KPIs, and balances the traffic load over the network topology.

Our contribution is three-fold:

1. we first assign a route to each flow, based on the traffic load and the reliability constraints. In particular, we adopt a multi-path approach: different flows from the same source may use different paths;
2. we propose Kausa, a KPI-aware scheduling algorithm supporting fragmentation. We provide a backtracking technique that enhances its performance;
3. we demonstrate with simulations the performance of our approach in terms of SLA satisfaction and allocation efficiency.

2   Related Work

2.1   Technical Background: A General Vision of 6TiSCH

FTDMA enhances the multi-channel technology, and reduces interference [10]. Several packets are multiplexed on independent channels and time slots. The nodes are synchronized so that they share an FTDMA schedule [8].

The IETF 6TiSCH Working Group runs an IP stack over IEEE802.15.4e TSCH networks [11]. In TSCH networks, the FTDMA schedules are build on periodical set of time slots, named *slotframes*. At each slot, every channel is mapped to a new channel offset. This *channel hopping* mechanism enhances the reliability of FTDMA by spreading the failure probability among the channels [8]. 6TiSCH considers both centralized and distributed schedule construction [6]. Each node has a *Scheduling Function* [11] to manage its allocations.

6TiSCH provides the possibility to dedicate resource to a multi-hop flow. A *track* is a set of time-frequency blocks along a route, that may only be used for the transmissions of determined frames [11]. Each node forwards the fragments according to their track IDs.

2.2 SLA-aware Scheduling

In order to satisfy the SLAs, the operator needs to manage and schedule the client traffic [2]. The schedule must consider the flow-level KPIs as constraints while maximizing the lifetime of the network.

Decentralized or distributed algorithms enable reactivity (e.g. in case of mobility, or node failure) and limit the control overhead by allowing local decisions. In the D-SAR [12] and CFDS [5] algorithms, the source nodes request QoS whenever they have traffic to transmit. According to the required bandwidth and QoS of each source node, the allocation is built *en route* to the destination. The QoS demands are supposed dynamic, changing with each traffic burst. Neither D-SAR nor CFDS provide stable guarantees of QoS for a given flow, because they are designed for temporary reservations. Phung et al. propose a multi-channel schedule based on distributed reinforcement learning and adaptation to network changes [7]. Being local to each node, the autonomous reinforcement learning does not provide end-to-end guarantees, necessary for satisfying the SLAs.

TASA centrally allocates resource for a given traffic load on each node [6]. Based on 2-hop conflicting sets, TASA gives priority to the most loaded sub-parts of the given routing tree. TASA yields optimal schedule length in this case [6]. Compact schedules optimize both the energy consumption and the delay. Neither flow-level KPIs in terms of delay and delivery nor load balancing are considered.

Industrial deployments also have strong constraints on their traffic: Pöttner et al. [9] propose a scheduling method to gather data under time-critical delivery in an oil refinery. Typical expected delay in the data delivery is 3 s. They partition the network into small sets of nodes (no more than 24 nodes in practice). They provide both a schedule and the required transmission powers, for one given reliability constraint. In our contribution, we consider different KPIs and network-wide techniques that reduce the overall load.

Dobslaw et al. propose SchedEx [1], a scheduler extension that modifies a schedule in order to guarantee a minimal network level end-to-end reliability. The authors calculate the number of necessary retransmissions for all the packets, at each link of the routing tree. This expected number of retransmissions is defined according to the load of a radio link and its reliability. In other words, Schedex does not guarantee flow isolation with differentiated PDR requirements.

Our solution provides flow-level guarantees and favors traffic balancing on the network while efficiently allocating the time-frequency blocks.

3   Network Model and Properties

In this section we specify the network characteristics and the parameters considered by Kausa. Table 1 summarizes the parameters and notations we use, as well as the default values we used in the performance evaluation.

3.1   Node Model

We model the IoT network as a set of wireless nodes exchanging messages in a multi-hop way. We assume the topology is known and static. Each node has a rank value that represents its distance to a gateway. Each node has one half-duplex radio interface. It can store a limited number of frames in its buffer. We consider 3 distinct types of nodes (Fig. 1):

1. **the leaf nodes:** the sensor nodes that only generate the client traffic;
2. **the relays:** the intermediary routers that forward the client traffic;
3. **the gateways:** the final receivers that collect the client traffic.

Each leaf node runs one or several applications that generate messages of constant size (possibly fragmented into several frames). We only consider upward traffic, converging toward the gateways. Figure 1 shows a portion of topology with 3 leaf nodes, 3 relays and one gateway.

3.2   Radio Link Model

We use the Packet Error Rate (PER) to characterize the link quality. We assume that the PER is time-invariant on the scale of the scheduling [8]. Because channel hopping is used, we also consider the PER independent from the channel offset [8].

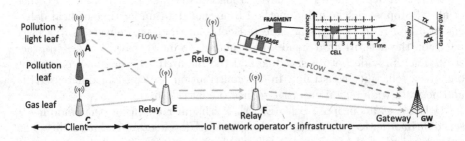

Fig. 1. Network topology

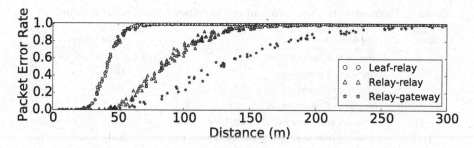

Fig. 2. Distribution of Packet Error Rates (PER) by role type. (Color Figure in Online)

The radio interference is heterogeneous: depending on the concentration of the nodes and the traffic density, the PER evolves [7]. We model an heterogeneous environment using a classical Path Loss model [4]. Noise and shadowing are normal random variables. Figure 2 shows the variations of the PER according to the distance, calculated for a large set of nodes. We assume the attenuation depends on the type of link because of their location and transmission power. For instance, the operator places the gateways on highpoints whereas the leaf nodes suffer bad radio conditions. The routers are located outdoor and use high transmission power (e.g. 3 dBm). We set the path-loss exponents and the reference distances in the typical ranges (Eq. 2.53 in [4]).

Using FTDMA, the medium is divided into time slots and frequency channels. We assume that co-channel interference is only limited to a few hops (e.g. 1 or 2). The algorithm allocates time-frequency blocks, named *cells*, within the FTDMA schedule (Fig. 1). The schedule is divided into periodical slotframes (set of time slots) that repeat in time, indefinitely.

Typically, during one cell, 2 neighboring nodes can exchange a frame of 127 B and its corresponding acknowledgement (Fig. 1). If necessary, an applicative message is fragmented into frames that separately transit on a single cell each.

3.3 Flow Model

For each application, a leaf node creates one *flow* of messages. The client traffic is converge-cast. The operator collects the traffic of each flow at a gateway (Fig. 1).

We create *tracks* assigning a given number of cells at each hop along a path. Each track is identified with a track ID. We associate each flow to a unique track. This way, each flow is isolated.

The scheduler provisions enough cells for the end-to-end transmission of the fragments of each message. The scheduler allocates additional over-provisioning cells in order to consider hop-by-hop retransmissions [3].

The operator only provides guarantees on the flows that respect the SLA specification [2]. We address the allocation of resource for periodical traffic patterns. For each flow, the source leaf node generates a given number of fixed-size

Table 1. Parameters of the network model and their values in the simulations

Variable	Explication	value	range
	Size of slotframe (slots)	1000	[100,1200]
	Number of channels	16	
	Number of leaves running each app	100	
	Number of relays, sinks	24, 2	
$n_{msg}(f)$	Num. of messages per slotframe for a flow f	1	[1, 12]
$n_{frag}(f)$	Num. of fragments per message (app.1, app.2)	2, 3	
$PDR_{msg}^{sla}(f)$	Expected end-to-end PDR (app.1)	0.80	[0.80, 0.98]
	Expected end-to-end PDR (app.2)	0.97	[0.97, 0.997]
$DEL_{msg}^{sla}(f)$	Expected end-to-end delay (slots) (app.1)	60	[30, 140
	Expected end-to-end delay (slots) (app.2)	90	[45, 210]
$pdr_{frag}^{sla}(f)$	Expected end-to-end PDR - fragments of f		
$per(i)$	Packet Error Rate (PER) on the i_{th} link of a path		
$n_{cell}(A)$	Number of allocated cells on a node A		
$n_{rtx}^{max(msg)}(f)$	Max. num. of retrans. per (hop, msg)	16	
$n_{rtx}^{max(frag)}(f)$	Max. num. of retrans. per (hop, frag)	8	
$n_{buffer}^{max}(f)$	Maximum buffer occupation	20	

messages at each slotframe. The leaf node buffers the corresponding fragments until their first transmission slot. During one slotframe, all the fragments must reach a gateway.

4   An Overview of Kausa

We consider a centralized scheduling algorithm that allocates FTDMA cells to a set of flows. Kausa takes as input parameters:

1. each flow's SLA parameters: the traffic profile and the expected KPIs;
2. the topology information: the PER of each link, and the rank of every node.
3. interference: the maximum hop distance after which we neglect interference.

In order to limit the final number of allocated cells, the algorithm considers, per hop, a maximum number of hop-by-hop retransmissions: either for each fragment ($n_{rtx}^{max(frag)}(f)$) or for each message ($n_{rtx}^{max(msg)}(f)$).

Figure 3 shows the allocation process. Kausa first builds a path for each flow (step 1). Each path is chosen in order to balance the load over the network while satisfying a minimum reliability. Kausa computes the number of cells needed at each hop to satisfy the PDR, according to the PER of each link (step 2).

If the chosen path does not suit the PDR or delay KPIs, a link-level backtracking takes place (Sect. 5.6). New paths are considered by recursively eliminating links from the possibilities. We avoid both the most loaded and the least reliable portions of the network. The recursion stops when the source has no more possible links to a next hop. The link-level backtracking favors the respect of the KPIs at the expense of the load-balancing.

The set of cells associated to a message at a given hop is named a *range*. At each hop, one *range* contains enough cells for the transmissions of the fragments of each message, as well as enough over-provisioned cells for the anticipated retransmissions. Kausa sequentially allocates the cells within each range into the FTDMA schedule (step 3). In Fig. 3, Kausa allocates flow $k + 1$ considering the cells already allocated for flow k and the previous flows.

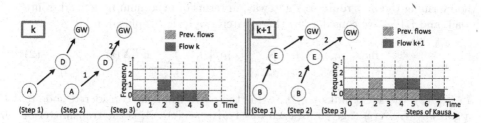

Fig. 3. Step-by-step overview of Kausa, for two flows k and $k + 1$.

In order to avoid buffer overflows, we limit the maximum amount of buffered fragments on each node, considering the least favorable scenario of retransmissions. Kausa accordingly anticipates or delays the allocations of a given message.

If no allocation is found for a given message, a flow-level backtracking takes place (Sect. 5.7). We successively remove the allocations of the previous flows, look for new paths, and try other allocations. The flow-level backtracking stops when it reaches the first allocated flow.

5 Detailing Kausa: An SLA-aware FTDMA Scheduler

5.1 Ordering the Flows

We order the flows in decreasing order of the load metric, so that Kausa first deals with the most greedy ones. We compute the load metric directly from the SLA parameters (Table 1). In Eq. (1), the load metric corresponds to the number of expected fragments during a slotframe, for a given flow f:

$$n_{msg}(f) \cdot n_{frag}(f) \cdot PDR_{msg}^{sla}(f) \tag{1}$$

In case two flows have similar values for this metric (e.g. less than 1 % difference), we consider the expected delay $DEL_{msg}^{sla}(f)$ as the second ordering criterion.

When several flows have the same KPIs (typically for one multi-source application, e.g. telemetering), they have the same load metric and delay. We use the source rank to decide between the flows (and in case of identical source rank the track ID). We give the furthest source in terms of rank the highest priority, since the route may be longer, and hence the load higher for the considered flow.

5.2 Building the Paths

We provide a load-balancing and KPI-aware multi-path mechanism. Our objective is to minimize the impact of each flow on the network in terms of allocation load. We construct the path for one flow, taking into account the previous allocations. We need to find valid next-hop nodes for each router in the path, in terms of traffic load and reliability.

We update the routing topology, starting from the gateways and using a vector-distance (e.g. Bellman-Ford) algorithm. We choose for each node the next-hop forming the best route to a gateway, in terms of maximum node load, route load, and ETX. We consider the three criteria in lexicographical order:

$$< \max_{i \in [1, l_{gw}]} n_{cell}(A_i), \sum_{i \in [1, l_{gw}]} n_{cell}(A_i), \sum_{i \in [1, l_{gw}]} ETX(i) > \tag{2}$$

In Eq. (2), the set of links $i \in [1, l_{gw}]$ is the route from the considered node to a gateway. $ETX(i)$ is given for each i as $1/(1 - per(i))$, A_i is the transmitter of i. Note that the sum of the ETX allows us to estimate the necessary number of timeslots for the allocation over a given path.

The next-hop of a node is a neighboring relay with a lower rank to avoid loop. This allows us to build robust path. We verify that the obtained path is sufficiently reliable while considering independent fragment transmissions [3]. Kausa disregards any path that requires, at any given link and for one fragment, more retransmissions than the maximum $n_{rtx}^{max(frag)}(f)$: thus, we avoid paths with too low a reliability. Equation (3) expresses the satisfaction of the fragment reliability constraint using the maximum number of retransmissions:

$$\prod_{l=1}^{l_{gw}} \left(1 - per(l)^{n_{rtx}^{max(frag)}(f)}\right) \geq pdr_{frag}^{sla}(f) \tag{3}$$

If Eq. (3) is not verified, the chosen link is not reliable enough: the path is not valid and we need to backtrack.

5.3 Computing the Hop-by-Hop Number of Allocations

We evaluate for each hop the minimum number of over-provisioning cells needed to satisfy the message PDR constraint. We use an inverse greedy algorithm starting with the max. number of retransmissions for a message, $n_{rtx}^{max(msg)}(f)$ [3].

The resulting number of allocated cells must respect the delay in case the cells are consecutively allocated. If this condition does not hold, the path cannot be valid and we need to backtrack.

5.4 Allocating the Messages

The path has been computed. We now have to allocate cells for each hop. The allocation is done message after message, for each ordered flow. We associate the

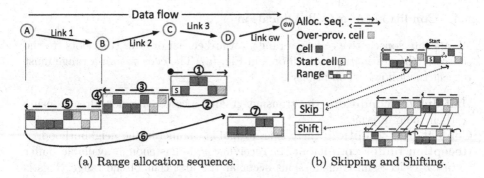

(a) Range allocation sequence.   (b) Skipping and Shifting.

Fig. 4. Kausa range allocation.

message with one range for each hop in the path (Fig. 4). Each range is a set of cells dedicated to the message on a given link.

The ranges are allocated in sequence. In order to avoid scheduling the next transmission before the last reception, the ranges of the same message are contiguous but do not overlap. Indeed, a set of ranges is valid if it respects the delay constraint: the difference between the first slot of the first range and the last slot of the last range is less than the delay (in slots). Figure 4a represents the allocation sequence for a given message, on a path from a leaf node A to a gateway. The numbered arrows show the allocation order. The sequence begins with the *starting range* at the link with the most allocated cells in the slotframe: e.g. at step 1, the range of Link 3 is the starting range.

1. **Allocating the cells within the starting range:** we first have to position the cells of the starting range. We iteratively try all the possible *start cells* by scanning the whole slotframe. We finally select the starting range which minimizes, for all the cells, the channel occupation. We terminate the scan early if we find a zero value for a range before finishing the slotframe.
2. **Allocating the cells for the previous hops:** we allocate the cells of the previous hops from last to first in upward direction. Next considered link is Link 2, steps 2 and 3, in Fig. 4a. We choose to minimize the buffering delay, starting from the cell closest from the next *start cell*. We repeat the same process (step 4 and 5) until we reach the first hop AB.
3. **Allocating the cells within the following hops:** the ranges of cells for the next hops are allocated consecutively to the last cell of the starting range (step 6). If the cell conditions cannot be verified or the delay constraint is not satisfied, scanning all the slotframe, the range is not valid.
4. **Modifying the starting range:** if the allocation fails on a given hop, or if the delay constraint is not verified, the allocation starts again for the same message with the second best starting range. We resume the search if it was not finished, and choose the new start cell accordingly.

If we cannot find any starting range, the allocation of the message fails and we need flow-level backtracking.

5.5 Conditions of a Cell Allocation

We allocate consecutive cells in a range, toward earlier or later time slots (to the source or the destination, step 5 or 7 in Fig. 4a). The cells within a range must respect the following conditions:

Half-duplex condition: the transmitter and receiver are not already using their radio interface;

Conflict-free condition: there is an available channel in the neighborhood;

Reception buffer condition: the receiving node has enough available buffer in case successful transmissions occur at the first cells of the range. This is the worst case scenario: the receiver's buffer is occupied early.

Transmission buffer condition: the sending node has enough available buffer for the not yet transmitted fragments in case successful transmissions occur at the end of the range. Among the cases that validate the SLA conditions, this is the worst case scenario: the transmitter's buffer is freed late.

If one of these conditions is not respected, we *skip* a cell (Fig. 4b) and try with the next ones. Note that the buffer conditions must be verified for the whole range, because as no fragment is transmitted in a skip, no memory is freed.

During the allocation of a given range, if the buffer conditions are no longer valid, we need to recursively *shift* the ranges to earlier in time until the problem is solved (Fig. 4b). For each range, we move the allocated cells and verify at each cell the conditions. Note that if we reach the slotframe size while shifting, the allocation is not valid and we need to modify the starting range.

5.6 The Link-Level Backtracking Procedure

Kausa builds an alternative path for a given flow (step 1 in Fig. 3):

1. when the routing algorithm does not find a suitable path for a given flow;
2. when the provision of cells fails on the path for a given flow.

One link in the path is blacklisted for the flow. Then, the new path is completed following the routing metric (Eq. 2) but avoiding the blacklisted links.

If the backtracking call is due to an impossibility to build a path or a non respect of the delay constraint, we eliminate the link with the worst PER, to build a more reliable path. In all other cases, we temporarily blacklist the link with the highest allocated load in the neighboring nodes, to build a less loaded path. If no valid path is found at this step, we definitely eliminate the link with the worst PER, and remove from the blacklist the temporarily added links.

We do recursive backtracking calls until either we find a valid path respecting the delay (success) or all source neighbors are blacklisted (failure). When the link-level backtracking fails, our algorithm drops this flow altogether and proceeds with the next one. Thus, the link-level backtracking does not cover all the possible paths, but progressively eliminates the links with the worst quality.

5.7   The Flow-Level Backtracking Procedure

A flow-level backtracking takes place when the allocation fails for a given message (step 3 fails in Fig. 3). We first save the state of the allocation.

If the link-level procedure on the flow fails, we backtrack on the links of the previous flow (e.g. in Fig. 3, if it fails on flow $k + 1$, we backtrack on step 1, flow k). When the backtracking procedure reaches the first allocated flow ($k = 0$) and fails, there is no solution under the assumptions we made. We resume Kausa with the previously saved allocation and proceed with next flow.

6   Performance Evaluation

6.1   Scenario and Simulation Setup

We run a Monte Carlo network simulator using Python. We compare Kausa with an extension of TASA, enabling the provision of cells for the retransmissions [3]. We run the algorithms on 16 randomly generated topologies and with 12 values of each of the 4 parameters (traffic intensity, slotframe size, expected PDR KPI, expected delay KPI). Table 1 summarizes the parameters and default values.

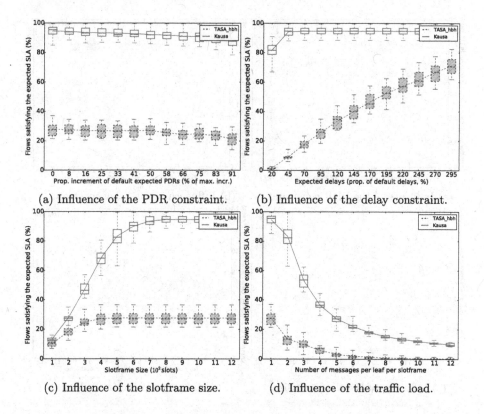

(a) Influence of the PDR constraint.   (b) Influence of the delay constraint.

(c) Influence of the slotframe size.   (d) Influence of the traffic load.

Fig. 5. Evaluation of the performance in terms of PDR satisfaction.

The simulations emphasize the impact of the variations of each parameter, the others kept at default. We use the same scenario as in [3] and our radio model (Sect. 3.2). We consider two applications with strong delay constraint (app.1) and with strong PDR constraint (app.2). For the two applications, the ranges of variations of the delay and PDR constraints are realized in proportion of the two default values (Table 1). They are expressed as a percentage in the figures.

The leaf nodes are uniformly spread in a rectangle of 400×200 m. The relays are placed on a triangle mesh (every approximately 70 m). The 2 gateways are placed at positions (100,100) and (300,100).

6.2 Results

Figure 5 shows that Kausa outperforms TASA with retransmissions, in terms of SLA satisfaction. With strong requirements (half the flows expect a PDR of 97 %) around 90 % of the flows validate their KPIs for Kausa (Fig. 5a, b), with a schedule length of 800 slots (Fig. 5c). The delay constraint is the main limiting factor in TASA: its performance increases almost linearly when relaxing the constraint (Fig. 5b). When the PDR constraint increases (Fig. 5a), the performance of

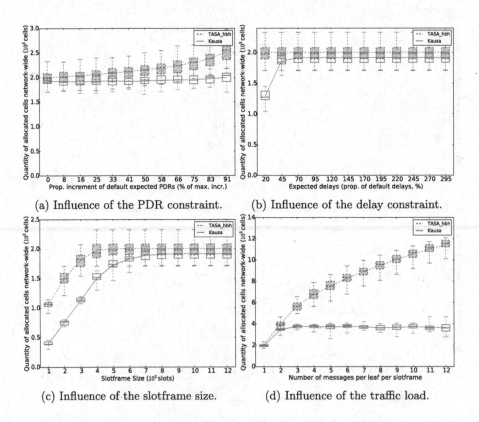

(a) Influence of the PDR constraint. (b) Influence of the delay constraint.

(c) Influence of the slotframe size. (d) Influence of the traffic load.

Fig. 6. Evaluation of the performance in terms of network resource usage.

Kausa slightly decreases: Kausa finds specific solutions depending on the topology. The two sinks are saturated when each leaf node generates more than 3 messages (Fig. 5d). Even when the intensity of traffic reaches the limit of capacity, Kausa allows some flows to be serviced (50 % with 3 messages).

In Fig. 6 the total amount of allocations on the whole network is similar for the two protocols (around 2000 cells for default traffic). In Fig. 6a, for high PDRs ($\geq$ 90 %, 50 % of maximum increase), the allocations increase in TASA while they remain stable in Kausa. Indeed, the backtracking process leaves flows unallocated in Kausa, whereas TASA serves all the traffic indistinctly. Figure 6b and c show that Kausa uses less allocations, whichever parameter is considered, because it finds paths with a better tradeoff between length and link quality. Figure 6d shows that Kausa does not provision cells in case of saturation, and does not waste bandwidth.

Finally, Fig. 7 shows that with default traffic, the maximum buffer occupation in Kausa remains around 10 fragments (half the maximum), versus about 40 fragments for TASA (Fig. 7a, b, and c). Figure 7d shows that with high traffic Kausa preserves the buffers better than TASA.

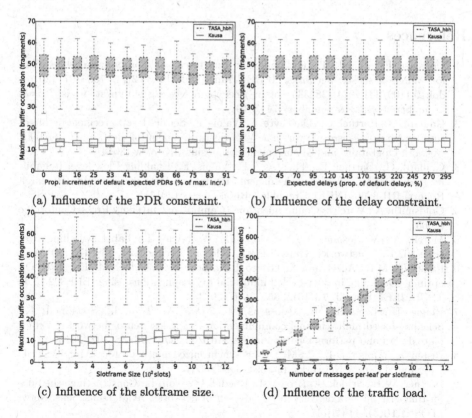

(a) Influence of the PDR constraint. (b) Influence of the delay constraint.

(c) Influence of the slotframe size. (d) Influence of the traffic load.

Fig. 7. Evaluation of the performance in terms of node buffer occupation.

7 Conclusion and Future Work

The number of devices that form the IoT grows, along with the requirements of quality of service for a diversity of applications. This makes the resource allocation for IoT networks more challenging in the smart cities.

In this work, we propose a solution that centrally allocates FTDMA resources under flow-level QoS constraints. Our algorithm balances the traffic load, to prolong the network lifetime, while satisfying delivery and delay constraints for a multi-flow scenario. A controlled backtracking algorithm allows reaching a satisfying solution in a reasonable time.

We simulate an urban environment, with a set of typical applications, their characteristics and requirements. We model a FTDMA technology, where the nodes have limited buffer capacities, and where the links have different qualities. In this context, Kausa performs better than TASA in terms of SLA satisfaction. We ensure flow-level QoS without creating buffer overflows.

In a future work we will study how to enhance Kausa by taking into account different queue management strategies on each router. We envision an adaptation of Kausa for dynamic or bursty traffics.

References

1. Dobslaw, F., Zhang, T., Gidlund, M.: End-to-end reliability-aware scheduling for Wireless sensor networks. IEEE Trans. Ind. Inf. **12**(2), 758–767 (2016). doi:10.1109/TII.2014.2382335. http://ieeexplore.ieee.org/stamp/stamp.jsp?tp=&arnumber=6987334&isnumber=7442910
2. Gaillard, G., Barthel, D., Theoleyre, F., Valois, F.: Service level agreement architecture for wireless sensor networks: a WSN operator's point of view. In: IEEE/IFIP Network Operations and Management Symposium, Krakow (2014)
3. Gaillard, G., Barthel, D., Theoleyre, F., Valois, F.: Enabling Flow-level Reliability on FTDMA Schedules with efficient Hop-by-hop Over-provisioning. (Research Report) RR-8866, INRIA Grenoble - Rhône-Alpes (2016)
4. Goldsmith, A.: Wireless Communications. Cambridge University Press, Cambridge (2005)
5. Morell, A., Vilajosana, X., Vicario, J.L., Watteyne, T.: Label switching over IEEE802.15.4e networks. Trans. Emerg. Telecom. Technol. **24**, 458–475 (2013)
6. Palattella, M.R., Accettura, N., Grieco, L.A., Boggia, G., Dohler, M., Engel, T.: On optimal scheduling in duty-cycled industrial IoT applications using IEEE802.15.4e TSCH. IEEE Sens. J. **13**(10), 3655–3666 (2013)
7. Phung, K., Lemmens, B., Goossens, M., Nowe, A., Tran, L., Steenhaut, K.: Schedule-based multi-channel communication in wireless sensor networks: a complete design and performance evaluation. Ad Hoc Netw. **26**, 88–102 (2015)
8. Pister, K., Doherty, L.: TSMP: time synchronized mesh protocol. In: IASTED Distributed Sensor Networks, pp. 391–398 (2008)
9. Pöttner, W.B., Seidel, H., Brown, J., Roedig, U., Wolf, L.: Constructing schedules for time-critical data delivery in wireless sensor networks. ACM Trans. Sens. Netw. (TOSN) **10**(3), 44 (2014)
10. Soua, R., Minet, P.: Multichannel assignment protocols in wireless sensor networks: a comprehensive survey. Pervasive Mob. Comput. **16**, 2–21 (2015)

11. Thubert, P., Watteyne, T., Palattella, M.R., Vilajosana, X., Wang, Q.: IETF 6tsch: combining ipv6 connectivity with industrial performance. In: IMIS, pp. 541–546 (2013)
12. Zand, P., Chatterjea, S., Ketema, J., Havinga, P.: A distributed scheduling algorithm for real-time (d-sar) industrial wireless sensor and actuator networks. In: 17th IEEE Conference on Emerging Technologies & Factory Automation (ETFA), pp. 1–4 (2012)

Theory and Communications

An Algorithm for k-Connectivity
Under Pure SINR Model

Flávio Assis[1,2]([✉])

[1] Department of Computer Science,
UFBA - Federal University of Bahia, Salvador, Brazil
`fassis@ufba.br`
[2] University of Liverpool, Liverpool, UK

Abstract. In this paper we describe a deterministic algorithm to compute a (vertex disjoint) k-connected subgraph ($k \geq 1$) of an undirected complete weighted Euclidean graph. The weight of the computed subgraph is within an $O(k \log n)$ factor of the weight of an optimum k-connected subgraph for a given graph, where n is the number of nodes. In particular, the algorithm was designed for *pure* SINR (Signal-to-Interference-plus-Noise Ratio) model. Variations of SINR models are currently considered the most appropriate ones to design and analyse algorithms for wireless networks when interference is taken into consideration. To the best of our knowledge, we describe the first algorithm to compute a k-connected vertex-disjoint subgraph under SINR. It has $O(\log g)$ runtime complexity, where g is the granularity of the network.

Keywords: k-Connectivity · SINR · Wireless networks · Distributed algorithms

1 Introduction

In this paper we describe a deterministic distributed algorithm to compute a *vertex disjoint k-connected subgraph* (or simply *k-connected subgraph*), $k \geq 1$, of a given complete undirected Euclidean graph modelling a wireless network. A *k-connected subgraph* is one that contains k vertex-disjoint paths between any two nodes of the original graph. We consider in this paper networks embedded in a two-dimensional Euclidean space, where the weight of each edge is equal to the (Euclidean) distance between its adjacent nodes.

Differently from previous work we assume in this paper an *interference model* for concurrent transmissions. The model assumed is pure *Signal-to-Interference-plus-Noise Ratio* (SINR). This model determines in which circumstances a message sent by a node can be received correctly by other nodes by taking into consideration how strong is the signal of a transmitting node at a receiver in relation to the sum of the signals of all other concurrent transmitters plus the

F. Assis—This paper was supported by the Royal Academy of Engineering under the Newton Research Collaboration Programme (grant number NRCP/1415/2/34). The work was also partially supported by CAPES/Brazil (grant number BEX 1836/14-5).

© Springer International Publishing Switzerland 2016
N. Mitton et al. (Eds.): ADHOC-NOW 2016, LNCS 9724, pp. 65–78, 2016.
DOI: 10.1007/978-3-319-40509-4_5

ambient noise. SINR models are also called *physical models* and are currently considered the most appropriate algorithmic models for wireless networks when interference is taken into consideration. The development of algorithms for variations of SINR models has attracted much attention recently (e.g. algorithms for broadcasting [5,10,11], local broadcasting [7,9], connected dominating set [23], among others).

The cost (or weight) of a graph is the sum of the weights of its edges. It is well known that the problem of computing a *minimum cost* k-connected subgraph is NP-hard [16]. The algorithm we describe in this paper computes a k-connected subgraph whose weight is within an $O(k \log n)$ factor of the weight of a minimum cost k-connected subgraph, where n is the number of nodes in the graph.

Our algorithm is designed for the particular case of *complete* graphs. Many wireless network scenarios can be modelled as complete graphs and can benefit from k-connected subgraphs. Wireless networks are being considered as an alternative to wired networks in critical application domains such as automotive [20,24]. The integration of many sensors using cables introduces extra complexity in the manufacturing process. Wireless connections among these nodes might contribute to decrease this complexity. Having k node-disjoint paths between any pair of nodes provides a level of resilience to node faults, as each message can be sent from a node to any other node through k non-overlapping paths. K-connectivity can be used to balance energy expenditure and workload among nodes by alternating the path through which messages are sent. Additionally, it can be used to control the transmission power of nodes while keeping robustness of the network, as an additional way of reducing energy expenditure.

Although each node can potentially communicate with any other node in our model, communication between nodes is performed through *SINR channels*. Communication between nodes is only possible when interference from other nodes is sufficiently low. Handling interference in this sense introduces a set of complex issues, which make the development and analysis of algorithms for the SINR model much more complicated than for traditional models.

Our contribution: In this paper we describe a *deterministic* algorithm for k-connectivity in complete graphs under *pure* SINR model. To the best of our knowledge, this is the first work that addresses this problem for $k \geq 2$ under this model. When $k = 1$, the problem is reduced to finding an approximate Minimum Spanning Tree (MST). Algorithms for approximating MST under a SINR model have been proposed in [1,13,15]. The algorithm proposed in [13,15] is randomized and is not based on pure SINR, but on a model that assumes that nodes might execute carrier sensing. The resulting model is thus a combination of SINR with a radio model. Pure SINR is assumed in [1]. The algorithm that we present in this paper is an extension of the one presented in [1] for the case of k-connectivity for $k \geq 2$, but restricted to the case of complete graphs.

The algorithm that we describe provides an $O(k \log n)$ approximation guarantee to an optimal k-connected subgraph. The algorithm is based on the NN-Scheme, proposed in [12,14]. However, the algorithm described in [12,14] is based on a model without interference. The adoption of the SINR model requires the use of different, specific and much more complex techniques to control the inter-ference among concurrently transmissions in the network.

Our algorithm has $O(\log g)$ time complexity, where g is the *granularity* of the network, as will be described in Sect. 3.

This paper is structured as follows. In Sect. 2 we discuss related work. In Sect. 3 we describe the assumed system model and notation. In Sect. 4 we describe the so-called NN-scheme, upon which our algorithm is based. In Sect. 5 we describe our algorithm. Section 6 concludes the paper.

2 Related Work

The k-vertex disjoint connectivity problem is NP-hard for $k \geq 2$, even in the case when edge weights satisfy the triangle inequality or in complete Euclidean graphs [16].

There is a large number of work on centralized approximation algorithms for this problem (for example, see the surveys in [16,18]). Among the most impor-tant results are [3,6,17,21]. In [17] Kortsarz and Nutov presented a centralized algorithm with an approximation guarantee of $O(\log k \cdot \min\{\sqrt{k}, \frac{n}{n-k} \log k\})$. In [6] Fackharoenphol and Laekhanukit presented an algorithm with an $O(\log^2 k)$-approximation. The approximation guarantee was improved by Nutov to $O(\log k \log \frac{n}{n-k})$ [21]. These results apply both to undirected and directed graphs. In [3] Cheriyan and Végh described the first algorithm that provides a constant approximation, but restricted to the case when the number of nodes is at least $k^3(k-1)+k$. The algorithm assumes as input an undirected k-connected graph with positive edge weights.

Many previous works considered distributed algorithms for k-connectivity as well, mainly in the area of Topology Control (e.g. [2,4,8,19,22,24]). All of these works, however, assume a model without interference. The main goal in [2,8,19,22], for example, is to describe algorithms that guarantee k-connectivity when nodes act *locally*, i.e. with knowledge about its neighbourhood in a constant number of hops. Each node chooses an appropriate set of neighbours and adjusts power accordingly. As a main goal in Topology Control, power adjustment results in less energy being spent by the nodes. In [4], the authors present distributed algorithms to determine whether a graph is k-connected or not.

Among the distributed algorithms, only the ones described in [8,12,14] pro-vide approximation guarantees. Hajiaghayi et al. describe in [8] a distributed approximation algorithm that outputs a k-vertex connected subgraph whose edge with maximum power approximates the power used by an optimal solution by an $O(k^{O(c)})$ factor, where c is the path-loss exponent. Khan *et al.* describe in [12,14] an algorithm for k-connectivity which is a distributed implementation of the NN-Scheme, as described in Sect. 4. Each node has a random rank and probes its neigh-bours in increasingly transmission powers until it has received k acknowledgement messages from its neighbours or has probed all of the neighbours. This algorithm gives an approximation ratio for the cost of the computed k-connected subgraph of $O(k \log n)$ for metric graphs, $O(k)$ for random graphs with nodes uniformly distributed in $[0,1]^2$ and $O(\log \frac{n}{k})$ for a complete graph with edge weights selected randomly from $[0,1]$ according to a uniform distribution.

Our algorithm is based on [12,14] and thus is based on the NN-Scheme as well. Our algorithm, however, is designed for the SINR model, where interferences among transmissions are taken into consideration. The algorithm in [12,14] is based on a model without interference. Adopting a SINR model implies the use of specific techniques to cope with the much more complex relationship between nodes transmissions.

The only previous related work based on the SINR model we are aware of are [1,13,15]. These works, however, compute approximations for minimum spanning trees (MST) under SINR. An MST is a k-connected subgraph for $k = 1$. As previously described, the algorithm described in [13,15] was designed for a model that combines SINR with a radio model, while our work assumes a pure SINR model. The algorithm we describe in this paper can be seen as an extension of [1] for $k \geq 2$. We are not aware of any other work for k-connectivity for $k \geq 2$ under the SINR model.

3   System Model and Notation

Network Model. We assume a network composed of n static nodes spread on an Euclidean plane and communicating by wireless medium. The network is modelled as an undirected complete weighted graph $G(V, E, w)$ with edge weight function $w : E \to \mathbb{R}^{\geq 0}$. For each $(u,v) \in E$, $w(u,v) = dist(u,v)$, i.e. the weight of each edge is the Euclidean distance between its adjacent nodes. For any graph H, we denote $w(H)$ the weight of H, i.e. the sum of the weights of its edges.

Each node v has a unique id from the set $\{1, 2, ..., n\}$, denoted $id(v)$, and a unique pair of x, y-coordinates. As we assume a model with messages of restricted size (see below), each coordinate is assumed to be of maximum $O(\log n)$ bits.

Each node is equipped with an omni-directional antenna and all nodes transmit with the same (maximum) power (so-called *uniform network*). We denote by r the *maximum transmission range*, i.e. the maximum distance from a transmitting node at which another node can still receive (successfully) the message.

Each node knows: its id; its x, y-coordinates in the plane; the number n of nodes in the network; and the *network granularity* g, defined as r divided by the minimum distance between any two nodes.

Communication and Interference. Communication between nodes is defined by the SINR model. In this model, there are three fixed parameters: path loss $\alpha > 2$, receiver sensibility $\beta \geq 1$ and ambient noise $\mathcal{N} > 0$. The $SINR(u, v, \mathcal{T})$ ratio for nodes u, v and a set of transmitting nodes $\mathcal{T}$ is defined as follows:

$$SINR(u, v, \mathcal{T}) = \frac{P_u \cdot dist(u,v)^{-\alpha}}{\mathcal{N} + \sum_{z \in \mathcal{T} \setminus \{u\}} P_z \cdot dist(z,v)^{-\alpha}} \tag{1}$$

A node v successfully receives a message from a node u in a round if $u \in \mathcal{T}$, $v \notin \mathcal{T}$, and:

$$SINR(u, v, \mathcal{T}) \geq \beta,$$

where $\mathcal{T}$ is the set of transmitting nodes at that round. Without loss of generality, we normalize the maximum transmission range to 1, i.e. $r = 1$.

Communication Graph. In practice, as nodes located at different parts of the network might transmit simultaneously, it is reasonable to assume that communication occurs only up to a range that is smaller than the maximum. According to SINR, a node v located at distance r from a node u can only receive successfully a message from u if u is the only node transmitting in the whole network. Thus we assume here (as it has become usual) a *communication graph* $C_G(V, E)$ of a network as consisting of all nodes and edges (u, v) such that $dist(u, v) \leq (1 - \epsilon)r = 1 - \epsilon$, where $0 < \epsilon < 1$ is a fixed model parameter. In particular, we assume $\epsilon = \Theta(1)$. Without loss of generality, $d_{min} < (1 - \epsilon)$. The communication graph is assumed to be connected and complete.

As defined in [11], a node u transmits *c-successfully* in a round t if u transmits a message in round t and this message is received by each node v in Euclidean distance from u smaller or equal to c.

Synchronization. The algorithm described in this paper works synchronously in *rounds*. In each round, each node can either transmit or receive a message. Additionally, we assume a *non-spontaneous wake-up* model, where the source node starts executing in the first round, while the other nodes start when they receive a message successfully for the first time. We do not assume knowledge of a global tick by the nodes.

Carrier Sensing. We assume a model *without carrier sensing*. A node has no other feedback from the wireless channel than receiving or not a message in a round.

Messages. Each message might contain at most an $O(\log n)$ number of bits. A message transmitted by a node in a round is received (according to SINR) at the end of that round. During the execution of the algorithm, a node might discard messages sent by nodes that are farther than a certain distance. We say that a *node processes a message* if the node does not discard it (i.e. the node receives the message and does some computation based on it).

Grids. As in [11] the algorithm described in this paper is based on divisions of the plane into grids. We use some of the notations from that paper as follows. A grid of square boxes of size $c \times c$ is denoted G_c. In a grid: all boxes are aligned with the coordinate axes; point $(0, 0)$ is a grid point; each box includes its left side without the top endpoint and its bottom side without the right endpoint and does not include its right and top sides. We say that (i, j) are the coordinates of the box with its bottom left corner located at $(c \cdot i, c \cdot j)$, for $i, j \in \mathbb{Z}$. For a station v located at position (x, y) we denote $box_c(v)$ the box in grid G_c where v is (i.e. $ic \leq x < (i + 1)c$ and $jc \leq y < (j + 1)c$) and $G_c(v)$ the coordinates of this box. We say that *a box b contains a node in grid G_c* if there is at least one node v such that $box_c(v) = b$. Finally, for tuples (i_1, i_2) and (j_1, j_2) the relation $(i_1, i_2) \equiv (j_1, j_2) \bmod d$, for $d \in \mathbb{N}$, denotes that $(|i_1 - j_1| \bmod d) = 0$ and $(|i_2 - j_2| \bmod d) = 0$.

Ranks. Each node has a *rank*. We denote by $rank(v)$ the rank of node v. We use the id of nodes as their rank, i.e. $rank(v) = id(v)$, and thus ranks are totally ordered. We assume that $rank(u) > rank(v)$, for nodes u and v, iff $id(u) > id(v)$, i.e. the higher the id, the higher the rank.

Knowledge of Nodes and Saturated Messages. In the algorithm that we describe in this paper, some nodes send messages containing data (id, position and rank) to other nodes. We say that a node u *knows a node v* at a certain instant iff u has received a message that contains data about v. Additionally, we say that a message is a *v-saturated message*, for some node v, iff the message contains data about $\min\{k, n - id(v)\}$ nodes with rank higher than $rank(v)$.

4   The NN Scheme for Computing k-Connected Subgraphs

Our algorithm is based on the NN (*Nearest Neighbour*) scheme presented in [12] to compute a k-connected subgraph of a complete graph. Each node has a *rank*, which can be defined in different ways, such as, for example, a random number or the coordinates of the node. In the NN-scheme each node *connects* to the k nearest neighbours with higher rank. Nodes that do not have k nodes with higher ranks in their vicinity connect to those existing nodes with higher ranks (for example, the node with the second highest rank will connect to the only existing node with higher rank, for any k). By *connecting a node v to a node u* we mean that edge (v, u) will be part of the subgraph.

In [12] the author shows that: (a) the NN-scheme generates a k-connected subgraph of a complete graph; and (b) in a complete weighted metric graph the subgraph generated using the NN scheme is an $O(k \log n)$ approximation of a minimum cost k-connected subgraph, irrespective of the way how ranks are defined. These results are expressed as the following proposition and theorem from [12]:

Proposition 1 *(Proposition 3.3.1 in [12]). Consider an enumeration of n nodes, $v_1, v_2, ..., v_n$, where v_i is the node with the ith rank and for any $j > i$, $rank(v_j) > rank(v_i)$. Let $H(V, E)$ be a graph on $V = \{v_1, v_2, ..., v_n\}$ with $n \geq k + 1$ so that every v_i has at least $\min\{k, n - i\}$ neighbors in $\{v_{i+1}, v_{i+2}, ..., v_n\}$. Then H is k-connected.*

As a corollary of the proposition above, the NN-scheme generates a k-connected subgraph in a complete graph, independently of how nodes are ranked.

Theorem 1 *(Theorem 3.4.1 in [12]). On a metric graph H of n nodes, for any arbitrary ranking of the nodes, the weight of the k-connected graph H_k constructed by the NN-scheme $w(H_k) = O(k \log n)w(MKG)$, where MKG is a minimum k-connected subgraph of H.*

5   Algorithm

5.1   Overview

Based on Theorem 1 and following the main idea of the algorithm for k-connectivity described in [12,14], our algorithm finds a k-connected subgraph by *connecting* each node to k close nodes (or to as many nodes are available) with higher rank. Differently from [12,14], however, our algorithm does not guarantee that a node will connect to *the closest* nodes with higher rank, but to *close enough* nodes. Additionally and most importantly, the algorithm we describe in this paper was developed for pure SINR model. The model assumed in [12,14] does not take transmission collision or interference into consideration. The development of algorithms for an SINR model requires the use of specific techniques to cope with the interference of concurrent transmissions. A transmission of a node might affect the transmissions of other nodes, even being located far apart.

5.2   Description of the Algorithm

In order to coordinate the transmissions of nodes under interference, we designed our algorithm as a variation of *GranLeaderElection* [11], a leader election algorithm for the case when nodes know the granularity of the network. This algorithm is based on *dilution* [10,11], where nodes far away enough from each other can transmit successfully within a certain range. We use the $DilutedTransmit(V, x, d)$ algorithm, as described in [11], specialized for the case when each node transmits its id and a set of $\langle id, position \rangle$ pairs. This is represented as Algorithm 1. The following fact applies to it:

Fact 1 *(Proposition 1 in [11]). Let V be a set of at most n stations such that there is at most one station in each box of G_x and $x \leq (1-\lambda)/\sqrt{2}$ for $0 < \lambda < 1$. Then, there exists a constant d_α such that each element of V transmits $(2\sqrt{2}x)$-successfully during $DilutedTransmit(V, x, \lceil d_\alpha/\lambda^{1/\alpha} \rceil)$.*

The algorithm to connect nodes is represented as Algorithm 2, named *ConnToKNodesWithHigherRank*. It is a fully distributed algorithm, described here with input parameters V, g and ϵ. This algorithm sets the value of $KN(v)$, for each node $v \in V$. The $KN(v)$ set represents the set of (maximum k) ids of nodes to which v connects. The values in $KN(v)$ for all nodes induce a k-connected subgraph whose weight approximates the weight of a minimum weight k-connected subgraph.

Algorithm 1: DilutedTransmit(V, x, d, HR) (adapt. from [11])

1 **foreach** $a, b \in [0, d-1]^2$ **do**
2 $\quad V_d \leftarrow \{v \in V \mid G_x(v) \equiv (a,b) \bmod d\}$
3 $\quad$ All elements $v \in V_d$ transmit a message containing its id and the current value of $HR(v)$

Algorithm 2: ConnToKNodesWithHigherRank(V, g, ϵ)

1 $x \leftarrow \max\{\frac{1-\epsilon}{2^i\sqrt{2}} \mid i \in \mathbb{N}, \frac{(1-\epsilon)}{2^i} \leq \frac{1}{g}\}$

2 **foreach** $v \in V$ **do**

3     $RP(v) \leftarrow \{\langle id(v), pos(v)\rangle\}$

4     $KN(v) \leftarrow \emptyset$

5 $V_a \leftarrow V$

6 **while** $x \leq \frac{(1-\epsilon)}{2\sqrt{2}}$ **do**

7     $\lambda \leftarrow (1 - 2\sqrt{2}x)$

8   **foreach** $v \in V_a$ **do**

9   Let $HR(v)$ be the set of the $\min\{k, |RP(v)|\}$ elements in $RP(v)$ with the highest ranks

10      $DilutedTransmit(V_a, x, d, HR)$, for $d = (d_\alpha/\lambda)^{1/\alpha}$

11   **foreach** $v \in V$ **do**

12   Let $M(v)$ be the set of messages received by v in the current stage from nodes within distance $2\sqrt{2}x$ from v

13       $RP(v) \leftarrow RP(v) \cup (\bigcup_{m \in M(v)} S(m))$, where $S(m)$ is the set of $\langle id, position\rangle$ pairs in m

14       $NR(v) \leftarrow \{\langle id, pos\rangle : (\langle id, pos\rangle \in RP(v)) \wedge (id > id(v))\}$

15     $V_a \leftarrow \{u \in V_a : u$ is the node with the highest rank in its box in $G_{2x}\}$

16     $x \leftarrow 2x$ { beginning of new round }

17 **foreach** $v \in V$ **do**

18    $KN(v) \leftarrow$ the $\min\{k, n - id(v)\}$ closest nodes from v in $NR(v)$

The main part of the algorithm is the *while*-loop in lines 6–16. Each execution of the body of this loop with a specific value of x determines a *stage*. The first stage begins in Line 7 and each new stage begins in Line 16, when the value of x is doubled. The last execution of Line 15, which makes $x > (1-\epsilon)/(2\sqrt{2})$, does not initiate a new stage, as the loop condition will be false (i.e. the last stage terminates after the execution of Line 15). Each stage is based on a division of the plane in a grid G_x, for the corresponding value of x.

The dissemination of messages is done by *leaders* of boxes of the grid G_x (for each value of x). During the execution of the algorithm, if a node is elected a leader, it continues *active*, i.e. transmitting message in the next stage. Otherwise, it becomes *inactive*, i.e. it only listens to messages.

Set V_a represents the set of currently active nodes. At the beginning of the first stage, each box of G_x might contain at most a single node, as $x\sqrt{2} \leq 1/g$ (Line 1) and $1/g$ is the minimum distance between any two nodes. The node in a box is the leader of the box. Thus, in the first stage, all nodes are active and leaders (Line 5). From one stage to the next one, the value of x is doubled (Line 16) and a new grid is used. For each value of x, each box of G_{2x} comprises four boxes of grid G_x. The leaders of these four boxes elect the one with the highest rank to become the leader of the G_{2x} box. For that, each of them transmits a message containing a set of $\langle id, position\rangle$ pairs using dilution. This step is represented by algorithm $DilutedTransmit(V_a, x, d, HR)$ (Line 10). Each leader collects the messages it receives from nodes within distance $2\sqrt{2}x$ (Line 12). The one with the highest rank becomes the leader for the next stage. The others become inactive (Line 15). The loop terminates when $x = (1 - \epsilon)/\sqrt{2}$ (Line 6), as the $(1 - \epsilon)$ transmission range has been achieved.

By Fact 1, all nodes will correctly receive messages sent by other nodes located at distance up to $2\sqrt{2}x$ (they may receive messages from nodes located at larger distances as well, but these messages are discarded). When a node u sends a message in a stage, this message contains the id and position of the nodes with the highest ranks (limited to a maximum of k nodes) in the box of G_x (for the value of x in the stage) where u is located. Node u is also the leader of this box. Messages received by a node v (within distance $2\sqrt{2}x$) are stored in the set $M(v)$ (Line 12). The sets of $\langle id, position \rangle$ pairs contained in these messages are kept in set $RP(v)$ (Line 13). Each node v keeps those $\langle id, position \rangle$ pairs from nodes with ranks higher than $rank(v)$ in set $NR(v)$ (Line 14). At the end of the algorithm, each node v connects to the closest nodes in $NR(v)$, up to a maximum of k such nodes (Lines 17–18).

Observe that the algorithm is consistent with the restriction on the size of messages, as coordinates and ids are each of $O(\log n)$ size and each message will have a maximum of a constant number of pairs of ids and coordinates.

5.3   Correctness and Complexity

The following facts, lemmas and theorem state the correctness and performance of *ConnToKNodesWithHigherRank*.

Lemma 1. *Algorithm* ConnToKNodesWithHigherRank *(V, g, e) terminates execution in $O(\log g)$ rounds.*

Proof. The number of rounds of the algorithm is the same as the number of rounds of algorithm *GranLeaderElection* [11]. The differences between these algorithms are on data processing (local to each node) and that more data are exchanged between nodes on each message during the diluted transmissions. These differences do not affect the time complexity. Thus from Theorem 1 in [11] we have that the time complexity of *ConnToKNodesWithHigherRank* is $O(\log g)$, as $z = (1 - \epsilon)/\sqrt{2}$ and $\epsilon = \Theta(1)$. ☐

The leader election procedure that we use in our algorithm is the same used in *GranLeaderElection* [11]. In each leader election, the node with the highest rank (highest id) becomes the new leader. Thus the fact below is a direct consequence of *Proposition 2* in [11].

Fact 2. *At the end of each stage, each box of G_{2x} that contains at least one node will have a single leader. The leader of a box (if there is one) has the highest rank in the box.*

Although there might be more than one leader at the end of the execution of the algorithm, all nodes will hear the transmissions of the leaders participating in the last stage.

Fact 3. *At the last stage, all nodes hear all sent messages.*

Proof. At the last stage, x will have value $(1-\epsilon)/2\sqrt{2}$ (Line 6). By Fact 1, all sent messages will be correctly received within distance $(1-\epsilon)$. As the communication graph is complete, all nodes will correctly receive the messages. □

Fact 4. *The message broadcast by each leader of a box b in a grid G_x will contain the id and position of the $\min\{k, n_b\}$ nodes with the highest ranks in b, where n_b denotes the number of nodes in b.*

Proof. The proof is done by induction on the stage index. In the first stage, each box of the grid contains a single node, which is also the leader of its box. All nodes belong to V_a (Line 5) and each node sends a message containing its own id and position (Lines 3, 9 and 10). As each node is the single node in its box, it is also the one with the highest rank.

Let us now assume that the fact is true from stage 1 to stage i. We show that it is also true for stage $i + 1$. For any x, a box of G_{2x} is built from four boxes of grid G_x. In each stage, the leaders of these four boxes elect the leader of the G_{2x} box. Each of these leaders, say a leader in box b', sends a message containing the ids and positions of the $\min\{n_{b'}, k\}$ nodes in its box, where $n_{b'}$ is the number of nodes in b'. By the induction hypothesis, these are the $\min\{n_{b'}, k\}$ nodes with the highest ranks in the box. By Fact 1 these messages are received by all the leaders in the G_{2x} box. By combining the messages received from all the leaders of the four G_x boxes, the elected leader of the G_{2x} box will have the data about the k nodes with highest ranks in the G_{2x} box or the total number of nodes in this box, if there are less than k nodes in the box. If it is not the last stage, this new leader will select the data about the $\min\{n_b, k\}$ nodes with the highest ranks (Line 9) to send in its message in the next stage (Line 10). □

Fact 5. *If a node v receives a message from a leader l of a box b that is not a v-saturated message, then v will know all the nodes with rank higher than $rank(v)$ in b.*

Proof. By Fact 4 the leader of a box will send a message with the $\min\{k, n_b\}$ nodes with the highest ranks in its box, where n_b is the number of nodes in the box. As the message sent by l is not v-saturated (recall the definition of a v-saturated message in Sect. 3): either (a) the number of nodes in the message is less than k; or (b) the number of nodes in the message is k but there is at least a node whose rank is less than $rank(v)$. In case (a) the message contains all nodes in the box. Thus, node v will know in particular all nodes with rank higher than $rank(v)$. In case (b), the message contains all nodes with rank higher than $rank(v)$ that exist in the box, as the message contains the k nodes with the highest rank in the box and there is at least one node whose rank is less than $rank(v)$. □

Fact 6. *At the end of stage i, if a node v has not received any v-saturated message, then v knows all the nodes with rank higher than $rank(v)$ that are at distant at most x_i from v, where x_i is the value of variable x in stage i.*

Proof. By Fact 1 during stage i all nodes receive messages correctly from nodes within distance $2\sqrt{2}x_i$. All boxes adjacent to $box(v)$ in grid G_{x_i} are completely within a $2\sqrt{2}x_i$ range from v. The radius of the largest circle centered at v that is completely inside the area covered by $box(v)$ and its adjacent boxes is x_i (this is the smallest distance between a point in $box(v)$ and the border of this area).

By Fact 2, each non-empty box will have a leader. Suppose first that v is not the leader of $box(v)$. Node v will receive a message from the leader of $box(v)$ and of the adjacent boxes. As no leader sends a v-saturated message, by Fact 5 v will know all the nodes in $box(v)$ and in adjacent boxes that have a rank higher than $rank(v)$. As the circle centered at v with radius x_i is completely inside the area formed by $box(v)$ and its adjacent boxes, the lemma applies. If v is the leader of $box(v)$ the lemma applies as well, as the only difference in this case is that v will not receive a message from itself. But it will know the nodes with higher rank in $box(v)$. Node v has the highest rank in $box(v)$. As it has never received a v-saturated message, by Fact 5 it has known the nodes with higher ranks in the adjacent boxes since the first round. □

Fact 7. *At the end of* ConnToKNodesWithHigherRank, *each node v knows a set of nodes S of cardinality $\min\{k, n - id(v)\}$ such that the sum of the weights of the edges connecting v to each element of S is less than or equal to a constant times the sum of the weights of the edges connecting v to the $\min\{k, n - id(v)\}$ closest nodes with rank higher than $rank(v)$.*

Proof. We consider two cases, as follows. During the execution of *ConnToKNodesWithHigherRank*, node v: (a) does not receive any v-saturated message; and (b) node v receives at least one v-saturated message.

Case (a): by Fact 3, v will receive during the last stage a message from all leaders. The area covered by the boxes of all the leaders cover the whole network, as each non-empty box contains a leader. By Fact 5, as v does not receive any v-saturated message, it will know all nodes in the network whose rank is higher than $rank(v)$. In this case the set S might be the set of the $\min\{k, n - id(v)\}$ closest nodes with rank higher than $rank(v)$. Thus the lemma applies.

Case (b): Let i be the first stage in which v receives a v-saturated message. Assume that $i > 1$ (for $i = 1$, you just have to consider that $x_{i-1} = 0$ in the following). By Fact 6, node v knows all nodes with higher rank in the range x_{i-1}. If there are nodes with higher rank in this range, they will be the closest ones. At the end of stage i, v knows at least $\min\{k, n - id(v)\}$ nodes with higher rank, as it receives a v-saturated message in this stage. Let C' be the set of nodes that are in the set of $\min\{k, n - id(v)\}$ closest nodes to v but whose distances to v are longer than x_{i-1}. The nodes whose ranks are higher than $rank(v)$ that are within distance x_{i-1} from v are already known by v. Let us now associate each node in C' with a node in the v-saturated message. This association can be done arbitrarily and it is always possible, as there are at least $\min\{k, n - id(v)\}$ nodes in the message. Let c' be a node in C' and z its associated node in the message.

Let us call l a leader that has sent the v-saturated message to v. Then we have $dist(v, l) \leq 2\sqrt{2}x_i$ (as v processes only messages from leaders within this distance - Line 12 in Algorithm 2). Let us consider any particular pair of such c' and z nodes. As z is in the same box as l in stage i, $dist(l, z) \leq \sqrt{2}x_i$. By triangle inequality, $dist(v, z) \leq 2\sqrt{2}x_i + \sqrt{2}x_i = 3\sqrt{2}x_i$. By the definition of C', we have $dist(v, c') > x_{i-1}$. As $x_i = 2x_{i-1}$, we have $dist(v, z) \leq 6\sqrt{2}x_{i-1} \leq 6\sqrt{2}dist(v, c')$. Thus, the distance between v and z is only by a constant factor the distance between v and the corresponding node c', which is an arbitrary node in the set of $\min\{k, n - id(v)\}$ closest nodes with rank higher than $rank(v)$. This is valid for all nodes in set C'.

From the discussion above, node v will have the following nodes available to connect: the closest nodes with higher rank within distance x_{i-1}; and those nodes located at a distance greater than x_{i-1} and made known to v in a v-saturated message. Additionally, the set of nodes with rank higher than v has at least $\min\{k, n - id(v)\}$ nodes. Node v will connect to the closest $\min\{k, n - id(v)\}$ nodes among this set of nodes.

Let us call C the set of $\min\{k, n - id(v)\}$ closest nodes with rank higher than $rank(v)$ (amongst all nodes in the network) and Z the set of nodes to which v connects. Thus we have: $\sum_{z \in Z} dist(v, z) \leq 6\sqrt{2} \sum_{c \in C} dist(v, c)$, which proves the lemma.   □

Let $K(V, E_K)$ be the graph induced by the values of $KN(v)$, i.e. V is the set of all nodes and an edge $(u, v) \in E_K$ iff either $u \in KN(v)$ or $v \in KN(u)$.

Lemma 2. *Let G be an input graph for* ConnToKNodesWithHigherRank. *When this algorithm terminates, $K(V, E_K)$ is k-connected. Additionally, $w(K) = O(k \log n) \cdot w(MKSG)$, where $MKSG$ is a minimum cost k-connected subgraph of G.*

Proof. The fact that *ConnToKNodesWithHigherRank* generates a k-connected subgraph is a direct consequence of Proposition 1, which states that the graph induced when each node connects to $\min\{k, n - id(v)\}$ nodes with higher rank is k-connected. This condition is satisfied after the execution of *ConnToKNodesWithHigherRank*, as stated by Fact 7.

The approximation ratio is proved as follows. By Theorem 1, connecting nodes according to the NN-scheme guarantees an $O(k \log n)$ approximation to the cost of $MKSG$. In the NN-scheme, nodes connect to the k *nearest* nodes with higher rank. In our algorithm, Fact 7 states that the sum of the weights of the edges connecting nodes in our algorithm is less than or equal to the sum of the weights of the edges connecting nodes in the NN-scheme times a constant factor. Thus Let $X(v)$ be the set of nodes to which a node v is connected in *ConnToKNodesWithHigherRank* and $C(v)$ the set of $\min\{k, n - id(v)\}$ closest nodes with rank higher than $rank(v)$. As Fact 7 applies to all nodes, we have:

$$\sum_{v \in V} \sum_{x \in X(v)} w(v, x) \leq b \cdot \sum_{v \in V} \sum_{c \in C(v)} w(v, c)$$

where b is a constant. Let H_k denote the graph induced by the NN-scheme on G. Therefore $w(K) = O(w(H_k)) = O(k \cdot \log n) \cdot w(MKSG)$.   □

The theorem below summarizes the time complexity of the algorithm and the approximation ratio of the computed subgraph.

Theorem 2. *Let $G(V, E, w)$ be a weighted undirected graph modelling a wireless network, where nodes know the network granularity g and communicate according to pure SINR model. For constant parameters $\alpha > 2$ and $\epsilon < 1/2$, a k-connected subgraph K ($k \geq 1$) can be computed over G deterministically within $O(\log g)$ rounds. Additionally, $w(K) = O(k \log n) \cdot w(MKSG)$, where $MKSG$ is a minimum cost k-connected subgraph over G, when $w(u, v) = dist(u, v)$.*

Proof. This theorem is a direct corollary of Lemmas 1 and 2.   □

6   Conclusion

We described an efficient deterministic algorithm for computing a k-connected subgraph of an undirected weighted Euclidean complete graph G whose weight is within an $O(k \log n)$ factor from the weight of an optimum k-connected subgraph. The algorithm was developed assuming a pure SINR model. Variations of SINR models are considered the current best models for designing and analyzing algorithms for wireless networks when interference is taken into consideration.

Our algorithm is based on the assumption that the underlying graph is Euclidean and on uniform power assignment. Interesting directions for further research are the investigation of algorithms for k-connectivity on non-metric spaces and on non-uniform power assignment models.

Acknowledgement. The author thanks Prof. Dariusz Kowalski (University of Liverpool, UK) for the many discussions and suggestions related to the topic of this paper.

References

1. Assis, F.: A deterministic and a randomized algorithm for approximating minimum spanning tree under the SINR model. In: Proceeding of the 8th IFIP Wireless and Mobile Networking Conference - WMNC (2015)
2. Bahramgiri, M., Hajiaghayi, M., Mirrokni, V.S.: Fault-tolerant and 3-dimensional distributed topology control algorithms in wireless multi-hop networks. Wirel. Netw. **12**, 179–188 (2006)
3. Cheryian, J., Végh, L.A.: Approximating minimum-cost k-node connected subgraphs via independence-free graphs. In: Proceedings of the IEEE 54th Annual Symposium on Foundations of Computer Science (FOCS) (2013)
4. Cornejo, A., Lynch, N.: Fault-tolerance through k-connectivity.In: Proceedings of IEEE International Conference on Robotics and Automation (ICRA): Workshop on Network Science and Systems Issues in Multi-Robot Autonomy (2010)
5. Daum, S., Gilbert, S., Kuhn, F., Newport, C.: Broadcast in the ad hoc SINR model. In: Afek, Y. (ed.) DISC 2013. LNCS, vol. 8205, pp. 358–372. Springer, Heidelberg (2013)
6. Fakcharoenphol, J., Laekhanukit, B.: An $o(\log^2 k)$-approximation algorithm for the k-vertex connected spanning subgraph problem. SIAM J. Comput. **41**(5), 1095–1109 (2012)
7. Fuchs, F., Wagner, D.: Arbitrary transmission power in the SINR model: local broadcasting, coloring and MIS. Technical report arXiv:1402.4994v2 (2014)

8. Hajiaghayi, M., Immorlica, N., Mirrokni, V.S.: Power optimization in fault-tolerant topology control algorithms for wireless multi-hop networks. IEEE/ACM Trans. Networking 15(6), 1345–1358 (2007)
9. Halldorsson, M.M., Mitra, P.: Towards tight bounds for local broadcasting. In: FOMC (2012)
10. Jurdzinski, T., Kowalski, D.R., Rozanski, M., Stachowiak, G.: Distributed randomized broadcasting in wireless networks under the SINR model. In: Afek, Y. (ed.) DISC 2013. LNCS, vol. 8205, pp. 373–387. Springer, Heidelberg (2013)
11. Jurdzinski, T., Kowalski, D.R., Stachowiak, G.: Distributed deterministic broadcasting in uniform-power ad hoc wireless networks. In: Gąsieniec, L., Wolter, F. (eds.) FCT 2013. LNCS, vol. 8070, pp. 195–209. Springer, Heidelberg (2013)
12. Khan, A.M.: Distributed Approximation Algorithms for Minimum Spanning Trees and Other Related Problems with Applications to Wireless Ad Hoc Networks. Ph.d thesis, Purdue University (2007)
13. Khan, M., Kumar, V.S.A., Pandurangan, G., Pei, G.: A fast distributed approximation algorithm for minimum spanning trees in the SINR model. Technical report arXiv:1206.1113v1, June 2012
14. Khan, M., Pandurangan, G., Anil Kumar, V.S.: A simple randomized scheme for constructing low-weight k-connected spanning subgraphs with applications to distributed algorithms. Theor. Comput. Sci. 385(1–3), 101–114 (2007)
15. Khan, M., Pandurangan, G., Pei, G., Vullikanti, A.K.S.: Brief announcement: a fast distributed approximation algorithm for minimum spanning trees in the SINR model. In: Aguilera, M.K. (ed.) DISC 2012. LNCS, vol. 7611, pp. 409–410. Springer, Heidelberg (2012)
16. Khuller, S.: Approximation algorithms for finding highly connected subgraphs. In: Hochbaum, D.S. (ed.) Approximation Algorithms for NP-Hard Problems. PWS Publishing Company, Boston (1995). Chapter 6
17. Kortsarz, G., Nutov, Z.: Approximating k-node connected subgraphs via critical paths. SIAM J. Comput. 35(1), 247–257 (2005)
18. Kortsarz, G., Nutov, Z.: Approximating minimum-cost connectivity problems. In: Gonzalez, T.F. (ed.) Handbook of Approximation Algorithms and Metaheuristics. Chapman and Hall CRC, Boca Raton (2007). Chapter 58
19. Li, N., Hou, J.C.: Localized fault-tolerant topology control in wireless ad hoc networks. IEEE Trans. Parallel Distrib. Syst. 17(4), 307–320 (2006)
20. Nolte, T., Hansson, H., Lo Bello, L.: Wireless automotive communications. In: Proceedings of the 4th International Workshop on Real Time Networks (RTN) (2005)
21. Nutov, Z.: Approximating minimum-cost edge-covers of crossing biset-families. Technical report arXiv:1207.4366v1 (2012)
22. Saha, I., Sambasivan, L.K., Patro, R.K., Ghosh, S.K.: Distributed fault-tolerant topology control in static and mobile wireless sensor networks. In: Proceedings of the 2nd International Conference on Communication Systems Software and Middleware, COMSWARE (2007)
23. Scheideler, C., Richa, A.W., Santi, P.: An O(log n) dominating set protocol for wireless ad-hoc networks under the physical interference model. In: MobiHoc (2008)
24. Thallner, B., Moser, H., Schmid, U.: Topology control for fault-tolerant communication in wireless ad hoc networks. Wirel. Netw. 16(2), 387–404 (2010)

An Original Correction Method for Indoor Ultra Wide Band Ranging-Based Localisation System

Nezo Ibrahim Fofana[(⊠)], Adrien van den Bossche, Réjane Dalcé,
and Thierry Val

IRIT, Université Fédérale de Toulouse, CNRS, INPT, UPS,
UT1, UT2J, Toulouse-Blagnac, France
{nezo-ibrahim.fofana,adrien.van-den-bossche,
rejane.dalce-val}@irit.fr

Abstract. During this decade, Wireless Sensor Networks (WSNs) brought an increasing interest in the industrial and research world. One of their applications is indoor localization. The ranging, i.e. the distance evaluation mechanism between nodes, is required to determine the position of the nodes. The research work presented in this article aims to use Ultra Wide Band (UWB) radio links to achieve an efficient ranging, based on Time of Flight (ToF) measurement. A good solution consists in integrating ranging traffic into the usual network messages. However, the ToF ranging process is based on information exchanges which are temporally constrained. Once this information is encapsulated into the usual messages, the temporal constraint cannot be honoured, resulting in important ranging errors due to clock drifts. To mitigate these errors, we have introduced an original dynamic correction technique which enables a precision of twenty centimetres allowing the inclusion of ranging traffic in usual traffic.

Keywords: Indoor localisation · Time of Flight · UWB · Ranging · TWR · SDS-TWR · Prototyping · Testbed

1 Introduction

The industrial world is a major application area for mobile localisation systems. Spatial and temporal mappings are useful for many applications. GPS is generally used for outdoor localisation, but suffers from a significant power consumption and a reduced performance in indoor environments. Indoor localisation and positioning based on other technologies is therefore required. They can take advantage of Wireless Sensor Networks (WSNs) and other wireless communication systems increasingly used in the industrial domain for the exchange of data from the sensors. The ranging, i.e. the distance evaluation mechanism between nodes, is essential to locate the nodes. Localisation based on range-free methods, as illustrated by the well-known DV-hop algorithm [1, 2] is simple to implement and is based on hypotheses of cellular connectivity between nodes. However, it is not very accurate. Range-based techniques can improve the localisation accuracy. Most wireless nodes can provide an indication of the power level of the received frames, but this method suffers from several drawbacks [3]

© Springer International Publishing Switzerland 2016
N. Mitton et al. (Eds.): ADHOC-NOW 2016, LNCS 9724, pp. 79–92, 2016.
DOI: 10.1007/978-3-319-40509-4_6

and is not reliable because of the influence of the antenna's polarisation and the existence of Non Line Of Sight (NLOS) paths. One of the best solutions is to rely on the Time of Flight (ToF) [4] between sender node and receiver node. This technique requires a precise signal timestamping at the physical layer, which is made possible through a radio transmission technology like Ultra-Wide Band (UWB) or even Chirp Spread Spectrum (CSS). In this sense, the DWM1000 [5] is a very interesting solution. As it complies with the IEEE 802.15.4 standard, it provides picoseconds-level precision on the timing of reception and transmission of the R-MARKER (Ranging Marker) bit of every frame.

While the IEEE 802.15.4 standard offers a normalisation of message exchanges between nodes to achieve the ToF measurement such as TWR (Two Way Ranging), it does not specify how to integrate the ranging service into the protocol stack. Nevertheless, we clearly see the benefits of a fully integrated ranging protocol in data exchange, that is to say providing the ability to perform a distance measurement between two nodes when these nodes exchange information without ranging-dedicated frames. Our objective is to design an "*opportunistic ranging*", where ranging messages are encapsulated into usual messages (data, acknowledgment…). The final objective of our work is to provide such a service, included in the protocol stack, as transparent as possible, minimising protocol overhead, while staying accurate on the distance measurement. To achieve this service, one of the difficulties is due to the time constrained characteristic of ranging messages. Messages will be delayed, because of the Medium Access Control (MAC) process, introducing discrepancies because the nodes' transceivers use different crystals, which introduce ranging errors. This paper deals with this last point, introducing an original correction method for the ranging system.

The remainder of the article is organised as follows: after this introduction, we begin with a brief state of the art. Then we re-examined the well-known problem of clock drift, because the delay in ranging message exchanges is an important aspect of our problem; We then present measurements performed on a real testbed which confirm the reality of the problem. Next, we introduce our original correction method and evaluate its performance on the same testbed, before concluding and presenting our perspectives.

2 Related Work

Many algorithms and localisation systems have been developed for indoor WSNs. The algorithms proposed in the literature can be classified into two categories: the *range-free* type and the *range-based* type. The *range-free* family uses the connectivity and the number of hops from source to destination in order to estimate the node's position. The second category is based on measurements between nodes that can be converted into distance or angle. This type of localisation proceeds in two phases: the ranging phase and the position computation phase. In the ranging phase, either the Time Difference of Arrival (TDOA) [6], the Angle of Arrival (AOA) [7] or the Received Signal Strength Indicator (RSSI) [3] is used to obtain the distance between two nodes. With this information, each node can compute its own 3D-coordinates using classical localisation algorithms such as trilateration or triangulation. The RSSI technique uses signal propagation in order to convert the received signal strength into distance. It doesn't require additional hardware because transceivers are usually equipped with electronic circuits allowing to associate each received frame to a power level. However, this technique suffers [3] usually from several drawbacks related to high variations in radio propagation, the mobility of nodes, or even antennas direction. The evaluation of distance between two nodes is altered, and the resulting location is then less precise. Ranging accuracy can be improved to respond to requirements of the location. We achieve this by proposing a dynamic correction exploiting the protocols based on ToF and using the properties of the UWB physical layer as specified by the IEEE 802.15.4 standard [8], specifically the RMARKER bit which is not available for DSSS based physical layers. This PHY header bit (Fig. 1) serves as a reference for accurate timestamping of a ranging frame (Fig. 2). The existing UWB transceivers are capable of producing a picoseconds-level timestamp for both incoming and outgoing frames.

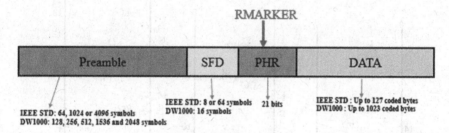

Fig. 1. Frame format: IEEE STD and DWM1000 compliant

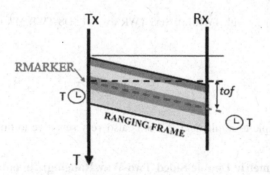

Fig. 2. RMARKER in ranging frame

In fact, temporal information provided by nodes in a UWB network is more reliable for ranging [9] than information extracted from the received power level. Ranging methods based on the ToF use a set of messages to determine the ToF. During transmission and reception of frames, nodes timestamp those moments on the physical layer. These timestamps are used to calculate the ToF and therefore the distance between transmitter and receiver. Multiple protocol variants are possible and are presented in the following sections.

2.1 Ranging Protocols

TWR (Two Way Ranging). TWR consists in a sequence of three messages (Fig. 3a). As indicated before, nodes A and B timestamp both outgoing and incoming frames. Node A extracts the timestamps collected by B from DATA_REPLY frame and applies the following equation.

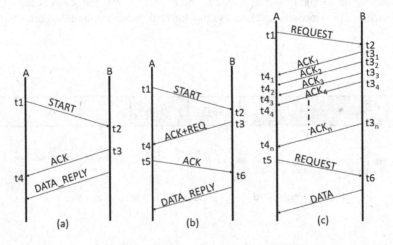

Fig. 3. A ranging with TWR (a), SDS-TWR (b) and SDS-TWR-MA (c) protocols.

$$ToF_{TWR} = \frac{(t_4 - t_1) - (t_3 - t_2)}{2} \tag{1}$$

While it is simple to implement, TWR is also very sensitive to timing inaccuracies due to clocks drift.

SDS-TWR (Symmetric Double-Sided Two-Way Ranging). In order to compensate the weakness of TWR, [10] proposes to symmetrise the protocol by adding a message to TWR basic ranging. This symmetry allows SDS-TWR (Fig. 3b) to reduce the effect of differences in the clocks of two nodes involved, in turn reducing the ranging error. The ToF is computed with Eq. (2) in this case:

$$ToF_{SDS-TWR} = \frac{(t_4 - t_1) - (t_3 - t_2) + (t_6 - t_3) - (t_5 - t_4)}{4} \tag{2}$$

SDS-TWR-MA (Symmetric Double-Sided Two-Way Ranging Multiple Acknowledgements) The SDS-TWR-MA was introduced in [11]. The purpose of this protocol is to reduce the inaccuracy in the original SDS-TWR. Since the clock difference varies over time, many estimates of distance are collected and averaged to reduce the impact of random disturbances. To obtain these "k" estimates, SDS-TWR must be executed "k" times. The multiple execution was used in [12]. In order to reduce the overhead due to multiple executions, the authors have developed a version of SDS-TWR with multiple acknowledgments called SDS-TWR-MA, where the number of frames, therefore timestamps is increased. For each exchanged frame (REQUEST, DATA, ACK), the nodes timestamp the departure and arrival time of the message. At the end of the measurement phase, an additional frame containing all the timestamps is sent to the node which initiated the protocol. This solution however has the disadvantage to have a large number of dedicated messages.

Our objective is to finely characterise these different protocols, and to propose an improvement in order to incorporate ranging traffic in the usual network traffic such as data, ack and beacon messages. This work had advantageously used a hardware and software platform developed by our team.

2.2 Testbeds for Evaluating Protocols Performance in Real Conditions

Some research projects have helped to design and deploy real open environments, allowing users to evaluate the performance of proposed protocols in real environment. The term testbed is commonly used to refer to these platforms, which may include a large number of nodes, allowing scalability tests.

One of the first public large-scale testbeds in France is the SensLab project (INRIA). An important step towards the Internet of Things was the deployment of the FIT/IoT-LAB [13] platform in 2012. One of its rivals, in terms of size, is the SmartSantander platform [14]: each of these testbeds hosts several thousands of nodes.

While most mentioned testbeds are based on the physical layer specified in IEEE 802.15.4-2006, 868 MHz and 2.4 GHz, we found a few public testbeds which propose less common modes of transmission such as UWB. Yet today, many ambitious projects still use the DSSS PHY, such as UWB LoRA Fabian [15] or Freescale [16] for example. Therefore, we resorted to the OpenWiNo platform [17] for this study. A presentation of this testbed will be provided in Sect. 4. The version [5] used in this study implements the IEEE 802.15.4 UWB specification.

3 Problem Statement

Ranging protocols previously presented suffer from a common problem: they all need dedicated messages for ranging. Especially, those exchanges should be performed on the shortest possible duration, which greatly constrains the message scheduling and may impact the Medium Access Control layer. Moreover, the ToF estimation is based on the use of two clocks, one on each node, assumed to be identical. TWR protocol

(Eq. 1) considers that timestamps t_2 and t_3 are based on the same crystal as t_1 and t_4, whereas on a real system, the two nodes are distinct. There is a difference between clocks, which introduces a bias in the ToF since $\{t_1, t_4\}$ and $\{t_2, t_3\}$ are not based on the same clock source. We will present an error model which focuses on this problem later in this article. If we increase the inter-message delay, this difference introduces a bias in the ToF, because of the accumulation of the drift's impact. SDS-TWR, imposing symmetry in the ranging protocol, precisely compensates the error introduced by the clock difference. However, this symmetry is made possible by an additional message, which increases the solution's cost (overhead). As such, if one of the two protocols should be the basis for a work of integration into an existing traffic, TWR would be less constrained because it requires only two messages for the materialisation of the timestamps t_1, t_2, t_3 and t_4, followed by a message containing t_2 and t_3 sent to the node at the initiative of ranging.

In order to integrate ranging traffic in the usual network traffic (data, beacons, ack, etc.), it is necessary to evaluate the real impact of clock drift on ranging measurement. In fact, network traffic will be encapsulated in messages which will be more or less delayed depending on the MAC protocols. In the case of CSMA/CA, the *backoff* significantly delays the transmission of messages. If ranging traffic is included in a series of beacons, the delay between beacons may be several tens of milliseconds. The real impact of this delay on the performance of ranging must be evaluated, preliminary to our work.

Finally, if SDS-TWR introduces a symmetry in order to compensate drift related errors introduced by the drift, this symmetry implies a fine control of transmission time: it seems impossible, in the context of an encapsulation of ranging traffic in the usual network traffic, to maintain this symmetry, since transmission instants will depend on the MAC protocols, and medium access method. A non-symmetrical approach as TWR is therefore preferred.

4 Preliminary Measurements Based on Testbed

In contrast to many studies which are based on theoretical assumptions, formal calculations, and more or less macroscopic simulations, our goal here is to finely characterize the real world based on testbed. We achieve this through our fast prototyping platform for protocols OpenWiNo [17]. A few public testbeds today are implementing UWB radios. We first implemented the well-known protocols TWR and SDS-TWR, to validate our environment and, on the other hand, to compare their performance. We used UWB transceivers developed by DecaWave [18]. Secondly, we evaluated the impact of the clock drift through experiments.

Testbed Description. OpenWiNo is an open source protocol development environment for WSN and the device layer of the IoT. It allows fast prototyping of original protocols, in C-language, for execution on a testbed of real nodes called "WiNos" (Wireless Nodes). The WiNos are developed using an Open Hardware approach. This allows a great versatility on the hardware; it is very simple, for example, to change a WiNo's physical layer: it is only a matter of replacing the transceiver and associated

Table 1. Features of developed WiNos

	WiNoRF22	TeensyWiNo	DecaWiNo	WiNoVW
CPU/RAM/ Flash	ARM Cortex M4 (32bit) 72MHz, 64kB RAM, 256kB Flash (PJRC Teensy 3.1)			
Transceiver (library Arduino)	HopeRF RFM22b : 200-900MHz, 1-125kbps, GFSK/FSK/OOK, +20dBm RadioHead		DWM1000 UWB IEEE 802.15.4 DecaDuino	Various VirtualWire
Sensors	temperature, light	idem + pressure, acceleration, compass, gyroscope	temperature, light	
Usage	WSN, IoT		IoT with ranging, indoor localisation	Very Low rate on medium
Availability	DIY	snootlab.com	DIY (Do It Yourself)	

library (Table 1). In this study, we used the DecaWiNo [5], a WiNo built on UWB transceivers developed by DecaWave, compliant with the IEEE 805.15.4 2011 standard, and an ARM CORTEX M4 (Teensy 3.1 Arduino-compliant board: Freescale MK20DX256VLH7 rated at 72 MHz; 64 kb RAM and 256 kb flash memories6). In order to use the physical layer efficiently, a dedicated library has been designed by our research team: this library is now available online [19].

Figure 4 shows three types of WiNo (c): WiNoRF22 (a) TeensyWiNo (b) and DecaWiNo (c). The WiNos are integrated into the Arduino ecosystem, which allow the researcher to easily add hardware components and/or software (sensors and actuators, advanced processing algorithms, interaction devices, library…) in order to prototype a complete solution.

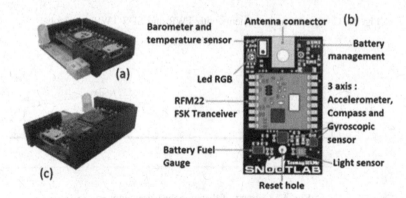

Fig. 4. The WiNos nodes

4.1 Implementation and Comparison of TWR and SDS-TWR

In a first step, in order to have real temporal characteristics, we performed a comparative metrology between TWR and SDS-TWR on our testbed. This preliminary experiment aims to validate the implementation of hardware and software components, but also to share the raw results from our testbed with the scientific community: the provided data may for example be used to develop models of nodes desynchronisation over time or test synchronisation solutions. The experiment was configured as follows: the maximum distance between the nodes is 5 ms (Figs. 5 and 6). Samples are collected every 0.5 m. The measurement protocol is as follows: a new ranging session starts every 200 ms and runs for about 30 s. The measurements were performed with a positioning of the nodes placed 15 cm above the aluminium rail with the PVC bracket. Through experiments, we verified that the impact of the rail on the ranging quality is negligible. In all experiments presented in the paper, we consider that the nodes are in Line Of Sight (LOS), i.e. without obstacle. The two nodes are powered with 1.8 m (6') USB cables.

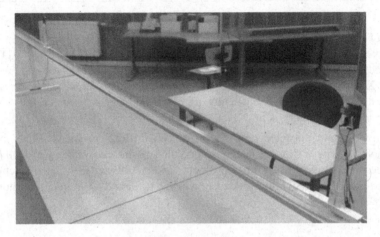

Fig. 5. Context of measurements with TWR and SDS-TWR protocols.

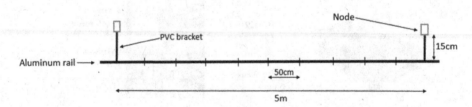

Fig. 6. Context of distance measurements

Ranging with TWR. The ranging data obtained by the TWR protocol is available in Table 2 and represented in Fig. 7. As we can see, the distance measured in TWR is very close to the actual distance: the absolute error is limited (16 cm) and the low standard deviation shows that the results are reproducible. Moreover, the results match the specification provided by the transceiver's manufacturer, which indicates an

Table 2. Error summary in TWR (in meters)

Actual distance	Dist. TWR	Average error	Max error	Min error	Standard deviation
0,5	0,339	−0,160	−0,11	−0,24	0,029745754
1	0,881	−0,118	−0,05	−0,17	0,02427616
1,5	1,417	−0,082	−0,03	−0,14	0,023683855
2	1,993	−0,006	0,05	−0,06	0,024150096
2,5	2,480	−0,019	0,03	−0,07	0,017567371
3,5	3,481	−0,018	0,03	−0,07	0,022417601
5	5,046	0,046	0,08	0	0,016043844

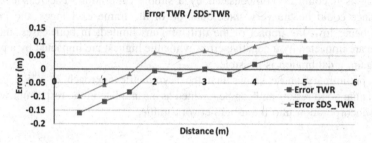

Fig. 7. Representation of SDS-TWR /TWR error as a function of the distance (Color figure online)

accuracy of ten centimetres. This step allows us to conclude positively about the proper implementation of TWR protocol in our environment.

Ranging with SDS-TWR. The ranging data is available in Table 3 and represented in Fig. 7. The experimental conditions are the same as those presented previously with the TWR protocol. We make the same observation, the distance measured by SDS-TWR is very close to the actual distance, which confirms the correct implementation of the protocol in our environment.

Table 3. Error summary in SDS-TWR (in meters)

Actual distance	Dist. SDS_TWR	Average error	Max error	Min error	Standard deviation
0,5	0,4	−0,099	−0,05	−0,15	0,02623868
1	0,942	−0,057	−0,02	−0,09	0,01638611
1,5	1,482	−0,017	0,03	−0,07	0,02390231
2	2,061	0,061	0,11	0,01	0,02122458
2,5	2,546	0,046	0,08	0,01	0,01724975
3	3,067	0,067	0,11	0,03	0,02023639
5	5,107	0,107	0,15	0,07	0,01559689

4.2 Discussion

The experiments showed that the mean error on both SDS-TWR and TWR does not exceed 16 cm and has a low standard deviation (~ 0.02). This implies a high reproducibility of the experiments: therefore, collecting a hundred samples per distance is not required to obtain accurate results. This is an encouraging trait since high variability would have required many executions of the ranging protocol in order to reach a stable distance estimation and would have hampered the encapsulation of the ranging protocol in usual network traffic. This characteristic will allow the reduction of the number of measurements.

We also noticed that there was little difference between TWR and SDS-TWR in terms of ranging error, although SDS-TWR has been designed to reduce the errors related to clock differences in TWR. The difference between the two curves of Fig. 7 appears as is it could be compensated by a simple calibration. The reason for this resemblance could be the very short duration of the frame exchanges: the protocol duration being low, the impact of the drift remains limited. In both cases, the short distances are impacted by a negative error, while the highest are impacted by a positive error. Again, a calibration is possible.

This preliminary phase shows that the asymmetric approach of TWR protocol presents quite good results and can be retained as a candidate in the rest of our work, to study its incorporation into the usual network traffic.

5 Artificial Delays and TWR Performance

In this section, we will evaluate the ranging accuracy in the presence of artificial delays introduced between the START and ACK messages in TWR (Fig. 8). In fact, when the ranging messages will be encapsulated in the usual traffic of the network, unavoidable delays will be introduced in particular by the medium access method. Therefore, we will investigate the consequences on the accuracy of a delay introduced in TWR.

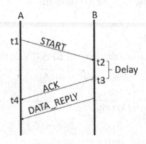

Fig. 8. The TWR protocol with a delay between exchanges

In order to get closer to the conditions of real radio communications, we introduced an artificial delay between t_2 and t_3 on node B and increased the timeout value on node

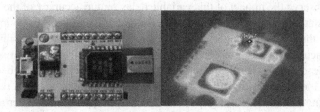

Fig. 9. Node DW1000: image in the visible and infrared

A. This will allow the difference between the clocks to increase, degrading the ranging performance.

A new experiment was performed on 11 positions regularly spaced 0.5 meter between nodes A and B (from 0.5 m to 5.5 m). For each point, we set a delay of 1, 2, 3, 4, 5, 6, 7, 8, 9, 10, 16 and 21 ms. Thirty measurements were performed for each delay value, yielding 240 samples per position. A new ranging session begins every 200 ms. The results for inter-node distance of 2 m are presented on Fig. 10.

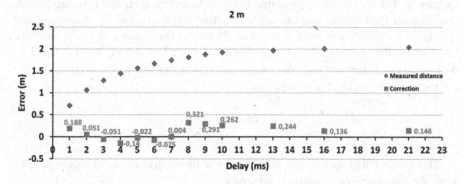

Fig. 10. Representation of the error and correction as a function of delay (Color figure online)

Table 4. Summary table of distances corrected.

Distance / Delay	0,5 m	1,5 m	2,5 m	3,5 m	4,5 m	5,5 m
1 ms	-0,26	0,06	0,15	0,06	0,24	0,23
2 ms	-0,32	-0,04	-0,01	0,04	0,21	0,31
3 ms	-0,29	-0,07	0,05	0,00	0,24	0,27
4 ms	-0,37	-0,02	0,17	0,02	0,22	0,38
5 ms	-0,17	-0,09	0,05	-0,10	0,36	0,17
6 ms	-0,30	-0,12	0,11	0,11	0,34	0,23
7 ms	-0,27	-0,04	0,03	0,05	0,04	0,24
8 ms	-0,23	-0,12	0,19	0,05	0,16	-0,01
13 ms	-0,16	0,06	0,24	0,12	0,32	0,06
16 ms	-0,11	0,04	0,27	0,18	0,38	-0,02
21 ms	-0,17	0,20	0,19	0,22	0,35	0,28

We can observe the impact of this variable delay on the accuracy of the ranging: the performance deteriorates with the increase in the delay. Large delays also imply that the nodes stay in the RX state longer: since reception is very energy-consuming [18], this leads to an increase in the transceiver's temperature. This impacts the resonance frequency of the quartz ad further degrades performance. Figure 9 has been captured using a laser thermometer and shows that, the longer the node stays in reception, the higher the temperature. While it is possible to alter the protocol in order to avoid this situation, it remains a general case which must be taken into account.

6 Proposed Error Mitigation Method

In order to compensate the error introduced by the increase of the delays in TWR, we propose a correction method based on the measurement of the clocks' relative drifts.

6.1 Principle of the Correction

Let's consider nodes A and B, A being the initiator of the ranging protocol. The equation of ToF in TWR was presented in Eq. 1. In reality, t_4-t_1 and t_3-t_2 are periods characterised by different clocks, running at frequencies f_A and f_B. Assuming that $f_A = f_B$, ToF estimation is correct. However, in reality, the clocks will not be identical. Let's rewrite the equation while taking into account the fact that f_A does not equal f_B. With k the coefficient differentiating these two clocks defined as $k = \frac{f_B}{f_A}; k \approx 1$.

The general formula for ToF becomes:

$$ToF_{Corrected} = \frac{(t_4 - t_1) - k * (t_3 - t_2)}{2} \tag{3}$$

This allows compensating the error introduced by the drift of two clocks, especially when the time between t_2 and t_3 is important.

The DW1000 on which our testbed is based features a *ClockOffset* functionality: by analyzing the corrections made by the PLL to decode the signal, it provides an estimate of the difference between the transmitting and receiving clocks, expressed in parts per million (ppm). Through the DecaDuino library, the protocols in the upper layer have access to this information. In our case, node A applies the correction as formulated in Eq. 4. This single measurement allows the mitigation of both drift and temperature related errors.

$$ToF_{Corrected} = \frac{(t_4 - t_1) - (1 + ClockOffset * 10^{-6}) * (t_3 - t_2)}{2} \tag{4}$$

6.2 Evaluation

In the last round of measurements, the *ClockOffset* if used to enhance ranging performance. The experiment is taking place in the same conditions as previously. Figure 10 shows the absolute error before (in red) and after correction is applied, as a function of the delay. These results correspond to an inter-node distance of 2 m. Table 4 summarizes all results.

It is worth mentioning that the ranging error has decreased considerably once the correction was applied: for delays up to 70 ms, the ranging error remains under 20 cm, which brings us back to the values initially measured for the time constrained version of the protocol. The error mitigation scheme we propose allows the same performance while relaxing the time constraint. This makes the inclusion of the ranging traffic in the network's communication flow a feasible task.

7 Conclusion

In this article, we addressed a fundamental issue in the field of WSN-based localisation. Improving the ranging performance will allow the creation of innovative services in the IoT context. We introduced the context of our work and the benefits of using TOF compared to RSSI. We introduced different ranging protocols encountered in the literature. In order to offer a localisation service compliant with the protocol stack, the ranging protocols should be integrated into the usual traffic (DATA, ACK, BEACONS). For this, they need to be delay-tolerant, which is not the case in theory: this sensitivity to delay was introduced in Sect. 3 and confirmed through measurements on a real testbed in Sect. 5. In order to reach our goal, we introduced an original and efficient correction of ranging error, based on clock drift estimation. We proposed an implementation of this correction on our testbed and evaluated the correction pragmatically. The results are satisfying and show that after correction, the ranging messages can be exchanged without any temporal constraint, with an error of about twenty centimeters in the worst case. Therefore, the ranging protocol will be integrated in the common messages in the network.

The results obtained so far led us to consider a few potential research topic. For example in the near future, a ranging protocol based on the cost-effective use of beacons and data frames will be investigated. The objective will be to create a continuous localisation service which will run in the background and take advantage of existing communication. Another perspective is the improvement of the ranging precision in Non Line-Of-Sight (NLOS) situations: in this context, we will evaluate the performance of the UWB technology which is said to be robust to multipath. Finally, we wish to give access to a subset of our testbed to the research community by making a set of 10 nodes remotely available through a web interface.

References

1. Gui, L., Val, T., Wei, A., Dalcé, R.: Improvement of range-free localisation technology by a novel DV-Hop protocol in wireless sensor networks. Ad Hoc Netw. J. (2014)
2. Guozhi, S., Dayuan, T.: Two novel DV-Hop localisation algorithms for randomly deployed wireless sensor networks. Int. J. Distrib. Sens. Netw. **2015**, 1 (2015)
3. Wang, P., Zhang, X., Sun, G., Xu, L., Xu, J.: Adaptive time delay estimation algorithm for indoor near-field electromagnetic ranging. Int. J. Commun. Syst. (2016)
4. Dalcé, R., van den Bossche, A., Val, T.: An experimental performance study of an original ranging protocol based on an IEEE 802.15.4a UWB testbed. In: IEEE International Conference on Ultra-Wideband, Paris (2014)
5. van den Bossche, A., Dalcé, R., Fofana, N.I., Val, T.: DecaDuino: an open framework for wireless time-of-flight ranging systems. In: Wireless Days, Toulouse (2016)
6. Shenghong, L., Hedley, M., Collings, I.B., Humphrey, D.: TDOA-based localisation for semi-static targets in NLOS. IEEE Wireless Commun. Lett. **4**(5), 513–516 (2015)
7. Kulakowski, P., Vales-Alonso, J., Egea-López, E., Ludwin, W., García-Haro, J.: Angle-of-Arrival localisation based on antenna arrays for wireless sensor networks. Comput. Electr. Eng. **36**(6), 1181–1186 (2010)
8. IEEE Standard for Local and metropolitan area networks, Part 15.4: Low-Rate Wireless Personal Area Networks (2011)
9. Microsoft Indoor Localisation Competition, IPSN (2015). http://research.microsoft.com/en-us/events/indoorloccompetition2015. Accessed 2016
10. Gentile, C.: Distributed sensor location through linear programming with triangle inequality constraints. In: IEEE Conference on Communications (2006)
11. Kim, H.: Double-sided two-way ranging algorithm to reduce ranging time. IEEE Commun. Lett. **13**(7), 486–488 (2009)
12. Wang, D., Kannan, R., Wei, L., Tray, B.: Time of flight based two way ranging for real time locating systems. In: 10th IEEE Conference on Robotics Automation and Mechatronics, Singapore (2010)
13. Tonneau, A., Mitton, N., Vandaele, J.: A Survey on (mobile) wireless sensor network experimentation testbeds. DCOSS (2014)
14. http://www.smartsantander.eu. Accessed April 2016
15. http://www.labfab.fr/portofolio/lora-fabia. Accessed April 2016
16. www.nxp.com. Accessed April 2016
17. van den Bossche, A., Dalce, R., Val T.: OpenWiNo: an open hardware and software framework for fast-prototyping. In: IoT International Conference on Telecommunications. Thessaloniki, Greece (2016)
18. http://www.decawave.com. Accessed April 2016
19. https://www.irit.fr/∼Adrien.Van-Den-Bossche/DecaWiNo. Accessed April 2016

Physics-Based Swarm Intelligence for Disaster Relief Communications

Laurent Reynaud[1,2(✉)] and Isabelle Guérin-Lassous[2]

[1] Orange Labs, Lannion, France
laurent.reynaud@orange.com
[2] Université de Lyon / LIP (ENS Lyon, CNRS, UCBL, INRIA), Lyon, France
Isabelle.Guerin-Lassous@ens-lyon.fr

Abstract. This study explores how a swarm of aerial mobile vehicles can provide network connectivity and meet the stringent requirements of public protection and disaster relief operations. In this context, we design a physics-based controlled mobility strategy, which we name the extended Virtual Force Protocol (VFPe), allowing self-propelled nodes, and in particular here unmanned aerial vehicles, to fly autonomously and cooperatively. In this way, ground devices scattered on the operation site may establish communications through the wireless multi-hop communication routes formed by the network of aerial nodes. We further investigate through simulations the behavior of the VFPe protocol, notably focusing on the way node location information is disseminated into the network as well as on the impact of the number of exploration nodes on the overall network performance.

Keywords: Controlled mobility · Physics-based swarm intelligence · Virtual forces · Unmanned aerial vehicles · Disaster relief communications

1 Introduction

In the wake of disaster, communication systems play a key role for the support of appropriate response operations. During such times, when existing terrestrial networks may be damaged or even completely impaired, Public Protection and Disaster Relief (PPDR) teams need to lean on a reliable emergency communications infrastructure to be able to restore essential services and more generally to quickly provide an adequate assistance to the affected population [2]. Regarding these requirements, multiple temporary network architectures relying on either terrestrial or satellite systems have been proposed [7]. All these architectures were tailored to meet the specific PPDR requirements, including the need to support a scalable network provision for challenging environments (e.g. in terms of temperature, hygrometry, landforms, obstacles, etc.), with robust equipment to be operated while the user is in motion, able to achieve a rapid deployment of a temporary infrastructure and services to address network outages in the affected zone [1]. In this regard, the European Electronic Communications Committee (ECC) defines

© Springer International Publishing Switzerland 2016
N. Mitton et al. (Eds.): ADHOC-NOW 2016, LNCS 9724, pp. 93–107, 2016.
DOI: 10.1007/978-3-319-40509-4_7

broadband PPDR temporary additional capacity as the means to provide additional network coverage at the scene of the incident with equipment such as ad hoc networks [2]. Over the last few years, airborne networking has also gained momentum to roll out rapidly deployable PPDR communication systems. In fact, aerial platforms, which can be designed to fly and operate at different altitudes, are increasingly valued for the multiple advantages they can offer in the context of disaster relief communications [12]. Some of the expected benefits are favorable propagation conditions with frequent line-of-sight transmissions, low latency compared to satellite equipment, and communication payload modularity enabling mission versatility. Moreover, regulation bodies have started considering aerial platforms as a viable set of technologies to roll out an emergency response in the first hours after a disaster [5], with many kinds and sizes of aerial platforms. Basically, each platform type displays distinct features best suited to different applicative perspectives: for instance, low altitude tethered balloons have been largely investigated in the context of disaster relief [3,4,6] for their cost-efficiency, low complexity of operation and ability to rapidly lift a telecommunication payload and act as temporary cell towers as long as response operations are required. Other initiatives investigate the use of high altitude, long endurance platforms in the context of disaster relief communications [8]. Although still facing multiple technical challenges, such solutions should provide large network coverages from the low stratosphere during weeks or months.

In this work, we take a particular interest in low altitude platforms with high mobility dynamics, also known as unmanned aerial vehicles, for their aptitude to self-propel and quickly bring small telecommunication payloads where and when needed. Further, we consider autonomous and distributed mobility control mechanisms for these platforms: those mechanisms, when they also enable neighboring nodes to cooperatively adapt their respective trajectories and behavior via local information exchange, are known under the name of swarming (or flocking) strategies [9,10]. In this regard, we investigate how to design an efficient swarming strategy based on virtual force principles, with the objective to deploy steerable nodes that are able to form a temporary wireless network and support disaster relief communications. This paper is organized as follows: Sect. 2 presents the disaster relief scenario as well as the network topology that are considered in the context of this study. Section 3 details prominent works and principles related to virtual force-based mobility control mechanisms and explains the design choices made for the extended Virtual Force Protocol (VFPe), which builds upon our previous study [15] by generalizing the force system and enabling node location dissemination in the network via extended multi-purpose beacon messages. Further, we analyze in Sect. 4 the performance of VFPe through representative network simulations, and finally conclude in Sect. 5.

2   Scenario

Regarding disaster relief scenarios, the main chronology events generally encompass three distinct phases: preparedness, response and recovery. We propose a network architecture which targets the response stage in this time line, with the specific objective to offer as quickly as possible a reliable and efficient communication environment to the public protection staff, rescue teams and other end users located at the scene of the considered incident or emergency. To this end, we specify four types of network nodes, as Fig. 1 shows:

- *Traffic (T-type)* nodes impersonate consumer devices and subsequent traffic requirement on the disaster site. A pair (source, destination) is arbitrarily chosen from this set and in the subsequent simulations, we monitor the user traffic from source to destination.
- *Surveillance (S-type)* nodes roam the exploration zone Z_e, discover nodes in physical proximity and record their location for later use and dissemination.
- *Relay (R-type)* nodes are former S-type nodes which became intermediate nodes in a communication chain between a (source, destination) T-type pair. When no longer useful in the chain, those nodes can revert to the S-type.
- A *prospection (P-type)* node is physically located at the apex of a forming communication chain. It not only has the role of intermediate node like regular R-type nodes, but is also in charge to evaluate which neighboring S-type node should be changed into a R-type node to further extend the communication chain. When no longer legitimate in this role, a P-type node switches to the R-type or S-type, depending on the context.

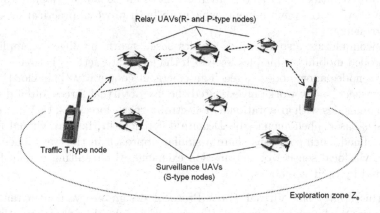

Fig. 1. Outline of the network equipment in the considered disaster relief scenario.

Regarding node movements, two mobility patterns can be distinguished:

- S- and T-type nodes roam the exploration zone Z_e in a similar fashion, at random, however T-type nodes possess a slower pedestrian-like velocity compared to the other nodes.
- R- and P-type nodes enforce a cooperative mobility strategy which we present and study in this work. With this mobility pattern, the considered R- and P-type nodes exert mutual virtual forces which result into node movement.

These mobility schemes are further detailed in the rest of this study, and nodes are in particular given specific mobility patterns in Sect. 4. Moreover, in terms of network deployment, we explain in the next section how nodes use the VFPe protocol to cooperatively create a communication chain between a user traffic (source, destination) pair through local information exchange.

3   Protocol Design

3.1   Related Works

With regard to swarming concepts applied to network communications, two prominent approaches have been pursued, based on either the principles of stigmergic collaboration or physicomimetics:

- Stigmergy, dating back to studies on social insects performed in the late 1950s [18], refers to the ability, for a swarm of nodes, to adopt an emergent behavior via cooperation and traces left in the environment, with the example of virtual pheromones. These messages intrinsically embed the location of their emission and may allow the creation of gradients based on pheromone additivity and decay [11]. Stigmergy can therefore be used to design path planning or obstacle avoidance strategies and more generally support collective node behavior. Yet, the implementation of pheromones requires either a centralized entity to act as the environment or an exchange system of incomplete views of the environment, which can both prove impractical in actual deployments.
- Physicomimetics, also known as physics-based swarm intelligence, applies to controlled mobility principles for which the local interactions between neighboring nodes allow nodes to reach an emergent cooperative behavior [13]. In this respect, many works investigated the use of virtual forces, often derived from analogies with gravitational or electromagnetic forces [13,15,17] or other physics-based phenomena [16]. Despite differences in the way virtual forces are defined, such solutions share a similar approach in the way nodes evaluate, via local sensing or information exchange, the resulting virtual forces exerted by their neighbors.

In this study, our Virtual Force Protocol design proposal also relies on the principles of physicomimetics. However, the general approach in the literature related to virtual forces presupposes swarms made of a large number of nodes, and incurs the formation of large-scale and steady topologies such as vast grids [13], lattices or hexagonal distributions [17]. Recent works investigate more dynamic and interest-driven topology formations [16] yet still address mesh topologies with many redundant communication links, best suited for mobile sensor networks. In contrast, we seek to address deployment scenarios in which the number of network nodes is constrained. In this challenging environment, our solution can steer a limited number of relay nodes to form suitable multihop routes between nodes in need to establish a communication. To meet this requirement, two classes of VFPe core mechanisms were designed: (i) a virtual force system exerted on all intermediary nodes of a communication chain, providing the means to control their mobility pattern and (ii) a beacon-based mechanism to create and maintain communication chains made of P- and R-type intermediary nodes between the user traffic endpoints. Beacons are also used to discover neighboring nodes and disseminate their positions, thereby allowing chain nodes to accurately match the desired topology. The following subsections detail both mechanisms.

3.2   Virtual Force-Based System

With VFPe, three types of virtual forces are applied on the network nodes. The first two forces are based on physicomimetics precepts [13], respectively impersonating physics-based attraction-repulsion and friction forces. As specified in our earlier works [14,15], both forces can be used to maintain neighboring nodes at a required distance by means of local distance-based interactions and without the need to interact with a centralized mobility planning entity.

An Attraction-repulsion Force f is defined so that any node N located in either the attractive or repulsive zone of another node P is under the influence of P's force f, collinear with vector $\boldsymbol{PN}$, as illustrated by Fig. 2: P's repulsive zone is the disc centered on P with a radius d_r, while P's attractive zone is the annular surface positioned at a distance $d\_f \leq d \leq d\_a$ from P. Further, f's intensity can be defined so that it depends on the distance between N and P, Fig. 2 outlining two exemplary intensity profiles for f. In the context of this work, we consider the simple piecewise-defined function such that $f = \|\boldsymbol{f}\| = I$ in the repulsive area, $f = -I$ in the attractive area and $f = 0$ otherwise.

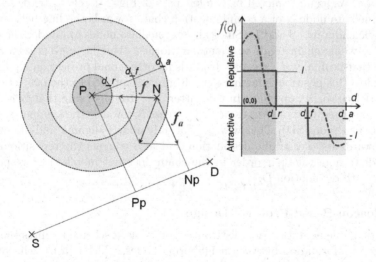

Fig. 2. Virtual forces f (attraction/repulsion) and f_a (alignment) exerted on a node.

A Friction Force f_{fr} is exerted on a node which is located in the friction zone of a nearby node. Retaining the previous notations, P's friction zone is the annular surface positioned at a distance $d\_r \leq d \leq d\_f$. If N were located in P's friction zone (which is not the case in the example of Fig. 2), a friction force f_{fr}, collinear with N's velocity vector, would be applied on N, so that $f_{fr} = \|f_{fr}\| = -Cx$. The use of a friction zone here is essential since it allows a neighboring node to decelerate and stop its motion within the boundaries of the considered zone. In this context, its absence would result into undesirable oscillating motion patterns for VFPe-enabled nodes, and be detrimental to the formation of the target chain topology.

It is worth highlighting that in [15], we conducted a in-depth study of the aforementioned parameters I and Cx and in particular sought optimal values. Consequently, key VFPe parameters in Sect. 4 were valued accordingly. However, the referenced study considered specific deployment environments in which the position of the user traffic destination is known and where communications chains are formed in a straight line. In contrast, as previously delineated by Sect. 2, our current work addresses more general deployment scenarios where no assumption is made on the knowledge regarding the location and movement patterns of the user traffic sources or destinations. In this challenged environment, the simple combination of attraction-repulsion and friction forces only ensures an acceptable distance between successive intermediate nodes in a communication chain, but does not guarantee that those chains are correctly directed towards the traffic endpoints. We therefore added a third component in the VFPe virtual force system.

An Alignment Force f_a is used to ensure that in a communication chain, successive intermediate nodes actually direct the chain towards the user traffic destination. As again depicted by the left part of Fig. 2, force f_a tends to steer an intermediate node N of a communication chain towards the line between the user traffic endpoints S and D. Since all intermediate nodes are under the effect of f_a, the whole chain tends to eventually form a straight line between S and D. Note that normally, f_a steers N towards its orthogonal projection Np on line (SD), unless this position is farther away from D, compared to the projection Pp of N's predecessor, P, on (SD). In this latter case, to make sure that N is closer from D compared to its predecessor P, f_a steers N towards the symmetric point of Np about Pp on (SD). That way, f_a not only allows aligning and orienting a chain towards its user traffic destination, but also triggers the repositioning of intermediate nodes if their order in the chain does not match their respective distance with destination D.

3.3   Beacon-Based Protocol Design

A realistic implementation of a distributed force-based scheme requires the local exchange of information between neighboring nodes, which can be achieved via the emission by each node at regular time intervals of a specific beacon message over the radio communication links. With VFPe, we designed a beacon containing information about the emitting node, in order to enable the creation and maintenance of communication chains between user traffic sources and destinations [15]. In essence, the beacon entries relate to the emitting node identifier and type, position and velocity vectors. Also, if the considered node is a relay in a chain, it contains information about successor, predecessor, destination identifiers and whether a new nearby node should be inserted in the chain as an additional relay. On this basis, VFPe is able to contextually insert S-type nodes into a chain, or on the contrary remove intermediary P- or R-type nodes, which then revert back to their S-type state.

Compared to [15], we extended the VFPe beacon specification in order to meet our PPDR scenario requirements where the position of user traffic endpoints is initially unknown, must be discovered, and varies with time. Our additions are twofold:

- Supplementary fields were added to account for both source and destination position discovery. A time-related field was also inserted to time-stamp the whole entry and be able to assess position errors with respect to time, as well as to allow the implementation of a VFPe scheme where beacons optionally use the freshest entries, as will be seen in the next section.
- As illustrated by Fig. 3, the VFPe structure was extended to allow multiple node entries, the first entry always relating to the beacon emitting node. Additional entries are used to more quickly disseminate network node position and status. The number of entries in the beacon as well as the way entries are sorted is tunable, and is investigated further in the rest of this work.

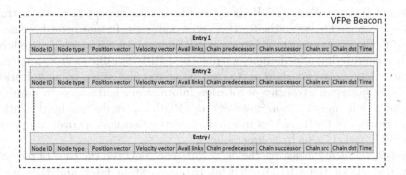

Fig. 3. Multi-entry structure of a VFPe beacon

4 Performance Evaluation

4.1 Simulation Parameters

As previously mentioned, VFPe entails both an exploration phase as well as a chain formation and maintenance phase. Since we were particularly interested to evaluate the impact of the exploration schemes on the overall VFPe performance, we designed the set of simulations accordingly. All (P, R, S and T)-type node mobility schemes were implemented within the release 3.23 of the ns-3 network simulator. Network nodes embed an IEEE 802.11b/g network interface card configured for High-Rate Direct Sequence Spread Spectrum (HR-DSSS) at 11 Mb/s. Moreover, the Optimized Link State Routing protocol (OLSR) [19] supports here multi-hop communication capability, and lossless radio propagation are assumed for the communication devices. Two T-type nodes constitute a user traffic (source, destination) pair and are initially scattered at random within the exploration area Z_e. Other nodes start with the S-type status and are initially located at the center of Z_e. Moreover, regarding the controlled mobility of the R-type and P-type nodes, VFPe is configured so that each node exerts a virtual force with key parameters valued as shown in Table 1. On this matter, it is worth noting that the description of these force-based parameters, as well as the justification for the chosen values, which allow an efficient use of the VFPe protocol, can be found in [15]. Likewise, mobility patterns are valued in Table 1, and the Random Waypoint Model (RWP) is mentioned when relevant.

Furthermore, each point of the simulated performance curves illustrated and analyzed in the rest of this section results from averaging 2000 independent simulations. The same applies for the measure of contact times between nearby S-type nodes mentioned hereafter. Additionally, the handling of error margins is such that confidence intervals are constructed at a confidence level of 95 %. Error bars are shown accordingly for each performance curve in the subsequent figures. Finally, results are analyzed based on the following performance metrics:

- *Packet Delivery Ratio* (PDR) is defined here with respect to the user traffic flow exchanged between the (S, D) T-type node pair. In that regard, we model this traffic with a constant bitrate (CBR) flow at 10 Kb/s, which is relevant when considering added-value, narrow band PPDR traffic such as predefined status messages and short messages, point-to-point voice communications, vehicle status and transfer of location information, and access to databases in small volumes [1]. More precisely, the PDR is here defined by the ratio of the number of CBR packets received by destination node D over the number of CBR packets sent by source node S.
- *End-to-end delay* measures the delay difference between the time of reception by node D of the CBR packets at the application layer and the time of emission of these packets by node S, still at the application layer.

4.2   Simulation Results

We first took interest in how the performance of the controlled mobility-based VFPe protocol evolves with the way VFPe beacon messages are disseminated by the S-type nodes, when roaming the exploration area Z_e. As described in Sect. 3, VFPe-enabled nodes regularly emit a beacon message, which contains a maximum of cs entries related to previously discovered nodes. Besides the first entry which is related to the emitting node itself, we purposely did not specify

Table 1. Main simulation parameters

Exploration zone	$Z_e = 1000$ m $\times$ 1000 m	
Mobility patterns	T-type	Position initially uniformly distributed on Z_e, RWP, velocity $\in [0.25, 1]$ m/s
	S-type	Position initially at center of Z_e, RWP, velocity $\in [5, 10]$ m/s
	P- and R-type	VFPe-based mobility, velocity $\in [0, 10]$ m/s
Nodes	Number of T-type nodes = 2	
	Number of (P+R+S)-type nodes = N, node mass = 1 kg	
	When fixed, $N = 15$. Otherwise, $1 \leq N \leq 30$.	
Network	802.11b/g, HR-DSSS at 11 Mb/s, radio range = 100 m, constant speed propagation delay model	
Routing	OLSR protocol with default parameter values [19]	
VFPe parameters forces	Beacon emission interval = 1 s	
	Interaction f	$d\_r = 50$ m, $d\_f = 75$ m, $d\_a = 100$ m, $Th\_dmin = 40$ m, $Th\_dmax = 75$ m, $I = 1$ N [15]
	Friction f_{fr}	$Cx = 2$ [15]
	Alignment f_a	$f_a = 2$ N if node is getting closer to the target realignment position, $f_a = 4$ N otherwise
User traffic	CBR flow at 10 Kb/s between the T-type node pair, CBR packet size = 100 bytes	

at the design stage how the remaining entries should be chosen and prioritized by the emitting node. Instead, we implemented 3 variants of the VFPe protocol, in terms of how VFPe-enabled nodes broadcast their beacon messages:

- *VFPe with random contacts*: according to this scheme, a VFPe-enabled node chooses the remaining entries at random, irrespectively of the time elapsed since the local information related to these nodes was first generated.

- *VFPe with fresh contacts*: this strategy, unlike the previous scheme, selects the remaining entries to insert in the VFPe beacon by prioritizing the freshest entries (i.e. which were added or updated last in the node local database).
- *VFPe with ideal node knowledge*: we also designed a VFPe extension in which the knowledge of other nodes' coordinates is not approximate and based on the disseminated VFPe beacons, but is assumed to be perfectly known at all times. We considered this ideal strategy to handle an upper range performance as a reference; however, this theoretical extension does not address how, and at what cost in terms of signalization overhead, these exact positions could be realistically retrieved in an actual network deployment.

In addition, we sought to complement the simulations by evaluating the performance of a RWP-only mobility scheme that does not make use of the VFPe controlled mobility. No VFPe beacons are in this case disseminated and therefore, S-type nodes are never turned into relay nodes. As a consequence, communication chains between the T-type (S,D) pair are never constructed and nodes, whether S- or T-type nodes, only rely on a RWP mobility pattern and on the underlying Mobile Ad Hoc Networks (MANET) OLSR routing protocol.

Incidence of the Contact Size. As previously mentioned, we define the contact size as cs, the maximum number of entries contained by a VFPe beacon. In this first series of simulations, we studied the impact of the number of node information exchanged during a contact between nodes in direct range able to exchange VFPe beacon messages. To this end, we carried out simulations with different values $1 \leq cs \leq 20$ and measured the PDR of the CBR flow between the (S, D) T-type node pair, as well as the average end-to-end delay taken by this flow. As illustrated by Fig. 4 (left), the RWP-only and VFPe-cs-ideal schemes give a low and high performance range with a respective PDR of about 4 % and 39 %, which both remain constant with cs. At this stage in the analysis, it must be stressed that in the context of the considered network scenario, a maximum PDR value below 40 % does not imply here a performance issue and is the expected consequence of the incompressible time needed for unmanned vehicles to move and for the desired chain topology to be established and allow the transmission of user traffic.

Regarding the VFPe schemes with beacon messages containing either random entries (VFPe-cs-random) or fresh entries (VFPe-cs-fresh), it can be observed that the PDR is about 14 % for $cs = 1$ (i.e. when nodes only disseminate their own information in the VFPe beacons) and increases with cs, converging towards a similar PDR value around 34 %. Further, the random scheme exhibits a significantly steeper increase, since the PDR is already approaching 32 % for $cs = 2$, while the PDR of the fresh scheme is still below 22 %. Although this behavior may seem counter-intuitive at first, it can however be explained with the scenario assumptions. As an illustration, with our parameter values and setting $cs = 5$, we measured for the random scheme an average contact time of 15.4 s between the nearby S-type nodes, which is much larger than the VFPe beacon emission interval. More generally, several successive beacons are likely to be exchanged during such contact times, and the random scheme will offer more diversity in terms of

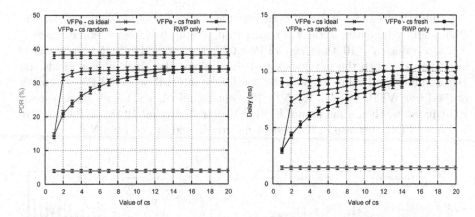

Fig. 4. PDR (left) and end-to-end delay (right) related to the CBR packets received by destination node D, with the number cs of node entries contained in beacon messages.

disseminated node entries. As a result, even with low cs values, nodes will handle more accurate coordinates with the random scheme and will be steered towards positions more likely to create the desired communication chains between the (S, D) T-type node pair.

In terms of end-to-end delay, Fig. 4 (right) shows the performance of the considered four schemes. The RWP-only scheme does not use VFPe beacons, hence exhibiting a delay which remains constant with cs. In contrast, the VFPe-cs-ideal scheme increases with cs despite a constant PDR on the whole range of cs, as previously seen. To explain this, it is worth highlighting that in the implementation of our 3 VFPe-enabled schemes, each node entry takes 36 bytes in a VFPe beacon message. Besides, although all simulations rely on a HR-DSSS modulation at 11 Mb/s, the VFPe frame payload is transmitted at a lower bitrate of 1 Mb/s, since the considered beacons are broadcast. As a result, each additional node entry (and therefore incrementation of cs) incurs an increase of about 0.29 ms, which in turns impacts the CBR queued traffic awaiting transmission on the wireless shared medium. Moreover, the delays for both VFPe-cs-random and VFPe-cs-random schemes are lower than 9.5 ms for the whole range of cs values and remain compatible with the low-latency requirements of interpersonal communications, for instance. In that light, the use of the VFPe-cs-random scheme may be preferred since the PDR approaches its maximum value with only a low contact size value (e.g. $cs = 4$), that is with a small-sized beacon.

Incidence of the Number of Nodes. For any given deployment scenario, the number of rolled-out systems has a strong impact on the workability of operations and on the overall cost of the solution. We therefore thought to assess how the performance of a controlled mobility scheme like VFPe evolves with the number of network nodes, and in particular with N, the initial number of S-type nodes. Figure 5 (left) gives a representation of the PDR results for each of the four considered mobility strategies, with $1 \leq N \leq 30$. In this set of simulations as well, the RWP-only scheme gives a reference, and a low performance range, regarding what can be expected from a simple MANET protocol here: overall, the PDR slightly increases with N, ranging from 3 % up to 4.5 %. Assuming here a constant value $cs = 10$, the three VFPe-enabled schemes give PDR results which are consistent with the previous set of simulations: the ideal scheme yields the best results for all values of N in the considered interval, while VFPe-cs-random outperforms VFPe-cs-fresh for $N \geq 10$. Besides, all three schemes show a steep PDR increase, then a maximum PDR value for N around 16 to 18, followed by a regular PDR decrease in the rest of the considered interval for N.

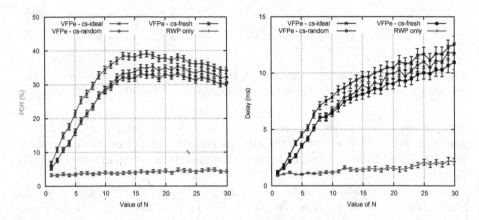

Fig. 5. PDR (left) and end-to-end delay (right) related to the CBR packets received by destination node D with N, the initial number of S-type nodes.

Moreover, Fig. 5 (right) allows stressing that the end-to-end delays for the three VFPe-enabled schemes, although still acceptable since they remain below 15 ms for all considered values of N, sharply increase with N. As a whole, this set of simulations shows that a limited set N_{max} of S-type nodes (such that $16 \leq N_{max} \leq 18$ with our assumptions) is needed to reach the best network performance. At this point, in order to properly apprehend those results, it is important to recall that the speed range of exploration S-type nodes, which is $[5, 10]$ m/s, exceeds the standard speed conditions normally found in pedestrian-type scenarios, where MANET protocols, and in particular link-state protocol like OLSR, behave best. As a result, when $N = 16$, enough S-nodes are initially available so that VFPe induces the best possible chain topologies. As a result, above that threshold for N, no additional S-type node is statistically expected to be converted into a relay (P, R)-type node to extend the communication chains, and the extra S-type nodes will be used by VFPe for exploration purposes only. When N increases above that threshold, the beneficial effects of adding S-type nodes in the network, and therefore increasing link diversity for the underlying multi-hop routing protocol, are severely hindered by the detrimental effects of high-velocity S-type nodes crossing the communication chains. In effect, the underlying MANET protocol will detect those transient and unstable links and try to exploit the corresponding routes, which is in turn likely to result into CBR packet losses, the time for OLSR to detect the link failures and to rebuild its routes accordingly. Further, an in-depth analysis of the simulation traces confirms the increasing amount of OLSR route losses due to the temporary presence of unneeded nearby S-type nodes, when N increases above the aforementioned threshold. In this regard, corrective steps could be taken to prevent the routing protocol from exploiting those suboptimal routes or even to repel S-type nodes (through an extra repulsion force effect, for instance) from an established communication chain. However, it is important to highlight that marginally improving the overall network performance via an increase of additional nodes is not desirable in the context of disaster relief deployment scenarios, for which we seek a limited number of deployed communication systems. With that requirement in mind, the implications from this analysis are threefold: first, VFPe allows reaching a satisfying network performance with a limited number N_{max} of nodes, then that this number N_{max} can be precisely defined, and lastly, that N_{max} remains sufficiently low ($16 \leq N_{max} \leq 18$ with the given assumptions) to offer the perspective of cost-efficient network deployments in real conditions.

5   Conclusion

In this work, we presented the VFPe protocol, which, by the use of virtual force principles, allows a swarm of mobile network nodes to fly cooperatively, to rapidly form wireless multi-hop communication routes and to efficiently transmit end-user traffic flows. After describing our network architecture proposal and explaining the design choices made for VFPe, notably in terms of force system and beacon-based information dissemination in the network, we analyzed a series of simulation results to assess the impact of the exploration schemes on the overall VFPe performance. We first analyzed how VFPe behaves with the way its beacon messages are disseminated by the network nodes in their exploration phase, and verified that our proposals for realistic VFPe scheme implementations compare satisfyingly with an ideal strategy. We then scrutinized the progression of VFPe performance with respect to the number of network nodes and identified, in the specific context of high mobility underpinning our applicative scenarios, that a maximum efficiency can be obtained with a limited set of nodes. Those results lead us to conclude that VFPe is able to provide an adapted solution for the rapid deployment of temporary networks, notably suited to the demanding context of disaster relief and cost-efficient emergency communications where in particular the amount of communication systems, and in this case the number of mobile network nodes, is constrained. Currently, we are planning to embed a VFPe implementation on quad-rotors to further evaluate the solution through experimentation. In the future, we intend to study the benefits of jointly using a virtual force-based protocol with disruption- and delay-tolerant schemes.

References

1. Report ITU-R M.2377: Radiocommunication Objectives and Requirements for Public Protection and Disaster Relief (2015)
2. CEPT ECC Report 199: Public protection and disaster relief spectrum requirements (2013)
3. Qiantori, A., Sutiono, A.B., Hariyanto, H., Suwa, H., Ohta, T.: An emergency medical communications system by low altitude platform at the early stages of a natural disaster in indonesia. J. Med. Syst. **34**, 1–12 (2010)
4. Reynaud, L., Rasheed, T., Kandeepan, S.: An integrated aerial telecommunications network that supports emergency traffic. In: 14th International Symposium on Wireless Personal Multimedia Communications (WPMC), Brest, France (2011)
5. The Role of Deployable Aerial Communications Architecture in Emergency Communications and Recommended Next Steps, White Paper, Federal Communications Commission (FCC), Washington, USA (2011)
6. Gomez, K.M., et al.: Aerial base stations with opportunistic links for next generation emergency communications. IEEE Commun. Mag. (2016)
7. Nelson, C.B., Steckler, B.D., Stamberger, J.A.: The evolution of hastily formed networks for disaster response: technologies, case studies, and future trends. In: 2011 IEEE Global Humanitarian Technology Conference (GHTC), pp. 467–475, Seattle, USA (2011)
8. Deaton, J.D.: High altitude platforms for disaster recovery: capabilities, strategies, and techniques for emergency telecommunications. EURASIP J. Wirel. Commun. Netw. **2008**(1), 1–8 (2008)
9. Varela, G., et al.: Swarm intelligence based approach for real time UAV team coordination in search operations. In: 3rd World Congress on Nature and Biologically Inspired Computing (NaBIC), Salamanca, Spain (2011)
10. Daniel, K., Rohde, S., Goddemeier, N., Wietfeld, C.: A communication aware steering strategy avoiding self-separation of flying robot swarms. In: 5th IEEE International Conference on Intelligent Systems (IS 2010). IEEE, London (2010)
11. Brambilla, M., Ferrante, E., Birattari, M., Dorigo, M.: Swarm robotics: a reviewfrom the swarm engineering perspective. Swarm Intell. **7**(1), 1–41 (2013)
12. Reynaud, L., Rasheed, T.: Deployable aerial communication networks: challenges for futuristic applications. In: 9th ACM Symposium on Performance Evaluation of Wireless Ad hoc, Sensor, and Ubiquitous Networks (PE-WASUN), pp. 9-16, Paphos, Cyprus (2012)
13. Spears, W.M., Spears, D.F., Hamann, J.C., Heil, R.: Distributed, physics-based control of swarms of vehicles. Auton. Robots **17**(2/3), 137–162 (2004)
14. Reynaud, L., Lassous, I.G., Calvar, J.-O.: Mobilité contrôlée pour la poursuite de frelons. In: 17èmes Rencontres Francophones sur les Aspects Algorithmiques des Télécommunications (ALGOTEL 2015), Beaune, France (2015)
15. Reynaud, L., Guérin Lassous, I.: Design of a force-based controlled mobility on aerial vehicles for pest management. Submitted to Elsevier Ad Hoc Networks Journal (2016)
16. Le, D.V., Oh, H., Yoon, S.: VirFID: a Virtual Force (VF)-based interest-driven moving phenomenon monitoring scheme using multiple mobile sensor nodes. Ad Hoc Netw. J. **27**, 112–132 (2015). doi:10.1016/j.adhoc.2014.12.002
17. Yu, X., Liu, N., Huang, W., Qian, X., Zhang, T.: A node deployment algorithm based on van der Waals force in wireless sensor networks. Int. J. Distrib. Sens. Netw. **2013** (2013)

18. Grassé, P.: La reconstruction du nid et les coordinations interindividuelles; la théorie de la stigmergie. Insectes Soc. **35**, 41–84 (1959)
19. Clausen, T., Jacquet, P.: RFC3626, Optimized Link State Routing Protocol (OLSR). Experimental, http://www.ietf.org/rfc/rfc3626.txt

PHY/MAC/Routing in Sensors/IoT

Routing and TDMA Joint Cross-Layer Design for Wireless Sensor Networks

Lemia Louail$^{(\boxtimes)}$ and Violeta Felea

FEMTO-ST Institute, University of Bourgogne-Franche-Comté, Besançon, France
{lemia.louail,violeta.felea}@femto-st.fr

Abstract. This paper presents a new cross-layer approach in which the network layer and the data link layer are combined in one new layer in order to correlate decisions on forwarding nodes involved in multi-hop communications in wireless sensor networks. More particularly, we propose a Routing and MAC joint protocol that takes the network as an input parameter and provides a routing tree and a collision-free (TDMA) schedule as output parameters while minimizing the latency of communications.

Simulation results show better performance of the proposed approach than classic routing and MAC protocols. The latency is improved from 17% up to 20% with shorter schedule lengths up to 9%.

Keywords: Wireless sensor networks · Cross-layer approaches · Routing protocol · TDMA scheduling

1 Introduction

Nowadays, Wireless Sensor Networks invaded many domains, from military to environment monitoring and even healthcare. This innovating technology enables connectivity between small devices (sensors) equipped with processing and communication capabilities. The sensors are deployed in an area in order to collect information and transmit it, through multi-hop communications, to a central location called the Base Station or the Sink.

To ensure the communication in such networks, the sensors use a simplified OSI (Open Systems Interconnection) stack composed of five layers [1]. In this protocol stack, each layer is responsible of ensuring particular functionalities, and protocols are implemented independently from one another. In this context, we are interested in two layers that are involved in communication decisions: the network layer responsible of finding routes between a sender and a receiver through the routing protocol and the data link layer responsible of coordinating communications in direct neighborhood of nodes through the MAC (Medium Access Control) protocol.

Traditionally, these two protocols are implemented independently, and minimizing the communication latency can be difficult when incoherent decisions are coming from these two protocols. Better performance is achieved when using cross-layer approaches [2,3].

© Springer International Publishing Switzerland 2016
N. Mitton et al. (Eds.): ADHOC-NOW 2016, LNCS 9724, pp. 111–123, 2016.
DOI: 10.1007/978-3-319-40509-4_8

In this paper, we aim to simultaneously determine a routing tree and a suitable collision-free (TDMA) scheduling for all nodes of the sensor network which minimize communication latency. We combine the network layer and the data link layer in one new layer providing one protocol that ensures functionalities of both layers. We use a traffic pattern that is uniformly distributed in which every node of the network captures environmental data and needs to send it regularly to the sink. This kind of scenario is used in many sensor network applications when monitoring an environment.

The remainder of this paper is structured in the following manner. Section 2 presents some existing cross-layer works where the network and the data link layers are combined into one layer. In Sect. 3 we present our contribution which is a new protocol combining routing with TDMA scheduling. Section 4 presents the simulation parameters that were used and the metrics we evaluated. In Sect. 5 we present the simulation results obtained in order to evaluate the performance of our approach. Section 6 presents the theoretical upper bound of the schedule length. We conclude in the last section.

2   State of the Art

Two types of cross-layer approaches concerning the routing and MAC protocols can be found in literature.

In the first type, one protocol is based on information of another protocol. MAR-WSN [4] is a MAC-aware routing protocol in which the next hop decision is made using information of the TDMA scheduling. Dissimilar to CoLaNet [5], which proposes a TDMA scheduling based on information of the routing tree. In the second type, the two layers can be combined in one layer and from the two former protocols emerges one new protocol.

In this paper, we are interested in the second type of cross-layer solutions. Three existing protocols fall in this category: AIMRP [6], RBF [7] and SIF [8].

AIMRP [6] is an address-light integrated MAC and routing protocol aiming to reduce the power consumption of the nodes in the network. It uses a tier-based routing scheme that is integrated with a contention-based MAC protocol similar to IEEE 802.11. Its main idea is to organize the nodes of the network into tiers around the sink, and to route the packets by progressively forwarding them to tiers closer to the sink. RTS and CTS packets are used to initiate the communication between the source and the next hop in the next tiers and to switch off the radio of all nodes within the communication range of the source, except for the next hop.

RBF [7] is a cross-layer integrated MAC/routing protocol based on the Received Signal Strength Indicator (RSSI) that is transmitted regularly by the sink. To choose a next hop, each node uses the RSSI parameter to ensure sending data in the direction of the sink, and the four-way RTS-CTS-DATA-ACK handshake to decide which nodes participate in the communication and which nodes can turn off their radio.

SIF [8] is a state-free and competition-based data delivery protocol aiming to reduce the communication overhead, the packet delivery ratio and the average packet delay. In SIF each node has a "distance-to-sink" information used to conduct the communication in the direction of the sink. A handshake system is associated to the distance-to-sink information to combine the tasks of routing and MAC via one protocol. A competition between receivers, at the transmission time, is used to eliminate the requirement of global and local state maintenance.

These three approaches use contention-based MAC protocols in which the neighbors of a sender (except for the next hop) can turn off their radios if they are not concerned with the communication and thus save their energy. This concerns all the nodes in the path leading to sink. But no information about the other nodes of the network is given. If a node is not in a routing path when can it be awake and when can it be switched to the sleep mode?

We propose to use contention-free MAC protocols jointly with routing protocols in a cross-layer approach. We intend to address node duty-cycle, both when it is involved in the communication and when it is not concerned with any. As a data flow pattern, we consider concurrent flows of packets generated at the same time (at the first slot) by all the nodes of the network. This pattern is mainly used in applications that need to continuously monitor an environment and where nodes have a regular frequency of sending information to the sink.

Figure 1 presents a sensor network composed of 8 sensors (A to H) using a particular TDMA.

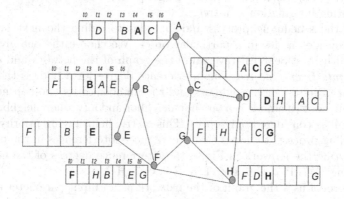

Fig. 1. A sensor network using a particular TDMA

The result of the TDMA protocol can be represented as a schedule. In this schedule, each node has one time slot to send information to its neighbors (marked in bold), the other time slots are either used to receive information from the neighbors (marked in italic), or they are not used at all (marked as empty). When a time slot is not used, the node can be switched to the sleep mode by turning off its radio and thus it saves energy.

In a TDMA schedule, each node knows exactly when it has to be awake and when it can go to sleep mode. This MAC protocol also ensures that there will be no collision caused either by the one-hop neighbors or induced by the two-hop neighbors through the hidden terminal problem. Finally, the communication latency is easily controlled within contention-free MAC protocols.

We note that no cross-layer combining routing with TDMA scheduling in a same layer was found in the literature.

3   IDeg-Routing&MAC

Our idea is to find, for each sensor node, a free slot in the schedule and a forwarding node; therefore, a contention-free MAC schedule and a routing tree are constructed simultaneously. This approach is called IDeg-Routing&MAC.

To model a TDMA schedule we consider a matrix $Schedule[N, l]$ where the number of lines N represents the number of nodes (each node has a line which is considered its schedule) and l is the length of the schedule. Each element of the matrix represents a slot. $Schedule[i, j] = i$ means that node i sends its data during slot j. $Schedule[i, j] = k$ means that node i receives data from its direct neighbor k at slot j. Finally, $Schedule[i, j] = \varnothing$ means that node i does not use the slot j for transmission neither for reception. We note that $1 \leq i \leq N$ ($1 \leq k \leq N$) and $1 \leq j \leq l$.

The routing pattern we considered is many-to-one where all nodes send information to the sink. Each node has one single parent as forwarding node, which models routing information as a tree.

During the scheduling process and in order to define the next hop for a node, we consider nodes in a particular order. We choose the one given by a BFS (Breadth First Search) traversal of the graph of the network and a metric called the *interference degree*. The interference degree for a node is the sum of the number of its one-hop neighbors and the number of its two-hop neighbors counted only once (if a node is a one-hop neighbor and a two-hop neighbor at the same time, it is considered only once). This metric helps giving priority, during the scheduling process, to nodes that create more interferences in the network.

Considering the network in Fig. 1, the interference degrees of the nodes are the following: A-7, B-7, C-6, D-6, E-5, F-7, G-7, H-7.

Our protocol uses the graph of the network as an input parameter and provides both the time slot schedule and the routing tree as output parameters. It works as follows:

1. Initially, the routing tree and the schedule are empty. The initial schedule length is equal to the maximum node degree in the graph incremented.
2. Construct a BFS tree applying a breadth first search on the graph of the network starting from the sink.
3. Find the leaf of the BFS tree with the highest interference degree in the network.
4. For this leaf:
    (a) Find its closest neighbor towards the sink in terms of distance, among all detected neighbors (based on graph information), and select it as a next hop.

(b) Add this couple (leaf, next hop) to the routing tree.
(c) Check if this leaf was chosen as a next hop:
  – If it was chosen as a next hop, consider the last slot among its children' slots and find a free slot after it, possibly in a circular research.
  – Otherwise, look for a free slot starting from the beginning of the schedule.
  – In both cases, if no free slot was found, a new slot is added at the end of the schedule and this leaf is scheduled on it.
5. Delete the current leaf from the BFS tree.
6. Repeat from step 3 until the BFS tree is empty.

At the end of this procedure, a TDMA schedule and a geographic routing tree are obtained. The slot allocation process concerns the transmission slot for a node. The other communication slots (in reception) can be induced afterwards based on the vicinity of every node given by the graph.

Figure 2 presents one BFS tree obtained for the graph in Fig. 1. This tree was constructed starting from the sink of the network and considering it as the root of the tree and then doing a BFS traversal of the graph modeling the network. The scheduling and the routing tree are implemented using this BFS tree and the interference degree that is computed for each node.

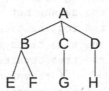

Fig. 2. A BFS tree of the graph in Fig. 1

Considering the BFS tree and the interference degrees of the nodes, the order in which the nodes will be considered is the following: H-G-F-D-C-E-B-A. We note that if two or more leaves have the same interference degree, one of them is chosen randomly.

The obtained scheduling is presented in Table 1, where the current node transmission slot is marked in bold and reception slots are in italic. This schedule provides an average latency of 5.14, while the random schedule in Fig. 1 has an average latency of 7. This average latency is computed as the sum of the latency of each node averaged by the number of nodes, 7.

The latency for a node depends on the data flow pattern and on the time slot of the nodes on the routing path to the sink. We consider that packets are generated with the same frequency at each node and are transmitted through the relay node to the sink during its transmission time slot. The reference time used to compute latency is considered the first slot, for all nodes.

We note that, because of the simplicity of the example, the obtained geographic routing tree is similar to the BFS tree.

Table 1. The obtained TDMA schedule

	slot1	slot2	slot3	slot4	slot5	slot6	slot7
node A	–	–	–	D	C	B	**A**
node B	–	–	F	E	–	**B**	A
node C	–	G	–	D	**C**	–	A
node D	H	–	–	**D**	C	–	A
node E	–	–	F	**E**	–	B	–
node F	H	G	**F**	E	–	B	–
node G	H	**G**	F	–	C	–	–
node H	**H**	G	F	D	–	–	–

Our approach IDeg-Routing&MAC needs to find successively, for every node, a free slot. A slot is considered free for the node if it is empty or if the nodes scheduled in the slot do not belong to its one-hop or two-hops neighbors and this to ensure that collisions do not occur.

The pseudo-code of IDeg-Routing&MAC is presented next.

```
algorithm IDeg-Routing-Joint-MAC (graph, routTree, schedule)
input   : graph          { the sensor network modeled as a graph }
output  : routTree        { the routing tree }
          schedule        { the time slots of each node }
local   : BFStree         { the tree obtained by applying a BFS
                            traversal on the graph }
          currentNode   { the current node }
          isSlotFound   { indication if a free slot was
                          found or not }
          nextHop       { the next hop of a node }
          lastSlotIndex { index of the last slot
                          of a node's children }
          freeSlotIndex { index of the free slot }
BFStree = computeBFS(graph)
while BFStree is not empty do
   currentNode  = findLeafHighIDeg(BFStree, graph)
   nextHop = findClosestNeighborToSink(currentNode, graph)
   add(currentNode, nextHop, routTree)
   if hasChildren(currentNode, routTree, schedule) then
      lastSlotIndex = findLastSlot(currentNode, routTree)
      isSlotFound = circularFindFreeSlot(lastSlotIndex, schedule)
      freeSlotIndex = getFreeSlotIndex(lastSlotIndex, schedule)
   else
      isSlotFound = findFreeSlot(1, schedule)
      freeSlotIndex = getFreeSlotIndex(1, schedule)
   endif
```

```
    if isSlotFound then
        allocate(currentNode, freeSlotIndex, schedule)
    else
        addNewSlot(schedule)
        freeSlotIndex = getNewSlotIndex(schedule)
        allocate(currentNode, freeSlotIndex, schedule)
    endif
    delete(currentNode, BFStree)
endwhile
```

4   Simulations Configuration

In this section, we present the configuration parameters used for the evaluation of the proposed algorithm through simulations and the evaluated metrics. We used a Java-based library called JUNG [10] that allows modeling, analyzing and visualization of networks as graphs.

5000 networks were generated randomly using $n = 100$ nodes, each node with a communication range $r = 25$ m. Nodes were deployed in a square area of size a^2. The connectivity model used is UDG (Unit Disk Graph [9]), in which two nodes are considered neighbors if they are within each other's communication range (r). Based on this model, the density of the generated networks is: $\delta = \pi * r^2 * n/a^2$. The deployment area size was changed to obtain different densities $\delta \in [4, 20]$.

Four metrics were used to evaluate the performances of our contribution:

- **Average latency:** The average latency was computed as the average of all total communication latencies from each node to the sink. This latency is the number of slots before the source node transmits (the reference slot being the first slot), and for every node in the routing path, except for the sink, one slot for transmission and the number of slots before its parent can transmit.
- **Average normalized latency:** The average normalized latency is computed as the average of per link latencies for all the nodes in the network. The latency per link for a node is computed as the total latency for the node divided by the number of hops in the routing path for the node to the sink.
- **Schedule length:** The schedule length is computed as the number of slots of the obtained schedule.
- **Duty cycle:** The duty cycle is computed as the ratio of the active period (estimated by the number of slots used either in transmission or in reception) to the total period (the schedule length). A lower duty cycle results in a lower energy consumption [15].

Results presented next are averages of the evaluated metrics over the 5000 generated networks.

5   Performance Evaluation

Since the existing cross-layer protocols based on joint decisions were all using contention-based MAC protocols, the comparison with our contribution which uses a contention-free MAC does not seem feasible. That is why, as our solution is the first routing joint TDMA schedule, we decided to compare it with a geographic routing and a random TDMA scheduling layered solution, that does not use cross-layering.

A Random TDMA is a schedule in which nodes are considered for allocation in a random order. For each node we look for a free slot in the schedule starting from the first slot. A new slot is added if no free slot is found.

Due to its stateless nature, geographic routing exhibits the lowest overhead. Early work in geographic routing considered only greedy forwarding [11] to decide to which node a packet should be forwarded. Every node knows the location of all of its neighbors and packets contain the location of the destination. Nodes forward packets to the closest neighbor to the destination. This is the routing protocol we used to compare our approach with. This kind of solution is very scalable but presents the problem of holes, i.e. nodes that are closer to the destination than any of their neighbors. As our goal is not to solve the problem of holes, networks presenting this problem were neglected in our simulations. Solutions (FACE [12], GOAFR+ [13]) derived from the greedy perimeter stateless routing [14] are proposed to avoid holes.

The left side of Fig. 3 shows that IDeg-Routing&MAC has better average latency than independent routing with random TDMA. First, because the BFS traversal of the graph gives a convenient order to nodes; nodes that are far from the sink will be leaves or at the bottom of the BFS tree, and nodes that are close to the sink will be in the upper part of the BFS tree. Then, considering nodes with the highest interference degree gives priority to nodes presenting the maximum constraints during the scheduling; these nodes can be leaves or internal nodes in the initial BFS tree.

The right side of Fig. 3 shows that the same remark stands for the average normalized latency. IDeg-Routing&MAC tries to find a slot for a node that is the closest to the slots of its children, therefore the average normalized latency which represents the per link latency is better than the normalized latency of the independent routing and MAC.

When IDeg-Routing&MAC looks for a free slot for a node, it does a circular research before adding a slot at the end of the schedule. This results in a shorter schedule length than the schedule of the independent routing and MAC as shown in Fig. 4.

As shown in Fig. 5, a shorter schedule length results in higher duty cycle. The duty cycle loss of our approach is higher only up to 7 % but this loss is balanced by the gain in the communication latency.

The previous results are averages of the obtained metrics for the 5000 generated networks and do not reflect data individually. We analyze the extent of variability of the obtained values in relation to the correspondent mean, on the basis of the coefficient of variation (Relative Standard Deviation). The coefficient of variation is expressed in percentage as $\frac{\sigma}{\mu}$ where σ is the standard deviation and μ is the mean. Table 2 shows the intervals for the coefficient of variation, irrespective of density. Three metrics are considered: the average latency,

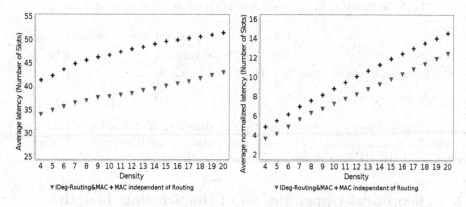

Fig. 3. Average latency (left) and average normalized latency (right) (Color figure online)

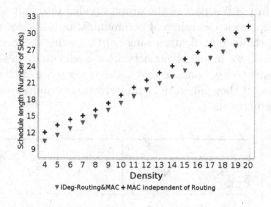

Fig. 4. Schedule length (Color figure online)

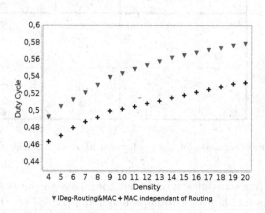

Fig. 5. Duty cycle (Color figure online)

Table 2. Intervals for the coefficient of variation irrespective of density

CV	Latency	Normalized latency	Schedule length
IDeg-Routing&MAC	$[7.5; 10.15]\%$	$[6.4; 10.6]\%$	$[3.8; 7.4]\%$
MAC independent of Routing	$[8.2; 11.3]\%$	$[7.0; 9.3]\%$	$[4.1; 7.1]\%$

the average normalized latency and the schedule length, depending on the analyzed allocation methods. All the obtained deviations are inferior to 11.3 % which shows that the absolute values are not very scattered.

6   Theoretical Upper Bound of the Schedule Length

We noticed from Fig. 4 that the schedule length depends on the density of the network with a multiplication factor that can be approximated theoretically. For that, we present in Fig. 6 a part of the graph divided into four $r \times r$ squares; in each square we represent the range of communication of nodes with a circle of radius r which is the communication range of the nodes according to the UDG [9] model with which we modeled our networks. Each square is also divided into four parts, each part $(r \times r)$ contains $\delta/4$ nodes (gray small circles in Fig. 6) that can induce collision if they generate communications simultaneously, where δ is the density of the network.

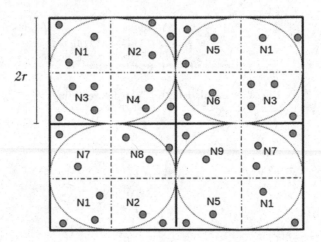

Fig. 6. UDG modeling and slot classes for slot allocation (Color figure online)

In contention-free schedules, a node can not be assigned a slot that is already used by one of its one-hop or two-hops neighbors. Based on the UDG model, this means that nodes in every square $(r \times r)$ cannot have the same slot as nodes in its direct adjacent squares and in the squares directly adjacent to the latter.

We obtain 9 different slot classes which are denoted from N1 to N9 in Fig. 6. A slot class is a set of slots that can be used by several nodes assuring that their communications are collision free. Every slot class needs at most $\delta/4$ different slots to be allocated to nodes inside the $r \times r$ square. This is explained by the fact that there are $\delta/4$ nodes in a $r \times r$ square, most of them in each other's communication range. Therefore, the length of the schedule is: $lg = 9*\delta/4$, which gives a multiplication factor of 2.25 applied to the network density in order to obtain the length of the schedule.

This reasoning was applied on a part of the network but can be extended to the hole network. The nine slot classes can be reused for the slot allocation necessary for every node.

The previous theoretical multiplication factor is an upper bound, but we can reduce even more the theoretical schedule length (diminishing the multiplication factor). This reasoning is based on a uniform distribution on nodes over the deployment area. When δ grows, the probability of having more nodes inside each square grows. In this case, the slots can be reused earlier, as shown in Fig. 7. On the left side of the figure we consider networks of densities $8 \leq \delta < 16$. In this case, the $r \times r$ squares contain, on average, one or two nodes. We can split each of these squares in two $r \times r/2$ rectangles (we illustrate here a vertical split, but an horizontal is similar). In this case, a maximum of 15 different slot classes are needed for scheduling all nodes of the network. In this case, multiplication factor equals $lg/\delta = 15/8$.

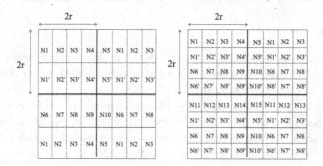

Fig. 7. Slot allocation for network densities between 8 and 16 (on the left) and superior to 16 (on the right)

Moreover, when density exceeds 16, split can be done on both dimensions (vertical and horizontal) such that each obtained square contains, on average, at least one node. In this case, 25 different slot classes are needed for scheduling all nodes of the network (right side of Fig. 7). Therefore, the multiplication factor equals $lg/\delta = 25/16$.

We estimate a theoretical upper bound for the schedule length which is:

$$lg = \begin{cases} 9/4^*\delta & \text{if } 4 \leq \delta < 8 \\ 15/8^*\delta & \text{if } 8 \leq \delta < 16 \\ 25/16^*\delta & \text{if } 16 \leq \delta \leq 20 \end{cases} \qquad (1)$$

These theoretical upper bounds are correlated with the results obtained in simulations as shown in Fig. 8.

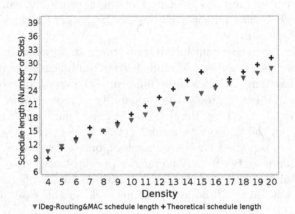

Fig. 8. Experimental schedule length vs Theoretical schedule length (Color figure online)

We remark that, for densities $4 \leq \delta < 8$ the multiplication factor 2.25 is the most adequate, the values of the theoretical schedule length being coherent with the values obtained experimentally.

For densities $8 \leq \delta < 16$ and $\delta \leq 16$, respectively, the theoretical schedule length is coherent with the experimentations at the beginning of each group of densities; then, the theoretical computation over-estimates the length of the schedule at the end of each group of densities. This shows that there are still nodes that can reuse slots already allocated which may decrease even more the multiplication factors 15/8 and 25/16.

7   Conclusion and Future Work

This paper focuses on the cross-layering approaches combining the network layer with the data link layer aiming to optimize communication latency in sensor networks. More particularly, it presents a new routing joint TDMA scheduling protocol in which a routing and a MAC protocol are combined in one new protocol.

Our approach IDeg-Routing&MAC is based on a BFS traversal of the graph modeling the network and the interference degree metric which gives priority, during the schedule process, to critical nodes presenting important interference in the network. Both a contention-free schedule for the communications between nodes and a geographic routing tree are obtained from this protocol.

Extensive simulations have shown that our approach improves up to 20 % the latency obtained with independent routing and MAC protocols. We also generated schedules with shorter lengths up to 9 %.

References

1. Akyildiz, I.F., Su, W., Sankarasubramaniam, Y., Cayirci, E.: Wireless sensor networks: a survey. Comput. Netw. **38**, 393–422 (2002)
2. Mendes, L.D., Rodrigues, J.J.: A survey on cross-layer solutions for wireless sensor networks. J. Netw. Comput. Appl. **34**(2), 523–534 (2011)
3. Melodia, T., Vuran, M.C., Pompili, D.: The state of the art in cross-layer design for wireless sensor networks. In: Cesana, M., Fratta, L. (eds.) Euro-NGI 2005. LNCS, vol. 3883, pp. 78–92. Springer, Heidelberg (2006)
4. Louail, L., Felea, V., Bernard, J., Guyennet, H.: MAC-aware routing in wireless sensor networks. In: IEEE BlackSeaCom 2015, Constanta, pp. 225–229 (2015)
5. Chou, C.F., Chuang, K.T.: CoLaNet: a cross-layer design of energy-efficient wireless sensor networks. In: IEEE Systems Communications, pp. 364–369 (2005)
6. Kulkarni, S., Iyer, A., Rosenberg, C.: An address-light, integrated MAC and routing protocol for wireless sensor networks. IEEE/ACM Trans. Netw. **14**(4), 793–806 (2006)
7. Awang, A., Lagrange, X., Ros Sanchez, D.: A cross-layer medium access control and routing protocol for wireless sensor networks. 10èmes Journées Doctorales en Informatique et Réseaux, Belfort, pp. 2–4 (2009)
8. Chen, D., Deng, J., Varshney, P.K.: A state-free data delivery protocol for multihop wireless sensor networks. In: WCNC, pp. 1818–1823 (2005)
9. Schmid, S., Wattenhofer, R.: Algorithmic models for sensor networks. In: 14th WPDRTS, pp. 450–459 (2006)
10. O'Madadhain, J., Fisher, D., Smyth, P., White, S., Boey, Y.: Analysis and visualization of network data using JUNG. J. Stat. Softw. **10**(2), 1–35 (2005)
11. Karp, B.: Geographic routing for wireless networks, Harvard University (2000)
12. Bose, P., Morin, P., Stojmenovic, I., Urrutia, J.: Routing with guaranteed delivery in ad hoc wireless networks. Wirel. Netw. **7**, 609–616 (2001)
13. Kuhn, F., Wattenhofer, R., Zhang, Y., Zollinger, A.: Geometric ad-hoc routing: of theory and practice. In: 22nd Annual Symposium on Principles of Distributed Computing, pp. 63–72 (2003)
14. Karp, B.N., Kung, H.T.: GPSR: greedy perimeter stateless routing for wireless networks. In: 6th International Conference on Mobile Computing and Networking, pp. 243–254, Boston (2000)
15. Zhuo, S., Song, Y.Q., Wang, Z., Wang, Z.: Queue-MAC: a queue-length aware hybrid CSMA/TDMA MAC protocol for providing dynamic adaptation to traffic and duty-cycle variation in wireless sensor networks. In: 9th IEEE WFCS, pp. 105–114, May 2012

Optimal Flow Aggregation for Global Energy Savings in Multi-hop Wireless Networks

Alexandre Laube[✉], Steven Martin, Dominique Quadri, and Khaldoun Alagha

LRI (Laboratoire de Recherche en Informatique), University Paris Sud, CNRS,
University Paris-Saclay, 91405 Orsay, France
{alexandre.laube,steven.martin,dominique.quadri,khaldoun.alagha}@lri.fr
http://www.lri.fr/

Abstract. Today, regardless of economic or environmental incentives, reducing energy consumption is of great importance when designing IT systems. In a multi-hop wireless network, taking into account energy usually consists in maximizing the lifetime of the network by uniformly distributing the traffic all over the nodes. In this paper, we propose a new routing paradigm which aggregates flows on a minimum number of nodes to maximize the number of nodes that can be turned off in the network. In this way, we intend to significantly reduce global energy consumption. To maintain the same quality of service while keeping the same network capacity, we combine flow aggregation with network coding. Our approach is particularly effective, in part because aggregation increases coding opportunities, as shown in the simulations in terms of global energy savings and network load compared to conventional routing algorithms.

Keywords: Multi-hop wireless networks · Global energy savings · Routing · Flow aggregation · Integer linear programming · Network coding

1 Introduction

Multi-hop wireless networks have performance issues especially in terms of bandwidth, delay and packet loss. These issues are mostly related to the wireless nature of the transmissions. Sharing the same communication medium implies the existence of interference between nodes. In such environment, a node within the interference area of another one can be prevented from receiving data properly. Thus, the capacity of these networks depends on the number of transmissions which can be performed simultaneously. In [1], the authors show that it decreases with increasing density.

Many researches have been conducted to address these issues, including Quality of Service (QoS) considerations. Related works propose methods to route flows more efficiently, taking into account interference [2,3], adapting the topology of the network by varying the transmission power of nodes [4,5] or coding packets in order to reduce the number of transmissions [6,7]. These works show

© Springer International Publishing Switzerland 2016
N. Mitton et al. (Eds.): ADHOC-NOW 2016, LNCS 9724, pp. 124–137, 2016.
DOI: 10.1007/978-3-319-40509-4_9

improvements in various metrics such as capacity, throughput, robustness, number of accepted flows and make such networks more popular. Such networks are now used for sensor and vehicular networks, to implement communication between robots and drones, to extend the coverage of an area or to quickly establish communications in an uncovered area in case of emergency. We can also find implementation of this technology in smartphones applications.

But another consideration has emerged in recent years, namely energy consumption. Indeed, in 2007, the ICT (Information and Communications Technologies) industry was responsible for 2 % of the global human made CO_2 emissions [8]. In 2013, this industry consumed 10 % of the total electricity produced worldwide [9]. This growth in the last decade can be explained by changes in use (e.g. smartphones have become commonplace), new applications that need more traffic and better accessibility (e.g. cloud computing, multimedia streaming, etc.). Thus adding energy awareness in communication protocols should be a priority concern to reverse this trend. This is largely the case in cellular networks, but the advent of smart cities, smart home, vehicular networks and the Internet of Things also requires effective solutions in the field of multi-hop wireless networks. Several works have already been done in this direction for mobile ad hoc and sensor networks. In [10], the authors study the ability of energy-aware protocols to provide QoS and show that they can be used without degrading too much performances. But these protocols make numerous assumptions about the nature of traffic, the power supply or the nodes' computing power in order to reduce the energy consumption in a local way. Typically, it is conventional to consider that nodes are powered with batteries. Hence, the energy consideration consists in maximizing the lifetime of the network (keeping the network connected as long as possible). A classical method is to distribute the load between nodes so that they deplete their battery power equivalently, in order to prevent a node from dying prematurely due to its overuse. However, even if this approach could reduce energy consumption locally, it cannot be optimal in terms of global energy savings of the network. In addition, nodes can be powered all the time, without batteries, which is the case we consider. Finally, as the energy consumption related to the wireless interface is just a portion of the overall node's power consumption, it is more efficient to completely turn off a node as much as possible.

This paper is focused in that direction. We propose to aggregate flows over a subset of nodes in order to minimize the number of nodes required to route a set of flows. Our solution consists in establishing an optimal routing mechanism that maximizes the global energy savings while respecting QoS constraints. Furthermore, by aggregating flows, we can strengthen the use of *network coding* [6], a technique designed to improve capacity, robustness and security of networks, by increasing coding opportunities. We use integer linear programming to modelize the problem. Branch-and-bound procedure is utilized to get the optimal solution, which will be used to evaluate future distributed and feasible approaches for the flow aggregation problem.

Our main contributions are the following:

1. Instead of minimizing the energy consumed by each node to maximize the lifetime of the network, in this paper we propose to minimize the global energy consumption by maximizing the number of nodes unused. Indeed, these nodes can be either switch off or put in a sleep mode. To do so, an innovative approach is proposed in wireless ad hoc networks, which aggregates the flows over a minimum number of nodes.

2. To maintain the same quality of service in terms of throughput, network coding is associated to aggregation. These two paradigms are complementary, as aggregating flows increases significantly coding opportunities.

3. To get an optimal solution, these two mechanisms are formulated through an integer linear program, solved using a branch-and-bound algorithm. We recall that this kind of algorithms is not a heuristic or approximating procedure, but an exact optimizing one that finds an optimal solution [11].

The remainder of this paper is organised as follows. Section 2 presents existing works related to energy consideration in multi-hop wireless networks and network coding. Section 3 recalls models used for multi-hop wireless networks and interference. Section 4 describes proposed solution based on flow aggregation that will be associated to network coding to enhance performances. Section 5 details simulation results. Section 6 concludes the paper.

2   Related Work

To address the problem of energy consumption in wireless networks, many studies have proposed to vary the transmission power of nodes (transmission power control). In multi-hop wireless networks, this is called topology control since varying the transmission power of a node modifies its transmission range and so its neighbourhood. A first approach consists of varying the nodes' transmission powers in order to obtain a 2-connected subgraph [12]. But this technique may need lots of computation and cannot be adapted in a distributed environment. Another approach considers the geometrical position of nodes instead of the global connectivity of the graph. In [13], the authors propose a cone-based protocol where a node sets its transmission power such that it has at least one neighbour in an angle of γ around it. This kind of mechanism doesn't take into account the traffic and only reduces the energy consumption per transmission. There is no global consideration on the energy consumption. To reduce the overall according to the existing flows, a solution is proposed in [5] where nodes can transmit using several power levels. It is efficient to minimize the transmission powers without compromising the provided QoS, but it does not attempt to turn off the nodes.

Some other works for ad hoc and sensor networks have addressed the issue of energy consumption at the network layer using energy as a cost in routing algorithms. In [10], an analysis of several energy-aware protocols shows that they are able to support QoS criteria quite well. Indeed, most of them increase the link stability and reduce packet losses. By reducing losses and limiting correction procedures (retransmission, redundancy, ...), they can improve the whole throughput.

These algorithms compute routes which uniformly distribute the traffic all over the network to avoid overusing some central nodes. Because most of sensor networks and mobile networks have nodes powered with battery, it is critical to keep them alive as long as possible, otherwise it may cause a partition of the network.

But distributing the load over a large number of nodes is not efficient in terms of global energy savings because it forces nodes to remain active, even if this is for transmitting a small amount of traffic. Switching off useless nodes enables to better use network resources and to reduce interferences. On the other hand, aggregating traffic may imply longer paths which can degrade the QoS of some flows. Hence, it would be wise to use techniques that can take advantage of aggregation.

Network coding (NC) is a recent method that allows intermediate nodes to combine packets before forwarding them in order to reduce the number of transmissions required to transport a flow. This principle can be illustrated with a simple Alice-Bob (A and B) communication scheme (Fig. 1). Indeed, let's be A and B two nodes who want to send a packet to each other using an intermediate node R (relay). Node A sends its packet to the relay and B does the same. If R uses a classical forwarding algorithm, it will use two transmissions to forward the packets (A's packet to B and B's packet to A). If the relay combines the packets, with a xor operation for example, it can broadcast the combination to A and B in one transmission. Then A (resp. B), having one part of the combination, can decode the packet sent by B (resp. A). Therefore, NC uses only three transmissions where four were needed with the classical method. This principle is used by Zhong et al. [6] who proposed a linear program to compute routes for a set of flows minimizing the number of transmissions. But this solution doesn't take interference into account. In [7], cliques constraints are added to improve the performance in terms of throughput reachable for each flow.

(a) Without network coding   (b) With network coding

Fig. 1. Network coding Alice and Bob scheme.

When transmitting with a wireless interface, the rate at which data are sent doesn't significantly impact the energy consumption of the interface [14]. Moreover, the authors conclude that it is, most of the time, more energy efficient to increase the sleep time of the interface, hence transmit data at the fastest rate available. When in idle mode (waiting to receive or transmit data), the interface consumes around height times more energy than in sleep mode. Considering that a simple energy model to calculate the energy consumption of any node i is $E_i = Es_i + Et_i(Tr_i) + Er_i(Rr_i)$, we can see the potential benefits of aggregation. Es_i is the static energy consumed when the node is on, its wireless interface remaining in idle mode. $Et_i(Tr_i)$ is the energy consumed by the node when it transmits data at rate Tr_i. $Er_i(Rr_i)$ is the energy used when receiving data at rate Rr_i. By aggregating flows, we increase the number of nodes that can switch the wireless interface in sleep mode or eventually switch themselves into a very low power mode.

3   Models Used

3.1   Network Model

We represent a multi-hop wireless network by a graph $G = (V, E)$ where V is the set of nodes of the network and E the set of links between those nodes. A couple (i, j), i and j belonging to V, is in E if there exists a link between nodes i and j. This link means that node i can send data to node j. More precisely, each node can transmit data to a certain distance. Thus, a link (i, j) exists if j is in the transmission range of node i.

3.2   Interference Model

Interference in wireless communications can be seen in two ways. The first one considers interference as noise. It takes into account the transmission power and the location of every node transmitting at the same time and computes a signal to interference ratio (SIR) for each transmission. A transmission is successfully received if its SIR is above a given threshold. The second way considers a transmission from i to j succeeded if the distance between these two nodes is lower than the distance between j and any other node transmitting at the same time as i. This approach prohibits interfering nodes to transmit at the same time. These two methods have been studied by Gupta and Kumar in [1].

We use a model based on a conflict graph [15] and cliques [16]. A conflict graph $CG = (V', E')$ models the interfering relationships between the links of the network, where V' is the set of links in G and E' is the set of edges in CG. There is an edge between two links (i, j) and (k, l) in CG if they interfere with each other. The conflict graph is defined from a set of interfering nodes. Considering two interfering nodes i and j, every outgoing link from i is connected with every adjacent link of j in CG. The set of interfering nodes can be built in two ways. A classical method considers that a node interferes with every node in its 2-hop neighborhood (its neighbors and their neighbors) [17]. The other method assumes that a node i interferes with a node j if j is in the interference area of i [1].

Cliques are the maximal subgraphs of CG and are sets of interfering links. Two links in the same clique must not be used at the same time otherwise the transmissions using those links will fail. Cliques can be used to derive necessary and sufficient constraints to accept or reject a flow in a multi-hop wireless network [16]. In wired networks, if the residual capacity of a link is lower than the rate required by a flow, we can't route that flow through this link without violating the QoS constraints. In a multi-hop wireless network, the residual capacity to take into account is not this of each link but this of each clique. Basically, cliques' utilization rates can be considered as a condition to define admission control. The utilization rate of any clique C is the sum of those of every link belonging to C. For example, if two links of C are used at 40 % and 60 % respectively, it is equal to 100 % and no more traffic can be routed through any link of C.

4   Solution

Our goal is to minimize the global energy consumption of ad hoc networks, which is the sum of energy consumed by the nodes. To do that, we want to maximize the number of unused nodes to be able either to switch them off or to put them in sleep mode. This section describes an innovative approach combining flow aggregation and network coding.

First, an optimal solution is proposed to aggregate flows (Sect. 4.1), in order to minimize the number of nodes used to route the flows while respecting QoS constraints in terms of throughput. Next, this solution is extended by adding network coding to enhance throughputs and increase the number of unused nodes (Sect. 4.2).

We use integer linear programming to modelize our problem. To get the optimum, this program is solved using a branch-and-bound algorithm, which is an approach developed for solving discrete and combinatorial optimization problems. The discrete optimization problems are those in which the decision variables assume values from a specified set. In our case, we will see that it is an integer set, leading to an integer programming problem. The combinatorial optimization problems, on the other hand, are to choose the best combination out of all feasible combinations. Most of these problems can be formulated as integer programs [18]. The essence of branch-and-bound approach is the following observation [18]: in the total enumeration tree, at any node, if it can be shown that the optimal solution cannot occur in any of its descendants, then there is no need to consider those descendent nodes. Hence, the tree can be "pruned" at this node. If enough branches are pruned in this way, the tree can be reduced to a computationally manageable size. Thus, the branch-and-bound approach is not a heuristic or approximating procedure, but an exact method that finds an optimal solution.

Figure 2 gives a simple example of our approach. Supposing a network composed of ten nodes and two flows a and b sending data with a rate of 10 units from 1 to 5 and from 10 to 6 respectively. A first routing algorithm (shortest path) will route flow a over the path $\{1, 2, 3, 4, 5\}$ and flow b over the path $\{10, 9, 8, 7, 6\}$. If every link has a capacity of 100 units, the two flows will be

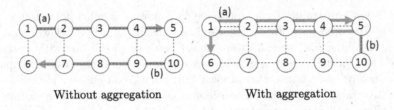

Fig. 2. Two flows that can be routed over parallel paths or aggregated together.

accepted and be able to increase their rate up to 20 units[1]. If their rate goes beyond 20 units, the cliques' utilization rates will exceed 100 %. Let's assume that flow b now follows path $\{10, 5, 4, 3, 2, 1, 6\}$. Flows a and b will be accepted if their maximum rate is lower than 14.28 units. Hence, by aggregating the flows we are still able to guarantee their QoS requirements. Besides, three nodes can be turned off, enabling 30 % energy savings. But the capacity of the network is decreased. This degradation is caused by flow b whose path is longer and the involved cliques.

This degradation can be avoided, since in nodes 2, 3 and 4, we can apply the Alice-Bob network coding scheme described above to save three transmissions for each packet sent by the sources (when data is waiting to be transmitted in the queue of each involved node). Rerouting flow b has added two transmissions: on links $(10, 5)$ and $(1, 6)$. Hence, with aggregation and network coding, we gain one transmission compared to the solution proposed by the first routing algorithm. So we can reach better performance in terms of energy consumption, but also in terms of capacity.

4.1   Aggregation

We propose RA (Routing with Aggregation) as an optimal solution for routing a set of flows minimizing the number of nodes used while respecting their QoS requirements in terms of capacity. This approach consists in solving the following optimization program, where our decision variables are x_i and x_{ij}^f which represent the state of node i (used or unused) and if link (i, j) is used to route flow f, respectively.

$$Min \sum_{i \in V} x_i$$

s.t.

(1) $\sum_{(i,j) \in c} \frac{d_{ij}}{C_{ij}} \leq 1, \forall c \in C$

(2) $\sum_{i \in V} x_{ij}^f - \sum_{k \in V} x_{jk}^f = 0, \forall f \in F, \forall j \in V - \{S_f, D_f\}$

(3) $\sum_{(i,S_f) \in E} x_{iS_f}^f = 0, \forall f \in F$

(4) $\sum_{(S_f,i) \in E} x_{S_f i}^f = 1, \forall f \in F$

---

[1] For sake of clarity, we don't detail computations and utilization rates of cliques. Details can be seen at https://www.lri.fr/~laube/RA/cliques.pdf.

(5) $\sum_{(D_f,i)\in E} x^f_{D_f i} = 0, \forall f \in F$

(6) $\sum_{(i,D_f)\in E} x^f_{iD_f} = 1, \forall f \in F$

(7) $\sum_{f\in F} x^f_{ij}.M_f = d_{ij}, \forall (i,j) \in E$

(8) $\sum_{(i,j)\in E} x^f_{ij} \leq 1, \forall f \in F, \forall i \in V - \{D_f\}$

(9) $\forall i \in V, x_i = \begin{cases} 0 \ if d_{ij} = d_{ji} = 0, \forall (i,j), (j,i) \in E \\ 1 \ otherwise \end{cases}$

with

$x_i \in \{0,1\}, \forall i \in V$

$x^f_{ij} \in \{0,1\}, \forall (i,j) \in E, \forall f \in F$

Here are the notations used in our formulation:

C: the set of cliques (derived from the conflict graph)

d_{ij}: the capacity reserved on link (i,j)

C_{ij}: the total capacity available on link (i,j)

F: the set of flows

(S_f, D_f): the source and the destination of flow f

M_f: the minimum capacity requested by flow f

x_i: equals to 1 if the node is used, 0 otherwise

x^f_{ij}: equals to 1 if (i,j) is used to route flow f, 0 otherwise

The first constraint is about the clique's utilization rates. It makes sure that the flows are routed with respect to the residual capacity of the cliques. Constraint (2) ensures the flow conservation which means that every intermediate node receiving a flow retransmits it to one of its neighbor. Constraints (3) and (5) avoid to route a flow such that it returns to its source or leaves its destination. Constraints (4) and (6) forces the source and the destination to be used as follows: source S_f transmits flow f on one of its outgoing links and destination D_f receives flow f from one of its ingoing links. Constraint (7) focuses on aggregation: the capacity reserved on a link is the sum of the capacity reserved by each flow on that link. Constraint (8) represents the single path constraint, that is a flow shouldn't be split in several paths. Finally, Constraint (9) is the condition that determines if a node is used or not. A node is used if it receives or sends traffic.

We can notice that this formulation is not linear due to logical constraint (9). Unfortunately, most of linear solvers need a linear formulation. This is why we use classical techniques in mathematical programming proposed by Glover [19] to linearize the above problem.

$$Min \sum_{i\in V} x_i$$

s.t.

(1) $\sum_{(i,j)\in c} \frac{d_{ij}}{C_{ij}} \leq 1, \forall c \in C$

(2) $\sum_{i\in V} x^f_{ij} - \sum_{k\in V} x^f_{jk} = 0, \forall f \in F, \forall j \in V - \{S_f, D_f\}$

(3) $\sum_{(i,S_f)\in E} x^f_{iS_f} = 0, \forall f \in F$

(4) $\sum_{(S_f,i)\in E} x^f_{S_f i} = 1, \forall f \in F$

(5) $\sum_{(D_f,i)\in E} x^f_{D_f i} = 0, \forall f \in F$

(6) $\sum_{(i,D_f)\in E} x^f_{iD_f} = 1, \forall f \in F$

(7) $\sum_{f\in F} x^f_{ij}.M_f = d_{ij}, \forall(i,j) \in E$

(8) $\sum_{(i,j)\in E} x^f_{ij} \le 1, \forall f \in F, \forall i \in V - \{D_f\}$

(9.1) $Max_j(C_{ij}).x_i - \sum_{(i,j)\in E} d_{ij} \ge 0, \forall i \in V$

(9.2) $Max_j(C_{ji}).x_i - \sum_{(j,i)\in E} d_{ji} \ge 0, \forall i \in V$

(9.3) $\sum_{j\in V}(d_{ij} + d_{ji}) - Min_{f\in F}(M_f).x_i \ge 0, \forall i \in V$

(10) $u^f_i - u^f_j + |V|.x^f_{ij} \le |V| - 1, \forall f \in F, \forall(i,j) \in E \ [i \ne S_f, j \ne S_f]$

with

$x_i \in \{0,1\}, \forall i \in V$

$x^f_{ij} \in \{0,1\}, \forall(i,j) \in E, \forall f \in F$

$u^f_i \in R^+, \forall f \in F, \forall i \in V - \{S_f\}$

Constraints (9.1), (9.2) and (9.3) are the linearization of logical constraint (9). Constraints (9.1) and (9.2) ensure that if a node is not used, no capacity is reserved on its adjacent links and Constraint (9.3) forces one of its links to reserve capacity if the node is used. Finally, we add Constraint (10) (from the Traveling Salesman Problem [20]) to prevent loop issues.

4.2   Network Coding

As mentioned before, aggregating flows increases coding opportunities. Applying network coding reduces the number of transmissions and thus reduces the load of the network. Hence, we can increase the capacity of the network. But instead of aggregating flows and then using opportunistic coding to reduce the load of the network, we propose RANC (Routing with Aggregation and Network Coding) that routes a set of flows while considering network coding to increase aggregation. We integrate in our solution the network coding from the Alice-Bob scheme presented in Sect. 3. In [7], the authors propose a solution that routes a set of flows, minimizing the number of transmission while taking into account cliques' constraints. When network coding is used, the utilization rate of cliques is decreased thanks to the transmissions saved. We use the same constraint and adapt it to our context. Constraint (1) in the previous program becomes:

(1) $\sum_{(i,j)\in c} \frac{d_{ij}}{C_{ij}} - \sum_{(i,j),(i,k)\in c, k<j} c_{kij}.\left(\frac{1}{2C_{ij}} + \frac{1}{2C_{ik}}\right) \le 1, \forall c \in C$

This constraint ensures that every clique's utilization rate is lower than 1, considering that the load is reduced thanks to network coding where c_{kij} is the portion of traffic on link (k,i) with next hop j that can be coded at node i. We do not consider transmission errors in this case.

The constraints on the amount of traffic that is coded at any node i can be written as:

(11) $c_{kij} - w_{kij} \leq 0, \forall (i,j), (k,i) \in E$
(12) $c_{kij} - w_{jik} \leq 0, \forall (i,j), (k,i) \in E$
(13) $c_{kij} - c_{jik} = 0, \forall (i,j), (k,i) \in E$

where variable w_{kij} denotes the total amount of traffic on link (k,i) destined to j. At node i, no more traffic can be coded from k to j that actually comes from k to j or from j to k. To maximize the coding, we can't add the maximization to the objective in the same way as in [6,7] due to the dependency between x_i and c_{kij}. To maximize the coding, we have to make sure that c_{kij} is equal to $min(w_{kij}, w_{jik})$. This is done by the two following constraints that set a lower bound to any c_{kij}:

(14) $c_{kij} - w_{kij} + C_{ij}.t_{kij} \geq 0, \forall (i,j), (k,i) \in E, t_{kij} \in \{0,1\}$
(15) $c_{kij} - w_{jik} + C_{ij}.(1 - t_{kij}) \geq 0, \forall (i,j), (k,i) \in E, t_{kij} \in \{0,1\}$

Finally, the following constraints are added to correlate the previous ones to our aggregation program:

(16) $(\sum_{f \in F} m_{ijk}^f) - w_{ijk} = 0, \forall (i,j), (j,k) \in E$
(17) $(\sum_{(j,k) \in E} m_{ijk}^f) - x_{ij}^f.M_f = 0, \forall (i,j) \in E, \forall f \in F$
(18) $(\sum_{(i,j) \in E} m_{ijk}^f) - x_{jk}^f.M_f = 0, \forall (j,k) \in E, \forall f \in F$

where m_{ijk}^f is the traffic of flow f on link (i,j), with k the next visited node by f.

5   Results

To highlight the pertinence of the proposed solution, we show by simulation performances of aggregation in terms of global energy consumption and network load (to emphasise the impact of network coding). By definition, the aggregation is constrained by respecting the QoS but our approach remains efficient in terms of energy saving.

In the following simulations, we have generated random connected graphs and counted the number of nodes unused when a set of n flows is inserted. Each flow has requirements in terms of throughput and its source and destination are selected randomly so that they are more than 1-hop neighbors and the capacities of cliques are not exceeded. These flows are inserted using the I2DIJSTRA routing algorithm [3] which routes flows over the largest path available between their source and destination. When a set of flows has been generated, the optimal solution is obtained with IBM ILOG CPLEX 12.6 [21].

Simulations compare our solutions with two other algorithms, I2DIJKSTRA and I2ILP [3], which are largest and shortest paths algorithms taking into account

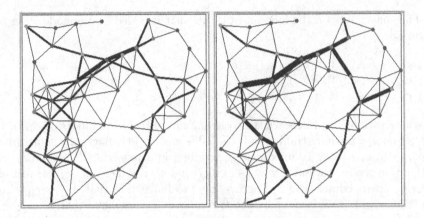

Fig. 3. Links used to route 10 flows with I2LP and RA algorithms (Color figure online).

inter-flow and intra-flow interferences. Performances are evaluated considering the number of nodes unused (nodes that can be turned off or put in a sleep mode) and the cliques' utilization rates. We consider 49-nodes random networks deployed in an area of 100 × 100 where nodes have a transmission range of 20 and links have a capacity of 100 units. Ten cases are studied, from 1 to 10 flows whose rate is comprised between 1 and 10 units. For each case, 100 sets of flows are generated. Figures show mean values of number of unused nodes and cliques' utilization rates.

Figure 2 represents a given case where 10 flows are routed using I2ILP (on the left) and our solution (on the right). The line thickness depends on the number of flows aggregated on it. When routing is performed with I2LP, the paths are more scattered and thus more nodes have to be used. This is due to the algorithm which routes a flow using the shortest path between the source and the destination without consideration of other flows. With the solution based on aggregation, we are able to route the same set of flows over a limited number of nodes while providing the same QoS. In our example, we can switch off around 30 % of nodes used with I2ILP.

Figure 3 shows how the number of unused nodes varies according to the number of flows generated. The decay of the function is expected because adding a flow will force its source and destination to be used. The gain is achieved on intermediate nodes and how routes pass through already used nodes, what is done with our solutions RA and RANC. Thus we are able to turn off more nodes than I2DIJKSTRA and I2LP for the same set of flows.

Figure 4 shows the impact of associating aggregation and network coding. The utilization rates of the most used clique are compared, especially when flows are aggregated with network coding (RANC) or without network coding (RA). In this last case, it is possible to apply network coding after having determined routes (RA+NC). As expected, when flow aggregation is performed, the load of the most used clique increases because more flows are routed over the same links. When network coding is applied on it (and when there is enough flows to combine), we are

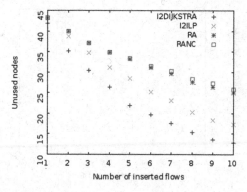

Fig. 4. Number of nodes that can be turned off when the number of flows varies.

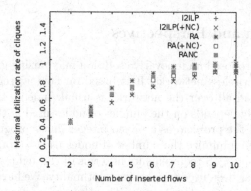

Fig. 5. Utilization rates of the most used clique when the number of flows varies.

able to reduce the load of this clique to a level comparable to that obtained with I2LP when network coding is used (I2ILP+NC). Hence, the number of used nodes is reduced significantly when aggregating flows and the degradation of the network's capacity is prevented when applying network coding. It is important to notice that applying network coding to increase flow aggregation (RANC) leads to a greater load of the most used clique compared to I2LP and RA+NC, but it is still better than aggregating flows without network coding.

To study not only the most loaded clique but also the global load of the network, Fig. 5 shows the average utilization rate of cliques according to the number of flows inserted in the network. Again, aggregating flows increases the load of the network, but associated to network coding, our solutions (RANC and RA+NC) give performances comparable to I2LP, even better in some cases. Therefore, applying flow aggregation and network coding is an effective approach to reduce the global energy consumption of the network by turning off or putting in sleep mode unused nodes, while preserving the capacity of the network and respecting the QoS requirements (Fig. 6).

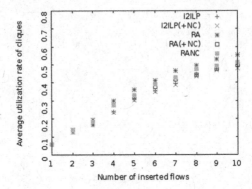

Fig. 6. Average utilization rates of the cliques when the number of flows varies.

6 Conclusion and Perspectives

In this paper, we proposed an innovative solution maximizing global energy savings in multi-hop wireless networks. It is based on flow aggregation instead of distributing the load all over the network, as usual. Indeed, the global energy consumption mainly depends on the number of nodes used in the network. This is why we modeled the problem as a mixed integer linear program with a set of linear constraints to minimize the number of nodes used to route a set of flows, while respecting their QoS requirements in terms of throughput. A branch-and-bound algorithm has been used to solve it to optimality. We have shown by simulation that the proposed solution can significantly save more energy compared to classical routing algorithms such as widest path or shortest path. Furthermore, flow aggregation creates coding opportunities, enhancing performances obtained with network coding. Associating both techniques decreases energy consumption and network load. In this way, we are able to aggregate more flows and then turn off or put in sleep mode more nodes. Finally, our solution may also benefit to the medium access layer. Indeed, by turning off nodes, we should be able to reduce access conflicts as well as interferences to improve capacity and reliability.

Our optimal solution gives an important bound which can be used as a comparison tool to develop distributed and feasible heuristics. Thus our future work will be to determine a metric that can be integrated in shortest path routing algorithms to achieve aggregation. The efficiency of this metric will be measured by comparison with our solution.

References

1. Kumar, P.R., Gupta, P.: The capacity of wireless networks. Trans. Inf. Theor. IEEE. **46**(2), 388–404 (2000)
2. Walrand, J., Jia, Z., Gupta, R., Varaiya, P.: Bandwidth guaranteed routing for ad hoc networks with interference consideration. In: IEEE Symposium on Computers and Communications, June 2005

3. Gawedzki, I., Benfattoum, Y., Martin, S., AlAgha, K.: I2ASWP: routing with intra-flow interference consideration in adhoc network. Research report, CNRS-University of Paris Sud-LRI, December 2010
4. Gomez, J., Campbell, A.T.: Variable-range transmission power control in wireless ad hoc networks. IEEE Trans. Mob. Comput. **6**, 87–99 (2006)
5. Martin, S., Al Agha, K., Pujolle, G.: Traffic-based topology control algorithm for energy savings in multi-hop wireless networks. Ann. Telecommun. **67**, 181–189 (2012). Springer
6. Zhong, Z., Ni, B., Santhapuri, N., Nelakuditi, S.: Routing with opportunistically coded exchanges in wireless mesh networks. In: Wireless Mesh Networks. IEEE, September 2006
7. Martin, S., Benfattoum, Y., AlAgha, K.: IROCX: interference-aware routing with opportunistically coded exchanges in wireless mesh networks. In: Wireless Communication and Networking Conference. IEEE, March 2011
8. Rossi, D., Bianzino, A.P., Chaudet, C., Rougier, J.-L.: A survey of green networking research. IEEE Commun. Surv. Tutor. **14**, 3–20 (2012)
9. Mills, M.P.: The cloud begings with coal, big data, big networks, big infrastructure, and big power. Digital Power Group, August 2013
10. Charu, V.A.: A quality of service analysis of energy aware routing protocols in mobile ad hoc networks. In: Contemporary Computing (IC3). IEEE, August 2013
11. Doig, A.G., Lang, A.H.: An automatic method for solving discrete programming problems. Econometrica **28**, 497–520 (1960)
12. Ramathan, R., Rosales-Hain, R.: Topology control of multihop wireless networks using power adjustment. In: Nineteenth Annual Joint Conference of the IEEE Computer and Communications Societies, vol. 2, p. 404 (2006)
13. Bahl, P., Wang, Y.M., Li, L., Halpern, J.Y., Wattenhofer, R.: A cone-based distributed topology-control algorithm for wireless multi-hop networks. In: Wirless Communication. IEEE, May 2004
14. Sheth, A., Halperin, D., Wetheral, D.: Demystifying 802.11n power consumption. In: Proceedings of the 2010 International Conference on Power Aware Computing and Systems, HotPower 2010. ACM (2010)
15. Padmanabhan, V., Jain, K., Padhye, J., Qiu, L.: Impact of interference on multi-hop wireless network performance. In: MobiCom ACM, September 2003
16. Musacchio, J., Gupta, R., Walrand, J.: Sufficient rate constraint for QoS flows in ad-hoc networks. Ad hoc Netw. **5**, 429–443 (2006). ACM
17. Luo, X., Chaporkar, P., Kar, K., Sakar, S.: Throughput and fairness guarantees through maximal scheduling in wireless networks. IEEE Trans. Inf. Theor. **54**, 572–594 (2008)
18. Wolsey, A.: Integer Programming. Wiley, New York (1998)
19. Glover, F.: Improved linear integer programming formulations of nonlinear integer problems. Manag. Sci. **22**(4), 455–460 (1975)
20. Tucker, A.W., Miller, C.E., Zemlin, R.A.: Integer programming formulation of traveling salesman problems. JACM **7**, 326–329 (1960)
21. IBM: Cplex optimizer. www.ibm.com/software/commerce/optimization/cplex-optimizer/

DTN/Opportunistic Networks

ONRECT: Scheduling Algorithm
for Opportunistic Networks

Saima Ali[✉] and Andreas Willig

University of Canterbury, Christchurch, New Zealand
saima.ali@pg.canterbury.ac.nz

Abstract. Opportunistic networks are characterized by having only intermittent connectivity on end-to-end paths, making successful delivery of messages very challenging. A key part of an opportunistic routing and forwarding scheme is message scheduling, where a node fixes the sequence in which it transmits messages for further relaying to its various neighbours. This paper presents a scheduling scheme which incorporates knowledge about the *remaining contact time* and compare its performance against baseline schemes from the literature. The results show that incorporating remaining contact time can indeed give substantial improvements in key performance indicators like the average delivery delay or the overhead ratio.

1 Introduction

Opportunistic networks belong to the broad category of delay tolerant networks (DTN) that are based on the *store-carry-and-forward* routing principle [1], where a node stores a message into its buffer, waits until it moves into the communication range of some other nodes and then replicates the message. These networks provide limited or intermittent connectivity and support the applications that can tolerate delays. Some example applications of the opportunistic networking principle can be found in habitat monitoring [2], underwater networks [3] and military networks [4]. They can have many applications in smart vehicular networks, social networking, agriculture and emergency networks.

The network environment in opportunistic networks is challenging due to only intermittent connectivity. A number of protocols have been proposed to improve the efficiency of these networks in terms of average delivery probability and to minimize the average delivery delay, either by improving routing or by providing more sophisticated message scheduling schemes. The routing protocols utilize a variety of methods, including probabilistic routing [5,6], network coding [7,8], controlled packet replication [9,10] and history-based learning [11,12] of routes, to improve performance. Despite the efforts invested in designing improved routing schemes, one of the crucial factors that could potentially improve performance has not yet received much attention: the issue of message scheduling, in which a node, upon encountering one or more other nodes, makes a conscious decision about the sequence in which to replicate messages to those

© Springer International Publishing Switzerland 2016
N. Mitton et al. (Eds.): ADHOC-NOW 2016, LNCS 9724, pp. 141–155, 2016.
DOI: 10.1007/978-3-319-40509-4_10

neighbours. There are only a few works that directly address message scheduling, for example the RAPID [13] protocol, GBSD protocol [12], and the Global History Based Prediction scheme [14].

An important overlooked aspect – and the one which this paper focuses on – is the "contact time duration", i.e. the time during which two nodes remain in each other's range. Although some studies mention *inter-contact time* (i.e. the time between the successive meeting of nodes [15]) to the best of our knowledge there has been only one protocol (called Opportunistic DTN Routing with Window-aware Adaptive Replication – ORWAR [16]) that involves the contact time based localized message scheduling and drop policy but does not explicitly consider the presence of several neighbours at once.

In this paper, we present a message scheduling scheme which, similarly to ORWAR, incorporates knowledge of the remaining contact time (RCT) left between the nodes. Furthermore, its scheduling algorithm explicitly addresses the case of having multiple neighbours at once, and the scheduling algorithm makes decisions that take into account both the "quality" of the neighbours (measuring how close they could bring a message to its destination) and the remaining contact time to them. We call this scheme **Opportunistic Networks Routing with REmaining Contact Time (ONRECT)**. The term "remaining contact time" refers to the (estimated) time that remains until the contact is finished, i.e. the nodes move out of radio range. The main aim of this paper is to assess the performance impact of ONRECTs' scheduling algorithm and to compare it against ORWAR with respect to the average delivery probability, the network overhead and the average delivery delay. Furthermore, as both algorithms use predictions of the remaining contact time, we compare the impact of ORWAR's estimator and the "ground truth" on their performance. Our scenario is simple, yet very large as compared to the other studies mentioned above and uses a mobility model geared towards modeling of mobile entities in an urban setting. It includes both mobile nodes and stationary nodes. We compare the ONRECT and ORWAR schemes also against another baseline scheme, the Spray-and-Wait scheme. Our results indicate that ONRECT can, in the presence of perfect remaining contact time information, outperform the other schemes for important performance indicators. At the same time, however, ONRECT shows a higher sensitivity to the quality of the available RCT information.

The rest of the paper is structured as follows: The next Sect. 2 provides some background on opportunistic networks, time calculation, message scheduling and routing. In Sect. 3, we outline our system model, and our proposed algorithms for time calculation and message scheduling are described in Sect. 4. The results of our performance analysis are given in Sect. 5, related work is briefly discussed in Sect. 6 and we give our conclusions in Sect. 7.

2   Background

We give a brief background on opportunistic routing and the ORWAR scheme, which we use as a baseline for comparison.

There are many wireless technologies that can be deployed for the implementation of opportunistic networks, for example the IEEE 802.11 WLAN standard [17]. Many of these standards can be used for the purposes of this work, as we do not make any strong assumptions about the services of the underlying wireless technology. Key properties of wireless standards which might influence the performance of the schemes considered in this paper include the data rate, transmission range, and the delay for neighbour discovery.[1]

In this work, we will consider IEEE 802.11 wireless technology. The IEEE 802.11 standard (also known as Wi-Fi) offers amongst other things a range of different physical layers which all use essentially the same MAC protocol. Out of the available physical layers we focus on IEEE 802.11g. This version operates in 2.4 GHz ISM band and supports data rates between 6 Mbit/s and 54 Mbit/s using OFDM transmission, and 2 Mbit/s when using direct-sequence spread spectrum (DSSS). To better highlight the impact of scheduling decisions when a node has many messages in its buffer and has only limited contact time with neighbouring nodes, we assume the slowest data rate of 2 Mbit/s. Furthermore, we assume that the transmission range is limited to 20 m.

Contact time is defined as the time duration from the time when the mobile nodes moves towards each other and build a communication link until the time when they move out of the range and the link is dropped [18]. At any time instant during the contact time, the *remaining contact time* can be defined as the time duration left until the both nodes lose contact with each other. It is important to mention here that the contact time and remaining contact time in general have different probability distributions.

Contact time is one of the factors that can potentially affect the performance of opportunistic networks. This duration depends upon how long the nodes remain in each other's communication range and how long the contact lasts. The contact duration defines the amount of data that can be transferred [15]. The studies in [19,20] show that the distribution of the contact duration for human mobility follows a power law distribution.

The remaining contact time has not been widely considered as a factor in opportunistic networking. The only work which uses this is idea [16], the *Opportunistic DTN Routing with Window-aware Adaptive Replication (ORWAR)* protocol.

The ORWAR scheme uses Spray-and-Wait (SnW) [9] as a baseline scheme to control the message replication. In SnW message replication, a source node *sprays* a number L of replicas of the message by replicating it to other nodes, so-called relay nodes. In plain SnW the source node picks $L - 1$ distinct relays and each relay keeps the message until it can deliver it directly to the final destination (wait state) or a timeout occurs. In binary SnW the source node replicates the message itself and an allowance of $L/2$ further replicas to the first relay it meets. Both the source and the first relay then hand over the message and allowances of $L/4$ replications to the next relays they meet. This process

---

[1] The discovery process often relies on periodically transmitted beacon packets and the discovery delay is then strongly influenced by the beacon period.

of splitting continues until the source or any relay has only an allowance of one message left. Again the relays try to deliver the message to the final destination or throw it away after a timeout. ORWAR is built on the binary SnW scheme.

The ORWAR scheme differentiates messages on the basis of a user-assigned priority. Unlike SnW, ORWAR adopts a different approach while choosing the initial number L of message copies. The messages with higher priority have a higher number of message copies as compared to lower priority messages. The messages are stored in a buffer according to the utility-per-bit ratio, which is defined as the ratio of the message priority and the message length in bits.

ORWAR estimates the contact time as follows:

$$t_{cw} = \frac{2 \cdot \min\{r_1, r_2\} \cdot \cos\alpha}{\|v\|} \tag{1}$$

where r_1 and r_2 are the transmission ranges of the two involved nodes, $\mathbf{v}$ is the difference vector of the velocity vectors of both nodes, $\|\mathbf{v}\|$ is its (euclidean) magnitude, and α is the angle between the nodes while they are in contact. Once the time is calculated, the messages with higher utility per bit ratio and a size that is small enough to be successfully transmitted during the contact time are replicated to the other node.

3   System Model

We consider a simple scenario which consists of 1,500 nodes, both mobile and stationary. All nodes are placed in a "playground" which resembles parts of Helsinki, Finland. The size of playground is $3500 \times 4500\ m^2$.

The nodes are divided into three groups of 500 nodes each. The first group comprises stationary nodes that serve as the message destinations. These nodes can represent for example fixed wireless access points in homes or in different buildings across the city. These nodes can also generate messages and can act as relays for the messages. The destination nodes are placed randomly and independently on the playground according to a uniform distribution. The remaining two groups of nodes are mobile, representing cars and trams moving with varying speeds, where cars have higher average speeds as compared to trams. The nodes speeds of the cars are chosen randomly from 10–50 km /hrs and the speed of the trams are chosen from 10–40 km/hr. The nodes in both of these groups can generate and relay messages but they cannot be the destination nodes.

Since the mobile nodes represent vehicles on roads, we use the *Shortest Path Map-Based movement Model (SPMBM)*. SPMBM initially puts the nodes at random locations on the roads. A node then selects a random destination on some road and calculates the shortest path to the destination (only on roads) using Dijkstra's shortest path algorithm. It then picks a speed from a uniform distribution between a designated minimum and maximum speed (which are specific for the type of mobile nodes the node under consideration belongs to). Every time the node reaches the next waypoint in the path (e.g. when entering a new road), it picks up a new speed from the uniform distribution and moves at that speed. Once the node reaches the destination, it picks a new destination point on the map and repeats the process [21].

It is assumed that nodes have unlimited buffer capacity, where they can store unlimited messages. The message format is shown in Fig. 1. It contains identifiers for the source and destination node of the message. We assume that the identifiers encode the geographical location of source and destination node. The message-id is unique for each new message generated by a particular source node. The Time-To-Live (TTL) field specifies the remaining time for which a message is valid before it is to be dropped. In this paper, however, we set the TTL value to a sufficiently large number to ensure that no timeout occurs within the simulated time of four hours. Finally, the size field specifies the size of the message. For reasons of simplicity messages are not fragmented.

We furthermore assume that all of the nodes are using the same wireless technology (IEEE 802.11g with 2 Mbit/s transmission rate) for message transmission. Each of the nodes in the network has one network interface and all nodes use the same transmit power. However, to focus on message scheduling and to keep simulation times bounded we have chosen not to add a MAC layer model to the ONE simulator.

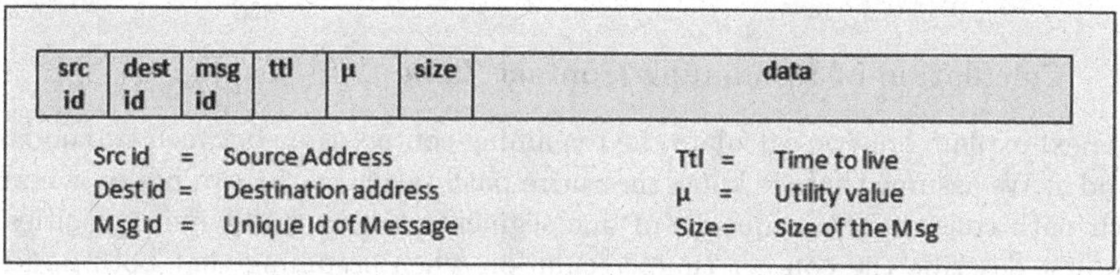

src id	dest id	msg id	ttl	μ	size	data

Src id	=	Source Address	Ttl	=	Time to live
Dest id	=	Destination address	μ	=	Utility value
Msg id	=	Unique Id of Message	Size	=	Size of the Msg

Fig. 1. Message Format

We consider the unit disk model [22] as channel model, i.e. a device has a certain transmission radius.

4   The ONRECT Scheme

The *ONRECT* scheme is a routing and scheduling algorithm for delay tolerant networks that uses controlled replication and knowledge of the remaining contact time left towards a chosen neighbour for message scheduling. The controlled replication strategy used is binary Spray and Wait [9]. Please note that in this paper we mostly do not attempt to *estimate* the remaining contact time but we use our knowledge of the paths taken by nodes to obtain the real remaing contact time. By this, we can assess the benefits of having remaining contact time information in isolation.

Each message in the network is assigned a utility value $u \in \{1, 2, 3\}$, where one refers to the lowest priority and three to the highest priority. The priorities are assigned to messages in round-robin fashion. This utility is used to prioritize the messages in the sending queue. Each node in the network maintains a sending message queue. The messages in the sending queue are arranged according to their utility-per-bit ratio, i.e. for message m we calculate u_m/s_m, where u_m is the priority of this message and s_m its size in bits. The messages that are generated by the node itself or received from other nodes are introduced into the sending queue at their respective place according to the utility-per-bit ratio, ties are broken randomly. The message with the highest utility-per-bit ratio is at the top and is first to be forwarded/replicated.

Furthermore, a node maintains for each message m its current replication counter L_m as per the binary SnW scheme. The initial L_m value for a freshly generated message may depend on its priority, but for the purposes of this paper we choose the same value L_0 for all priorities.

In general a node can be connected with several other nodes at a given point in time. Therefore, each node maintains a table of all the nodes it is in contact with and the respective remaining contact time with those nodes. From the remaining contact time toward a neighbour node j, a node i can derive the amount of data that may be transferred during this time from the characteristics of the underlying wireless technology. Node i refers to this table every time it needs to make a scheduling decision.

4.1 Calculation of Remaining Contact Time

We next explain how we calculate the remaining contact time between two nodes i and j. We assume that we know the entire path taken by the two nodes, where each path consists of a sequence of line segments taken at different velocities. We first describe the contact time calculation when assuming that both nodes move on straight lines forever, and then explain how we take account of changes in direction and speed. For simplicity we restrict to the two-dimensional case, generalization to three dimensions is straightforward.

Two Nodes Moving on Straight Lines. Suppose we are given two nodes i and j with velocity vectors $\mathbf{v}_i$ and $\mathbf{v}_j$ and initial positions $\mathbf{x}_i = (x_{i,0}, y_{i,0})$ and $\mathbf{x}_j = (x_{j,0}, y_{j,0})$, respectively. We transform the problem to assume that node i is stationary and located at the origin (position $(0,0)$), whereas node j then moves with velocity vector $\mathbf{v} = (v_x, v_y) = \mathbf{v}_j - \mathbf{v}_i$ and starts at initial position $\mathbf{x}_0 = (x_0, y_0) = (x_{j,0} - x_{i,0}, y_{j,0} - y_{i,0})$. Let $\mathbf{p}(t)$ be the position of node j at time t, i.e.:

$$\mathbf{p}(t) = \begin{pmatrix} x_0 \\ y_0 \end{pmatrix} + t \cdot \begin{pmatrix} v_x \\ v_y \end{pmatrix}$$

then the distance of node j to node i is given by its distance to the origin, i.e.:

$$\|\mathbf{p}(t)\| = \sqrt{(x_0 + t \cdot v_x)^2 + (y_0 + t \cdot v_y)^2} \tag{2}$$

When $R > 0$ denotes the mutual transmission range of i and j, then we seek values of t for which $R = \|\mathbf{p}(t)\|$ holds – this leads to a quadratic equation.

When no solution exists, node i and j never get in range and their contact time is zero. When two solutions $0 < t_1 < t_2$ exist, then node j has started outside the range of i, then comes close enough to i for time $t_2 - t_1$ and then leaves again. When only one solution exists, two cases are possible: the first one is that j's starting position was outside i's range, then j's trajectory just touches the circle of radius R around i and the contact time is zero. The second case is when node j's starting position is located within the range of i and leaves it at some time $t > 0$, then the contact time is t. In these considerations we have implicitly assumed that $\|\mathbf{v}\| > 0$ holds. In the border case where $\|\mathbf{v}\| = 0$ and thus $\mathbf{v} = 0$, the contact time is either zero (when $\|\mathbf{x}_0\| > R$) or infinity (when $\|\mathbf{x}_0\| \leq R$).

Contact Time over Paths. In our setup two nodes normally don't move on straight lines, but follow a (shortest) path on a street network, i.e. they change direction when entering a new street, and change their speed. Suppose node i moves along a path $(\mathbf{x}_{i,1}, \mathbf{x}_{i,2})$, $(\mathbf{x}_{i,2}, \mathbf{x}_{i,3})$, $\ldots$, $(\mathbf{x}_{i,k-1}, \mathbf{x}_{i,k})$, and node j along path $(\mathbf{x}_{j,1}, \mathbf{x}_{j,2})$, $(\mathbf{x}_{j,2}, \mathbf{x}_{j,3})$, $\ldots$, $(\mathbf{x}_{i,m-1}, \mathbf{x}_{i,m})$, where each $(\mathbf{x}_{a,b}, \mathbf{x}_{a,b+1})$ denotes a line segment. On each of the line segments $(\mathbf{x}_{i,b}, \mathbf{x}_{i,b+1})$ on its path node i has speed $v_{i,b}$, similarly for node j. The times where node i changes directions (moving to the next line segment) are given by $t_{i,b}$, similarly for node j.

In short, our algorithm proceeds by sorting the times $t_{i,b}$ and $t_{j,b}$ in ascending order to get times $t_1, t_2, \ldots, t_{k+m}$. We then apply the "straight-line" computation described before to each of the time intervals $[t_i, t_{i+1}]$ (keeping track of the positions of both nodes) and sum up the results. When the "straight-line" contact time for some interval $[t_i, t_{i+1}]$ extends beyond t_{i+1}, then the contact time is clipped.

4.2   Vaccination Mechanism

The vaccination mechanism is used both in ORWAR and ONRECT, its main purpose is to stop further propagation of messages that have reached their final destination, to limit the effort spent by the network. Broadly, the information that a message has been received is flooded into the network and all existing copies of the message are then dropped.

In detail, each node maintains a list of messages id's m_d that have been delivered to their final destinations. Whenever two nodes come in contact with each other, they exchange their lists. Both of the nodes then update their sending message queues (m_q) using these lists by dropping any message that has been mentioned in m_d.

Once a message is delivered to the final destination node, an acknowledgment message is sent back to the sending node. The id of the message is added to the delivered message queue m_d at both sending and receiving queue.

4.3   ONRECT Algorithm

The ONRECT message scheduling decision depends on multiple factors: (i) remaining contact time (ii) the best neighbour to forward the message to (in particular, we only consider neighbours which have a destination that is within 700m of the message destination, and from these we pick the one that gets closest), and (iii) the position of the message in the queue (the message with higher utility-per-bit will have higher position).

The pseudo-code for the algorithm is given below:

```
=============================================
For each node i:
Variables
x_i, y_i    current location of the node
X, Y        Next waypoint of the node
v_i         node speed if it is not stationary
R_i         Wireless transmission range of node
m_s         size of a message m
L_k         number of message copies
s_max       maximum data that can be transferred during calculated
            remaining contact time
m_q         Sending message queue (message are stored in u_k/m_s order )
m_d         Message id's list that are delivered to the final destination
connTime    table to maintain the remaining time (in terms of maximum
            data that can be transferred during this time, stores the
            s_max value) with the other nodes
=============================================
A node i comes in contact with other node j:

send m_d to j      // send delivered messages list to j
receive m_d from j     // receive delivered messages list from j
update m_q      // delete all messages that have been delivered

/** Compute the remaining contact time and add in the
connTime table **/
connTime.add(calculateTime(nodeId))
=============================================
/** Assuming that this node is connected with multiple nodes **/
/** Direct Delivery Message **/

For each message m in m_q:
    if m_s < s_max of destination node:
    deliver m to the destination node
    if m.ack:
        Remove the messages from the m_q and add the message id in m_d
        update connTime:
        s_max = s_max - m_s
/** Rest of the messages **/
For each message in m_q:
    //Find the best neighbor
    bestNeighbor = neighbor(message m)
    if s_max > m_s ∧ L_k > 1:
        transmit m to bestNeighbor
        L_k = L_k/2
        in connTime for bestNeighbor:
        s_max = s_max - m_s

calculateTime(nodeId):
    get (x_i, y_i), (X, Y) and v_i
    compute remaining time
    calculates s_max
    return s_max
neighbor(message m):
```

```
List Nodes = getAllConnectedNodes()
foreach(Node otherNode:Nodes) {
   if(otherNode.finalDestination < m.destination){
      return otherNode

   }
}
```

The ONRECT algorithm shares many similarities with ORWAR, however, the key difference is in the scheduling algorithm: in ORWAR we consider one neighbor and then send all the message to it in order of their utility-per-bit ratio and as long as we can (i.e. as long as the remaining contact time does not expire), whereas in ONRECT we consider all possible neighbors for each message and pick the best one. ONRECT makes a choice of the best neighbor by choosing the neighbour whose current destination is closest to the message destination while ORWAR just transfers messages to any node that comes in range.

5   Performance Analysis

We have performed comprehensive simulations to evaluate the performance of the ONRECT algorithm in comparison with ORWAR and SnW. Different message sizes were considered for this purpose. For each message size a sufficient number of simulation replications have been carried out to obtain a relative confidence interval width of 5 % at a confidence level of 95 % for the delivery probability. Each replication considers a different placement of stationary and mobile nodes, message generation times and message destinations.

The results were compared with the results from the two other algorithms, binary Spray and Wait [9] and ORWAR [16]. These protocols are run with similar settings as ONRECT to compare the results. The values for the fixed parameters are summarized in Table 1.

Aside from the vaccination algorithm, there are two major differences between ONRECT and ORWAR: the contact time calculation (using Eq. 1 for ORWAR and the algorithm from Sect. 4.1 for ONRECT) and the neighbour selection scheme. To analyse the impact of the contact time calculation, we have introduced two variations of the considered schemes: the ONRECT- scheme is very similar to ONRECT but uses the estimated contact time information according to Eq. 1, whereas the ORWAR+ scheme is very similar to ORWAR but uses the contact time calculation from Sect. 4.1.

5.1   Simulation Setup

All algorithms were implemented in the Opportunistic Network Environment (ONE) simulator version 1.5 [21]. This simulator is specifically designed to simulate algorithms and protocols for opportunistic networks, and in fact ONE is a popular tool in this area. Many commonly known opportunistic network protocols and algorithms like Spray and Wait, Epidemic Routing, or PRoPHET are available as a part of ONE. The simulated time for each replication is 4 h.

Table 1. Simulation parameters settings for all the protocols

Parameter	Value
Number of Message Copies L_k	6
Time to Live	1000 min
Buffer space	infinite
Speed of the Cars	$10 - 50$ km/hr
Speed of the Trams	$10 - 40$ km/hr
Number of Host Groups	3
Number of Hosts per Group	500
Message generation interval	1–10 s
Number of wireless interfaces	1
Interface Transmit speed	2 Mbps
Maximum Message Size	10 KB

The experiments were conducted for three different message size ranges. Small sized messages are drawn randomly from the interval between 200 Bytes and 400 Bytes, medium sized messages have a size between 1 KB and 2 KB, and large sized messages are between 5 KB and 10 KB. The messages are generated every 5–10 s in the network following a uniform distribution.

5.2   Results

One of the most important performance measures in this work is the average delivery delay, defined as the delay (in seconds) between message generation and its first reception at the intended destination (computed only for those messages which actually have reached the destination). In Fig. 2a we show the average delivery delay for different message sizes.

There is a substantial gap in the average delivery delay of ORWAR and ONRECT. It is obvious that the ONRECT algorithm incurs the smallest delays as compared to ORWAR and SnW. ONRECT incurs nearly 42 % less delay than ORWAR and nearly 18 % less delay than SnW. To understand whether this performance difference is due to the availability of the real contact time to ONRECT or due to ONRECT's scheduling algorithm, we have also included the results for ONRECT− and ORWAR+. Interestingly, ONRECT offers still significantly better delivery delay than ORWAR+, so the availability of real contact time information does not fully explain the difference between ORWAR and ONRECT. To our surprise, however, it can be seen that ONRECT− performs significantly worse than ORWAR+, which suggests that the neighbour selection scheme of ONRECT is susceptible to the error in the remaining contact time calculation introduced by ORWAR's estimator. It is a subject for future work to better understand this behaviour.

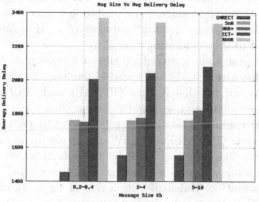

(a) Average Delivery Delay

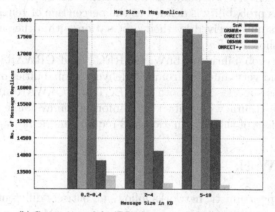

(b) Comparison of the 'Effort' in terms of Message Copies

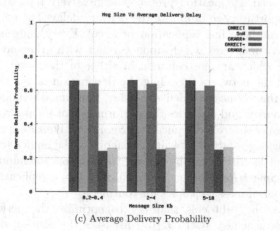

(c) Average Delivery Probability

Fig. 2. Graphs (Color figure online)

An important performance measure is the overall effort required in terms of the number of times a message is replicated between nodes for successful delivery. In Fig. 2b we report the total number of replications created by all messages generated during simulation time.

The results suggest that the contact time estimation of ORWAR (which does not account for turns) leads to a reduced number of message replications. A possible explanation for this is that ORWARs estimator tends to underestimate the contact time, and so this estimator allows fewer messages to be transmitted during contact. The difference between ORWAR and ONRECT− (and similarly between ORWAR+ and SnW on the one hand and ONRECT on the other hand) can likely be explained by the selection rule of ONRECT−, which only allows replications to neighbours that come within a certain range of the message destination. But achieving a deeper understanding of these differences is a possible subject of future work.

The average delivery probability is another important performance measure of this work. This probability is defined as the percentage of generated messages which are successfully received by their destination within the simulation time. The results are shown in Fig. 2c.

The delivery probabilities of SnW, ONRECT and ORWAR+ are at about the same level and very different from the delivery probabilities of ORWAR and ONRECT−. This suggests that both ONRECT and ORWAR are substantially influenced by the accuracy of contact time information available to them while, SnW is independent of this information and requires more message replications.

6   Related Work

Routing in the opportunistic networks follows an entirely different pattern when compared to the traditional wireless ad-hoc networks. Most common protocols are replication-based. Replication protocols aggressively replicate messages at every contact opportunity to minimize the delivery delay. The epidemic routing protocol [23] is the base line replication protocol. Every node in the network maintains a list of messages which upon contact with other nodes exchanges all the messages which are not common in either of their lists. The messages are spread to all the networks including the destination node. This protocol is the fastest in terms of message delivery delay, however, the flood of messages exhausts the resources and degrades the performance of the network. Some protocols use the idea of controlled replication. *Spray and Wait* [9] sprays only a few replicas of the message and then "waits" until one of the replica reaches the destination. Different heuristics can be defined to "control" the number of replicas in the network. *Smart Replication* [10] or "utility based replication" strategies use a fixed number of message copies for routing.

There are routing protocols that try to optimize the performance with improved routing schemes. RAPID [13] discusses the routing of messages as a resource allocation problem under the constraints of limited buffer and bandwidth.

GBSD & HSBD [12] presents two different policies for a joint scheduling and buffer management algorithm. The first policy, GBSD, gives an ideal theoretical framework that serves as a reference to measure the performance of the other policy HSBD. GBSD derives a per-message utility based on the global knowledge of the network. History based scheme (HSBD) adopts a learning history approach while exchanging information over the in-band channel for the estimation of parameters (number of replicas m and number of nodes that have seen the message n) required to derive the utility. Global History-Based Prediction (GHP) [14] is another framework similar to the HSBD. This framework implements four different modules for message scheduling and buffer management.

7   Conclusions

This work addresses the potential impact of the knowledge of remaining contact time on message scheduling in opportunistic networks. We have introduced the *Opportunistic Networks Routing with REmaining Contact Time (ONRECT)* scheme, which incorporates the information about contact time and the destinations of neighboured nodes and the messages into scheduling decisions. We compared its performance against the SnW scheme and the ORWAR scheme. The results show that although the average delivery probability of ONRECT and SnW are very close, ONRECT requires less effort as compared to SnW and achieves nearly the same average delivery probability when highly reliable contact time information is available. ONRECT selects a neighbor for the directional replication of the messages which means the messages are replicated in the direction of the destination of the message. In this way, the messages are better diffused in the network towards the message's destination. It is observed that the average delivery probability remains nearly constant with the increase in message size. However, with an increase in message size, the effort, in terms of number of message copies, also increases. This effort is not very prominent for ONRECT and SnW, but ORWAR shows a sharp increase. Another important observation is the small average delivery delay offered by ONRECT. The ONRECT's improved time estimation and neighbor selection results in a smaller average delivery delay as compared to other protocols.

There is some potential for future research. One important issue is to gain a better understanding of the apparent sensitivity of ONRECT and ORWAR with respect to the quality of the available contact time information, and to devise schemes which allow for more graceful performance degradation when the quality of the contact time estimate degrades. We furthermore believe that there is some potential in further optimizing the selection rule of ONRECT, and it is also interesting to assess the impact of vaccination.

References

1. Jain, S., Fall, K., Patra, R.: Routing in delay tolerant network. In: ACM SIGCOMM (2004)

2. Juang, P., Oki, H., Wang, Y., Martonosi, M., Peh, L.S., Rubenstein, D.: Energy-efficient computing for wildlife tracking: design tradeoffs and early experiences with zebranet. In: Proceeding ASPLOS 2002 (2002)
3. Heidemann, W., J., Ye, J. Wills, A. Syed, Y. Li.: Research challenges and applications for underwater sensor networking. In: Proceedings of the IEEE Wireless Communications and Networking Conference (2006)
4. Disruption tolerant networking. www.darpa.mi
5. Burgess, J., Gallagher, B., Jensen, D., Levine, B.N.: MaxProp: routing for vehicle-based disruption-tolerant networks. In: IEEE Infocom
6. Lindgren, A., Doria, A., Schelen, O.: Probabilistic routing in intermittently connected networks. In: SIGMOBILE Mobile Computing and Communication Review (2003)
7. Small, T., Haas, Z.: Resource and performance tradeoffs in delay-tolerant wireless networks. In: Proceeding ACM WDTN, pp. 260–267 (2005)
8. Wang, Y., Jain, S., Martonosi, M., Fall, F.: Erasure-coding based routing for opportunistic networks. In: ACM SIGCOMM Workshop on Delay Tolerant Networks (2005)
9. Spyropoulos, T., Psounis, K., Raghavendra, C.S.: Spray and wait: an routing in intermittently connected mobile networks. In: IEEE/ACM Transactions on Networking (2005)
10. Spyropoulos, T., Obraczka, K.: Routing in delay-tolerant networks comprising heterogeneous node populations. In: IEEE Transactions on Mobile Computing (2009)
11. Burns, B., Brock, O., Levine, B.N.: MV routing and capacity building in disruption tolerant networks. In: Proceedings of IEEE INFOCOM (2005)
12. Krifa, A., Barakat, C.: Message drop and scheduling in DTNs: theory and practice. IEEE Trans. Mobile Comput. 11(9), 1470–1483 (2012)
13. Balasubramanian, A., Neil Levine, B., Venkataramani, A.: DTN routing as a resource allocation problem. In: SIGCOMM 2007. ACM, Japan, August 2007
14. Elwhishi, A., Ho, P., Naik, K., Shihada, B.: A novel message scheduling framework for delay tolerant networks routing. IEEE Trans. Parallel Distrib. Syst. 24(5), 871–880 (2013)
15. Li, Y., Jin, D., Wang, Z., Zeng, L., Chen, S.: Exponential and power law distribution of contact duration in urban vehicular ad hoc networks. IEEE Signal Processing Letters 20(1), 110–113 (2013)
16. Sandulescu, G., Tehrani, S.N.: Opportunistic dtn routing with window-aware adaptive replication. In: Proceedings of the 4th Asian Conference on Internet Engineering, AINTEC 2008, pp. 103–112 (2008)
17. IEEE Computer Society, sponsored by the LAN/MAN Standards Committee, IEEE Standard for Information technology - Telecommunications, Information Exchange between Systems - Local, Metropolitan Area Networks - Specific Requirements - Part 11: Wireless LAN Medium Access Control (MAC) and Physical Layer (PHY) Specifications (2012)
18. Li, Y., Qian, M., J. D., L. Su, Z. L.: Adaptive optimal buffer management policies for realistic DTN. In: GLOBECOM (2009)
19. Wang, W., Srinivasan, V., Motani, M.: Adaptive contact probing mechanisms for delay tolerant applications. In: 13th ACM International Conference Mobile Computing and Networking (MOBICOM), pp. 230–241, September 2007
20. Chaintreau, A., Hui, P., Crowcroft, J., Diot, C., Gass, R.: Pocket switched networks: real-world mobility and its conse-quences for opportunistic forwarding. Univ. Technical report, Cambridge, Computer Lab, U.K. (2005)

21. Keränen, A., Ott, J., Kärkkäinen, T.: The ONE simulator for DTN protocol evaluation. In: SIMUTools 2009: Proceedings of the 2nd International Conference on Simulation Tools and Techniques. ICST, New York (2009)
22. Huson, M.L., Sen, A.: Broadcast scheduling algorithms for radio networks. In: Military Communications Conference, IEEE MILCOM, pp. 647–651 (1995)
23. Vahdat, A., Becker, D.: Epidemic routing for partially connected ad hoc networks. Duke University, Technical report, April 2000

Improving Message Delivery Performance in Opportunistic Networks Using a Forced-Stop Diffusion Scheme

Jorge Herrera-Tapia[✉], Enrique Hernández-Orallo, Andrés Tomas, Pietro Manzoni, Carlos T. Calafate, and Juan-Carlos Cano

Department of Computer Engineering, Universitat Politècnica de València, Camino de Vera S/N, 46022 Valencia, Spain
jorherta@doctor.upv.es, antodo@upv.es,
{ehernandez,pmanzoni,calafate,jucano}@disca.upv.es

Abstract. The performance of mobile opportunistic networks strongly depends on contact duration. If the contact lasts less than the required transmission times, some messages will not get delivered, and the whole diffusion scheme will be seriously affected.

In this paper we propose a new diffusion method, called Forced-Stop, that is based on controlling node mobility to guarantee a complete message transfer. Using the ONE simulator and realistic mobility traces, we compared our proposal with the classical Epidemic diffusion. We show that Forced-Stop improves the message delivery performance, increasing the delivery ratio up to 30 %, and reducing the latency of message delivery up to 40 %, with a limited impact on buffer utilisation and message relaying.

These results can be a relevant indication to the designers of opportunistic network applications that could integrate in their products strategies to inform the user about the need to temporarily stop in order to favor the overall data delivery.

Keywords: Adhoc opportunistic networks · Epidemic protocol · DTN · Data transfer · Wireless adhoc networks

1 Introduction

Opportunistic wireless ad-hoc networks [1,2] are an alternative to consider in environments where the wireless infrastructure has become inefficient due to the saturation of requests, or when no communication infrastructure is available. Instead of using the established Internet infrastructure, the communication in mobile opportunistic networks takes place upon the establishment of ephemeral contacts among mobile nodes using direct communication (i.e. Bluetooth or WiFi Direct), and storing the messages in these devices to achieve their full dissemination.

Considering an epidemic distribution, the duration of the contact time between nodes is probably the key factor in the dissemination of messages.

© Springer International Publishing Switzerland 2016
N. Mitton et al. (Eds.): ADHOC-NOW 2016, LNCS 9724, pp. 156–168, 2016.
DOI: 10.1007/978-3-319-40509-4_11

If the contact is too short, the nodes cannot complete the transfer operation, thereby increasing the probability of failure for message dissemination among the nodes in the network.

In this paper we propose a novel messaging diffusion approach, called *Forced-Stop*, that increases the effectiveness of the diffusion process by interacting with the user so as extend the contact duration by making the user wait until the message transmission is complete.

This scheme is already being adopted in some existing short-range messaging protocols such as *Apple iOS Airdrop* and *Google Android Copresence*.

We compare the performance of the basic epidemic diffusion and our Forced-Stop variation using the ONE (Opportunistic Network Environment) simulator [3]. This simulator was designed and built to specifically evaluate DTN protocols and applications, and focuses on the network layer without considering the details of lower layers such as MAC or physical. In our analysis we have selected a scenario based on a set of human geo-tagged traces obtained during fifteen days, collected by smartphones in the *National Chengchi University (NCCU)* campus [4]; the message generation patterns (frequency and size) are based on statistics related with social networking applications [5].

The experiments evaluated the dissemination performance of both solutions using different settings, like the buffer sizes or the messages TTL (Time To Live). The results show that the Forced-Stop approach clearly improves the message delivery performance by increasing the delivery success ratio and reducing the delivery time, although extending the message transmission time and introducing some extra overhead in terms of buffer utilisation.

This paper is organised as follows: an overview of related works addressing opportunistic networks and message diffusion is presented in Sect. 2. The description of our diffusion proposal, experiments and evaluation details are presented in Sects. 3 and 4, respectively. Finally, in Sect. 5, we present some conclusions and future work.

2   Related Works

Some authors [6] consider opportunistic networks as a subclass of Delay Tolerant Networks (DTNs) [7]. This model is being promoted by the "Internet Research Task Force", and we can find its specifications at http://www.dtnrg.org. Data transmission in DTNs is based on messages or *bundles*, which are received and then forwarded by nodes. This method is known in networking as *Store, Carry and Forward*, and it relies on the *Bundle Protocol (BP)* [8]. Unlike the Internet, in DTNs the information delivery time can increase beyond the minimum required because the communication channels and the data links are intermittent, a phenomenon related to the mobility of transmitters and/or receivers.

Various research works in the literature focussed specifically on message distribution in opportunistic networks. In the case of messages transmitted in social networks, the authors of [9] present a detailed analysis with findings from a large scale text messaging study of 70 university students in the United States during four months. In [10,11] the authors examine an utility-based cooperative data dissemination system where the utility of data is defined based on the social relationships between users. There are other proposals, such as [12–16], which evaluate the message dissemination behaviour of the Epidemic protocol by focusing on the mobility patterns of the nodes. In these works, the authors explain the relationship between factors such as speed, mobility model, density of nodes, and places. In addition, some of these authors propose their own mobility model to improve the diffusion in opportunistic networks. There are also some works [17–19] where the authors explain how to improve the dissemination process of epidemic protocols in order to save battery energy. In the case of buffer management, and the influence in the messages dissemination process, some authors [20–22] evaluate the use and optimisations of the buffer trough priority rules to deliver the messages, without performance loss in the information transmission process.

In the context of analytical researches, an analytical model based on Delay Differential Equations is proposed for the authors of [23] to evaluate the diffusion of messages in groups taking into account the transmission time of the messages. This model was validated through simulation. The authors of [24] introduce a mathematical approach for message diffusion in opportunistic networks using the Epidemic protocol.

Summing up, the results of previous papers, highlight the importance of the contact pattern and duration on the message diffusion performance.

3   Forced-Stop: A New Message Diffusion Approach

In this section we describe the operations associated to contact-based message diffusion, and we propose a new scheme called *Forced-Stop*. Moreover, we detail the modifications introduced to the ONE simulator in order to evaluate this type of protocols.

3.1   Contact Based Message Diffusion Schemes

The rationale of contact-based messaging is to establish short-range communications between mobile devices. A reference case study could be described as made of various mobile devices provided with a messaging application that notifies and shows to the user any received messages in the subscribed groups. The application is cooperative i.e., it must store the received messages and perform the diffusion of such messages to other nearby nodes. Each node has a limited buffer to store the messages received from other nodes. When two nodes establish a pairwise connection, they exchange any messages stored in their buffers and check whether some of the new messages are suitable for notification to the user.

Message spreading is based on epidemic diffusion, a concept similar to the spreading of infectious diseases. Basically, when an infected node (i.e., a node that has a message) contacts another node, it infects it (by transmitting the message). Epidemic routing obtains the minimum delivery delay at the expense of increased buffer usage and number of transmissions. A critical factor that determines whether a contact is long enough for a complete transfer of messages is the nodes behaviour when a contact is established. Two variants are possible:

- The nodes continue their movement. In this case, the completion of message transmissions will depend on how long they remain within in communication range. If this contact duration is smaller than the message transmission time, the transmission will fail.
- The nodes stop when they need to exchange information. In this case, the owner of each mobile device will control this exchange by stopping and waiting until the message transmission is completed.

Both approaches raise an interesting question: how much time do contacts last based on a particular mobility model? This clearly depends on average speed and communication range, as detailed in the evaluation section.

In this work we tested the spreading properties of the epidemic protocol by considering both variants of people behaviour to share information. We propose the latter variant to improve the efficiency in data distribution, and we called it Forced-Stop (FS).

3.2   ONE Simulator Modifications

The ONE simulator was designed specifically for evaluating the performance of contact-based dissemination protocols. Among its features are the possibility to generate synthetic traces based on six main DTN routing protocols, and some of their variants, namely: Epidemic, Spray and Wait, Prophet, First Contact, Direct Delivery, and Maxprop. It also provides some relevant mobility models such as Random Walk, Random WayPoint, Linear, and Grid. In addition, it enables using maps and routines based on common user patterns, such as Work and Office Day. Finally, it is easily extendable through the implementation of novel contact-based protocols and mobility models.

Figure 1 shows which modules in ONE were modified. First, in order to implement the Forced-Stop approach, we modified the *ActiveRouter, DTNHost,* and *Connection* Java classes. Devices are forced to stop every time they have a message to transmit by setting the speed of both nodes to 0 m/s. Any stopped device will resume its movement only when the transmission is actually finished.

Moreover, the original ONE message generator, in the *MessageEventGenerator* class, injects a new message using a random interval time. This random time is uniformly distributed from a range configured in the simulation parameters. In order to get a simulation closer to real human behaviour we modified this module to generate messages following a Poisson process, using an exponential random distribution.

Finally, although the ONE simulator produces a large variety of reports about the simulation process, there was no mechanism to obtain the buffer occupancy. We added a new report class that outputs the average and maximum buffer occupancy of all nodes for each step of the simulation. It also computes the maximum of the average buffer occupancy during the whole simulation.

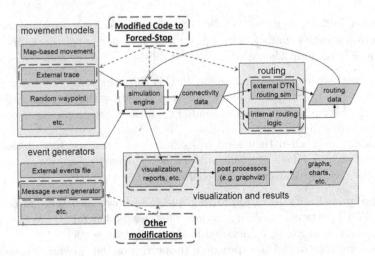

Fig. 1. The red boxes indicate the modified modules in the ONE simulator code. (The figure is based on the original from [3]).

4   Performance Evaluation

The goal of this section is to evaluate the performance of message dissemination under the Epidemic Diffusion and Forced-Stop approaches. The experiments were performed using the ONE simulator, and rely on real human mobility traces from the NCCU University campus. The workload, in terms of message generation is also based on realistic message patterns.

4.1   Description of Experiments

The experiments were performed using the ONE simulator with the modifications described in Sect. 3.2 using a real-life movement trace from an experiment at the NCCU University campus [4]. The *NCCU Traces* were collected using an Android app installed in the smartphones of students belonging to the National Chengchi University. A total of 115 students participated in the experiment. Their GPS data, application usage, Wi-Fi access points, and Bluetooth devices in proximity were recorded over a period of two weeks. Time is specified with a resolution of one second, and the position information is rounded to meters. Figure 2 shows a snapshot of ONE running with the corresponding graphical map information. We considered both behaviour models presented previously: classical Epidemic and Forced-Stop (FS).

A key aspect on these experiments is the workload generation. Our generation pattern of messages is related with social networking applications. We considered a typical multimedia messaging application where each user generates messages of different sizes, and shorter messages are far more common than larger ones. Three message sizes were considered: a short text message ($1\,kB$) every hour, a low-resolution picture or photo ($1\,MB$) every 18 h, and a short video or high

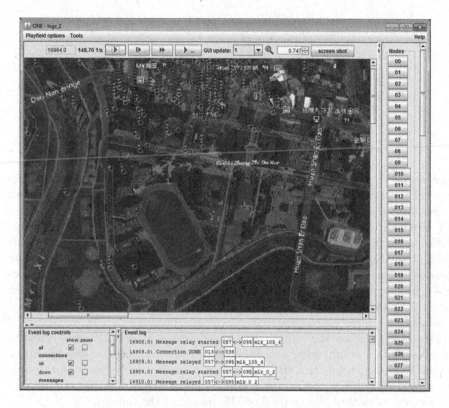

Fig. 2. The ONE simulator running with the NCCU traces.

resolution picture (10 *MB*) every 96 h. These frequencies are loosely based on [5], while sizes are approximations of the typical content produced by current mobile phone hardware. In order to obtain a realistic model of the user behaviour, the interval between messages is generated using an independent Poisson process for each user and message type. Note that this workload is not the same used in [4], so the results presented here can differ from the ones presented in that paper.

Communication range (r) was set to 7.5 m with bandwidth $Bw = 2.1\,Mb/s$, and setup time $T_s = 0.1\,s$. These values were selected based on Bluetooth 2.0, Class 2 specifications. Although the maximum range is 10 m, we assume a certain interference, and thereby we reduced the coverage value. Finally, the message Time To Live (TTL) was set to 12 h or 24 h, and buffer sizes varied from 50 MB to 1 GB. The main simulation parameters are summarised in Table 1.

4.2 Evaluation

In this section we detail the main experiments performed, and discuss the results. We are mainly interested in two performance parameters: the *delivery success ratio*, that is the ratio of nodes that receive the message, and the *delivery latency*,

Table 1. The main simulation parameters.

Parameter	Values
Buffer Size	50 MB, 100 MB, 200 MB, 1 GB
Message size	1 kB, 1 MB, 10 MB
Routing	Epidemic, Forced-Stop
Time to Live	12 h, 24 h

that is the time it takes for a message to reach its destination(s). Other key aspects are also studied, such as the overhead, buffer occupancy, and the amount of messages aborted/relayed.

Although there are several alternative diffusion protocols to the Epidemic diffusion (such as the PRoPHET, Spray and Wait, etc.) we compare our approach to the basic Epidemic routing, as it obtains the minimum delivery latency and greatest delivery success ratio. Using the same traces and simulator, the authors of [4] also evaluated the PRoPHET and Spray and Wait protocols, showing that the Epidemic approach obtains the best results at the expense of increased overhead.

We now focus on the dynamics of the diffusion process. Figure 3 shows a sample diffusion graph, using a medium sized (*1 MB*) message for dissemination. In this figure, each line is a transmission between two nodes showing the time when each transmission was started. Following these transmissions, the message finally arrives to all nodes (a total of 115, including the source).

We now focus on the *delivery success ratio*. Figure 4 provides a graphical representation of the average delivery success ratio at intervals of 1 h. This ratio was obtained by calculating the number of messages that are generated at a given hour h of a day: $msj_s[h]$, and the number of these messages that reach their destination $msj_r[h]$; so the hourly ratio is $msj_r[h]/msj_s[h]$. Due to space constrains, we include only two representative plots for the smallest and biggest buffer size (*50 MB* and *1 GB*) with different TTLs. In the first plot (Fig. 4a) with a TTL of 12 h, we can see that the delivery success ratios are related to user activities, i.e. at night people are sleeping, thus their motion is reduced and the delivery probability is reduced. At daytime the motion is restored, and the delivery probability goes back to previous levels. Nevertheless, when we increase the TTL to 24 h (and the buffer size), the results are quite different (see Fig. 4a). In this case, due to the longer life time of messages, the daily activities are not so evident, but we can clearly see the weekly activities (for example, days 3–4 and 10–11 are weekends, so message diffusion is reduced). Regarding the effectiveness, in all cases the trend is similar, showing that Forced-Stop has a higher delivery probability than Epidemic diffusion when the rest of parameters remain the same.

Figure 5a shows the average delivery success ratio of all the messages, depending on different buffer sizes for the four combinations of TTL and message spreading approaches simulated. In this case, the ratio is obtained as the quo-

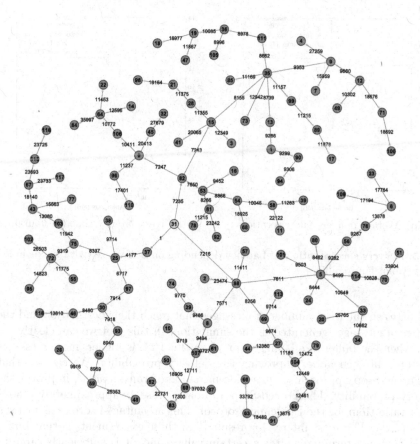

Fig. 3. Message diffusion.

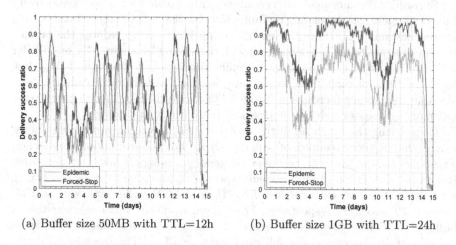

(a) Buffer size 50MB with TTL=12h   (b) Buffer size 1GB with TTL=24h

Fig. 4. Average delivery success ratio by hour. (Color figure online)

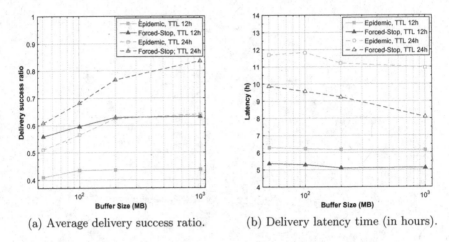

(a) Average delivery success ratio.   (b) Delivery latency time (in hours).

Fig. 5. Delivery success ratio and Latency depending on buffer size (with x-axis in log scale) (Color figure online)

tient between the total number of messages that reach their destination and the number of messages generated in the simulation. In this plot we can clearly see that, when the buffer size is bigger, or when the TTL is higher, more messages are stored in each node, improving the delivery probability. We can see that the Forced-Stop approach, as it avoids incomplete transmissions, improves the delivery probability. The Forced-Stop approach presents approximately a 30 % higher ratio than for the Epidemic protocol. This advantage is even higher for a TTL of 24 h. The most interesting result from these experiments is that buffer size is not so determinant after a certain value, since it is sufficiently large to store most of the generated messages.

Regarding the message delivery latency, Fig. 5b shows the average delivery time of all messages depending on buffer size, for the two different dissemination approaches and TTL times. In general, using Forced-Stop reduces the latency time in contrast to the Epidemic protocol, decreasing by about 20 % or 30 % for a TTL of 12 or 24 h. The impact of buffer size is not as dramatic, being close to 5 % or 10 % for each TTL, as it is the case for the delivery probability. Note also that the latency increases with the TTL, due to the improved delivery ratio, i.e., more messages reach the destination, but with greater latency.

Considering the overhead of the protocols, Fig. 6 compares the buffer occupancy (that is, the percentage of buffer used), for both dissemination approaches. These plots show the average buffer occupancy among all nodes at each time step in the simulation. As in previous experiment, and, due to space constrains, we only include two extreme cases, 50 MB and 1 GB buffer size, with a TTL value of 12 and 24 h, respectively. From these results it becomes clear that a buffer of 1 GB is large enough to store all messages even for a large TTL, while a buffer of 50 MB gets easily full even for a smaller TTL. In order to determine the required buffer size, we evaluate the maximum buffer occupancy for

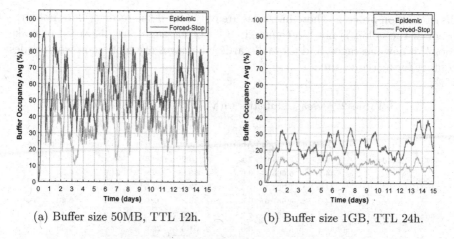

(a) Buffer size 50MB, TTL 12h.   (b) Buffer size 1GB, TTL 24h.

Fig. 6. Buffer occupancy over simulation time. (Color figure online)

the whole simulation. Figure 7a plots this maximum buffer occupancy for different TTL values and dissemination approaches. As expected, buffer occupancy is higher for the Forced-Stop than for the Epidemic diffusion scheme because, as the delivery probability is higher, more messages remain alive in the whole network.

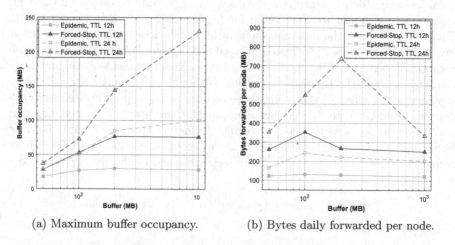

(a) Maximum buffer occupancy.   (b) Bytes daily forwarded per node.

Fig. 7. Overhead results: buffer occupancy and forwarded bytes (Color figure onlines)

Another interesting metric, related to protocol overhead, is the amount of information forwarded per node. Similarly to the previous analysis, the results in Fig. 7b are grouped by TTL. As the delivery probability increases, the amount of data delivered increases until the buffer size is big enough to hold most messages (200 MB for a 12 h TTL, and 1 GB for 24 h). This effect is caused by the dissemination process, where messages are only sent if the receiver does not have them. If the buffer is not large enough, a lot of messages are discarded to make room for new messages, and those discarded messages are sent again when nodes contact for a second time.

Table 2. Message statistics with a TTL of 12 h (x 1,000).

Buffer size	Protocol	Created	Relayed	Aborted	Dropped	Removed	Delivered	Delivered probability
50 MB	Epidemic	44.2	2,154.9	7.1	1,480.9	696.5	18.1	.0.41
	Forced − Stop	44.1	6.065.2	0.0	2,665.8	3,409.7	24.6	0.56
100 MB	Epidemic	44.2	1,793.1	7.8	1,677.6	136,5	19.2	0.44
	Forced − Stop	44.0	4,967.6	0.0	2,886.8	2,089.8	26,1	0.60
200 MB	Epidemic	44.3	1,690.9	7,7	1,712.0	0.0	19.3	0.44
	Forced − Stop	44.4	3,272.1	0.0	3,151.6	128.2	27.9	0.63
1 GB	Epidemic	43.9	1,701.5	7.5	1,722.2	0.0	19.2	0.44
	Forced − Stop	44.3	3,166.8	0.0	3,174.3	0.0	28.1	0.63

Finally, we present the results obtained related to the message dynamics, that is, the number of messages relayed, dropped and aborted. When mandatory stops are not enforced, transmissions will depend on the duration of contacts. If a transmission cannot be completed for the duration of a given contact, it is considered an *aborted* transmission. Also, a message could be *dropped* if the TTL expires, and *removed* to make room for new messages when the buffer is full. Tables 2 and 3 show the message count for all simulations considered, also showing the number of messages created, relayed and delivered. As expected, there are no aborted messages using the Forced-Stop approach this number comparatively small when using the Epidemic approach. Although this number is not large by any means, the effect of not losing those transmissions is quite noticeable in the delivery probability. In the Force-Stop approach there are more relayed, dropped and removed messages than in the regular Epidemic approach, simply because buffer occupancy is higher. As there are more messages in the buffers, more messages are transmitted, and so more message could be dropped or removed.

Table 3. Message statistics with a TTL of 24h (x 1,000).

Buffer size	Protocol	Created	Relayed	Aborted	Dropped	Removed	Delivered	Delivery probability
50 MB	Epidemic	44.2	4,088.9	7.5	1,525.2	2,561.4	22.6	0.51
	Forced − Stop	44.5	13,522.1	0.0	2,604.5	10,881.0	27,0	0.61
100 MB	Epidemic	44.4	3,846.4	9,5	1,854.7	1,979.7	25.1	0.60
	Forced − Stop	44.0	13,048.5	0.0	3,064.2	9,906.6	29.9	0.68
200 MB	Epidemic	44.2	2,969.3	9.3	2,385.3	564.8	27.7	0.63
	Forced − Stop	44.1	11,137.0	0.0	3,581.6	7,444.6	33.8	0.77
1 GB	Epidemic	44.2	2,541.8	9.2	2,513.1	0.0	28.3	0.64
	Forced − Stop	44.3	4,183.0	0.0	4,065.8	0.0	37.0	0.84

5   Conclusions and Future Work

This paper presented a new diffusion method, called Forced-Stop, based on controlling nodes mobility to achieve complete message transfers in opportunistic networks. Using the ONE simulator and a realistic environment based on real human mobility traces, we compared our proposal with the classical Epidemic diffusion.

Our experiments showed that forcing devices to stop moving to complete the data delivery process can improve the performance of the whole diffusion process. Our diffusion model provides a higher delivery success ratio and lower delivery times at the expense of higher buffer occupation and longer transmission.

These results can be a relevant indication to the designers of opportunistic network applications that could integrate in their products strategies to inform the user about the need to temporarily stop in order to increase the overall data delivery. Our interest in this type of analysis, on the long term, is focused on the design of cross-layer content distribution strategy to improve information sharing in opportunistic networks, and to provide a clear insight on how to develop and deploy efficient cooperative applications.

Acknowledgment. This work was partially supported by the *Ministerio de Economía y Competitividad, Programa Estatal de Investigación, Desarrollo e Innovación Orientada a los Retos de la Sociedad, Proyectos I+D+I 2014*, Spain, under Grant TEC2014-52690-R, the *Generalitat Valenciana*, Spain, under Grant AICO/2015/108, the Secretaría Nacional de Educación Superior, Ciencia, Tecnología e Innovación del Ecuador (SENESCYT), and the Universidad Laica Eloy Alfaro de Manabí, Ecuador.

References

1. Pelusi, L., Passarella, A., Conti, M.: Opportunistic networking: data forwarding in disconnected mobile ad hoc networks. IEEE Commun. Mag. **44**(11), 134–141 (2006)
2. Ferretti, S.: Shaping opportunistic networks. Comput. Commun. **36**, 481–503 (2013)
3. Keränen, A., Ott, J., Kärkkäinen, T.: The ONE simulator for DTN protocol evaluation. In: Proceedings of the Second International ICST Conference on Simulation Tools and Techniques, Rome (2009)
4. Tsai, T.-C., Chan, H.-H.: NCCU Trace: social-network-aware mobility trace. IEEE Commun. Mag. **53**, 144–149 (2015)
5. AnAverage WhatsApp User Sends Messages per Month, 15 September 2015. http://www.statista.com/chart/1938/monthly-whatsapp-usage-per-user
6. Niu, J., Guo, J., Cai, Q., Sadeh, N., Guo, S.: predict and spread: an efficient routing algorithm for opportunistic networking. In: Wireless Communications and Networking Conference (WCNC), 2011 IEEE, pp. 498–503, Cancún (2011)
7. Thakur, G.S., Kumar, U., Helmy, A., Hsu, W.-J.: On the efficacy of mobility modeling for DTN evaluation: analysis of encounter statistics andspatio-temporal preferences. In: 7th International Wireless Communications and Mobile Computing Conference (IWCMC), pp. 510–515, Istanbul (2011)

8. Warthman, F.: Delay-and Disruption-Tolerant Networks (DTNs) a tutorial, version 2.0. In: The InterPlaNetary (IPN) Internet Project. InterPlanetary Networking Special Interest Group (IPNSIG) (2012)
9. Battestini, A., Setlur, V., Sohn, T.: A large scale study of text messaging use. In: 12th International Conference on Human Computer Interaction with Mobile Devices and Services MobileHCI, pp. 1–10, Lisbon (2010)
10. Förster, A., Garg, K., Nguyen, H.A., Giordano, S. On context awareness and social distance in human mobility traces. In: Third ACM International Workshop on Mobile Opportunistic Networks, pp. 5–12, Zürich (2012)
11. Boldrini, C., Conti, M., Passarella, A.: Modelling data dissemination in opportunistic networks. In: Proceedings of the Third ACM Workshop on Challenged Networks - CHANTS 2008, pp. 89–96, San Francisco (2008)
12. Natalizio, E., Loscrí, V.: Controlled mobility in mobile sensor networks: advantages, issues and challenges. Telecommun. Syst. **52**(4), 2411–2418 (2013)
13. Neena, V.V., Rajam, V.M.A.: Performance analysis of epidemic routing protocol for opportunistic networks in different mobility patterns. In: 2013 International Conference on Computer Communication and Informatics, pp. 1–5, Coimbatore (2013)
14. Mehta, N., Shah, M.: Performance evaluation of efficient routing protocols in delay tolerant network under different human mobility models. Int. J. Grid Distrib. Comput. **8**(1), 169–178 (2015)
15. Su, J., Chin, A., Popivanova, A., Goel, A., Lara, E.D.: User mobility for opportunistic ad-hoc networking. In: Sixth IEEE Workshop on Mobile Computing Systems and Applications (WMCSA 2004), Low Wood (2004)
16. Feng, Z., Chin, K.-W.: A unified study of epidemic routing protocols and their enhancements. In: IEEE 26th International Parallel and Distributed Processing Symposium Workshops PhD Forum (IPDPSW), pp. 1484–1493, Shanghai (2012)
17. Vardalis, D., Tsaoussidis, V.: Exploiting the potential of DTN for energy-efficient internetworking. J. Syst. Softw. **90**, 91–103 (2014)
18. Rango, F.D., Amelio, S., Fazio, P.: Epidemic strategies in delay tolerant networks from an energetic point of view: main issues and performance evaluation. J. Networks **10**(01), 4–14 (2015)
19. Herrera-Tapia, J., Manzoni, P., Hernández-Orallo, E., Calafate, C.T., Cano, J.-C.: Power consumption evaluation in vehicular opportunistic networks. In: IEEE 12th CCNC 2015 Workshops - VENITS, pp. 925–930, Las Vegas (2015)
20. Erramilli, V., Crovella, M.: Forwarding in opportunistic networks with resource constraints. In: Proceedings of the third ACM workshop on Challenged networks - CHANTS 2008, pp. 41–47, San Francisco (2008)
21. Fathima, G., Wahidabanu, R.: Buffer management for preferential delivery in opportunistic delay tolerant networks. Int. J. Wirel. Mob. Netw. (IJWMN) **3**, 15–28 (2011)
22. Pan, D., Ruan, Z., Zhou, N., Liu, X., Song, Z.: A comprehensive-integrated buffer management strategy for opportunistic networks. EURASIP J. Wirel. Commun. Netw. **2013**(1), 1–10 (2013)
23. Hernández-Orallo, E., Herrera-Tapia, J., Cano, J.-C., Calafate, C.T., Manzoni, P.: Evaluating the impact of data transfer time in contact-based messaging applications. IEEE Commun. Lett. **19**, 1814–1817 (2015)
24. de Abreu, C.S., Salles, R.M.: Modeling message diffusion in epidemical DTN. Ad Hoc Netw. **16**, 197–209 (2014)

CIMPL: A Public Safety Tool
Based on Opportunistic Communication

Waldir Moreira[1,2(✉)], Antonio Oliveira-Jr.[3], and Marcos Aurélio Batista[4]

[1] COPELABS, University Lusófona, Lisbon, Portugal
waldir.junior@ulusofona.pt
[2] PPGMO, Federal University of Goiás, Catalão, GO, Brazil
[3] INF, Federal University of Goiás, Goiânia, GO, Brazil
antonio@inf.ufg.br
[4] IBiotec, Federal University of Goiás, Catalão, GO, Brazil
marcos.batista@pq.cnpq.br

Abstract. Public safety plays an important role in what concerns the social welfare of citizens, thus impacting their quality of life. There are different mechanisms (e.g., police, firefighters) in place to help citizens feel safe, and the citizens themselves may have a more proactive posture (e.g., neighborhood watch). The proposed Android application, called CIMPL, targets the willingness of these model citizens in engaging in public safety efforts and the capabilities of their mobile personal devices to improve social welfare. CIMPL allows users to opportunistically exchange public safety content among their devices. This content, produced by users, identifies potential hazard situations, and is used to duly notify the authorities involved in the maintenance of public safety.

Keywords: Opportunistic communication · Public safety · Social welfare

1 Introduction

Public safety is an important aspect when it comes to guaranteeing social welfare. Along with well-known approaches (e.g., police presence, road maintenance task forces), society also counts with the cooperation of its citizens to improve safety measures: neighborhood watch[1] is based solely on the reports of residents on suspicious activities. This simple action has helped police to successfully reduce disorder and crimes such as burglary, drug dealing, among others [4].

Furthermore, given the fast-paced advances in technology, there are different approaches in society that make use of technology (e.g., drones watching over crowds, sensors providing temperature levels, cameras monitoring roads) to keep track of any action and situation that may disturb orderliness and/or put the life of citizens at risk. These approaches make use of different types of communication technologies (e.g., Bluetooth, Wi-Fi) and of any piece of data obtained

---

[1] http://www.ourwatch.org.uk/.

© Springer International Publishing Switzerland 2016
N. Mitton et al. (Eds.): ADHOC-NOW 2016, LNCS 9724, pp. 169–183, 2016.
DOI: 10.1007/978-3-319-40509-4_12

from different sources that not only can be employed to guarantee the safety of the public, but also be extended to other social welfare aspects (e.g., health, education) that affect the well-being and quality of life of citizens [5].

CIMPL[2] exploits (i) the proactiveness of exemplary citizens, who are willing to help in keeping their communities safe and sound, and (ii) their mobile personal devices' increased capabilities (e.g., processing, storage), available wireless communications technologies (e.g., Wi-Fi direct), and built-in sensors (i.e., location), to allow citizens to easily engage in public safety efforts to help respective authorities in identifying hazard situations, and in maintaining social welfare.

The paper is structured as follows. Section 2 provides a brief overview of few applications that exploit user device to provide/generate public safety information. Section 3 provides a detailed view of the implementation of CIMPL. Section 4 shows a performance evaluation of existing opportunistic dissemination solutions in an application scenario for CIMPL, along with a small-scale experiment to show the feasibility of the application. Finally, Sect. 5 provides our considerations and future work towards the improvement of CIMPL.

2   Related Work

Public safety measures can follow a reactive or a proactive approach. Reactive approaches begin upon the detection of a hazard situation (i.e., an incident has already taken place), while proactive approaches may take place immediately (or even before) hazard situations. Moreover, these measures can be mitigating (i.e., to lessen the effects of hazard incidents that already took place), or preventive (i.e., to avoid bad situations such as the aforementioned neighborhood watch).

Depending on the perspectives, a specific approach shall be employed [10], and the success of such approaches is directly related to how the respective authorities execute them, affecting the levels of public safety experienced in society [2].

There are few applications that serve the purpose of public safety. They usually target incidents that already happened and still may put in cause the life of citizens. Thus, such applications are reactive and/or mitigating. However, with proper configuration, they can be used in a proactive and preventive manner.

The MobilePatrol Public Safety application is reactive as it does not prevent incidents, and it is mitigating as it serves to notify authorities. However, most of its use report on a hazard situation that already has taken place, and the application serves public safety by keeping track of these situations and by providing users with such information. To be able to report incidents, users are required to create an account. By allowing the capturing of photos and videos, the app may be used in a proactive and preventive way.

Snappii's Public Safety application is also reactive and allows users to gather information about incidents and report them accordingly to authorities. The application falls into the mitigating perspective as it can also allow authorities

---

[2] Read simple, which comes from Portuguese CIdadão exeMPLar that means exemplary, model citizen and it is meant, as its name suggests, to be rather simple.

to quickly react upon the reported incident. The user is also required to create an account and supply a contact list for reporting incidents. The incident reports also allows for the addition of photos and location information.

Public Safety app of the Medical College of Wisconsin is reactive and mitigating, allowing its users to report incidents taking place in the campus. Incidents are reported by email and suspicious activities may be reported along with photos, which makes the app also proactive and preventive.

Despite of being proactive and preventive, these applications rely on Internet connectivity, list of contacts, account creation, and even detailed user input.

On the other hand, CIMPL is: (i) simple - all the user needs is to report by means of a photo (with automatic incorporated relevant information, such as location, time) to help improve public safety measures; (ii) proactive - the basic assumption is that all users are willing to contribute to public safety efforts; and (iii) preventive - by relying on the fact that engaged individuals proactively participate in the public safety efforts, authorities can take fast actions.

Moreover, the main requirement for the conception and functioning of CIMPL is that the exchange of public data takes place upon any contact opportunities among citizens' (i.e., users') devices without relying on costly infrastructure and/or Internet connectivity (although such infrastructure and connectivity may be exploited when feasible).

It is worth noting that there are other applications that focus solely on the opportunistic communication (nearbyPeople, OMiN), on the public safety information (Massachusetts Alerts, CodeRED Mobile Alert), or image analysis (Hudl Technique, Sports Camera Analysis Free). However, we refrain from a deep analysis at these solutions since: (i) separately, they do not bring any relevant knowledge for the development of CIMPL; (ii) some of them provide too basic mechanisms (e.g., simple data exchange; help users analyze images instead of analyzing such images themselves); and (iii) we are interested in solutions that actually bring these worlds together as intended with CIMPL.

There is, however, a recent application, Oi! [1], that allows for opportunistic dissemination based on how users socially engage. This is rather interesting and can further complement CIMPL since its users do have social interactions that can be considered to improve the dissemination of public safety data (the potential of social-based solutions are discussed on Sect. 4).

3   CIMPL Implementation

CIMPL provides an easy-to-use interface that allows for a seamless, opportunistic exchange of public safety data among users towards respective authorities. Figure 1 presents CIMPL's implementation that can be broken down into two distinct parts concerning the user interface activities and background services.

Regarding the user interface, CIMPL has different activities concerning the main interface presented to the user, the application information, the capture of a public safety incident (i.e., photo), and the tagging of such incident (i.e., identification of the responsible authority). These activities are MainActivity, AboutActivity, CameraActivity, and TagActivity as illustrated in Fig. 2.

The main user interface is the **CIMPL_MainActivity** (cf., Fig. 2a), which is displayed in the foreground when the user runs CIMPL. From this activity, CIMPL users can call the AboutActivity by pressing the information button on the top right corner of the screen. Both the CameraActivity and TagActivity are called sequentially after the user presses the "Be a nice citizen" button.

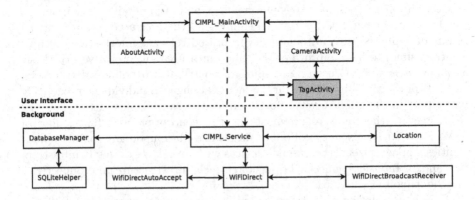

Fig. 1. High level design of the CIMPL application.

All the information about CIMPL (name, applicability and goals, the project to which it belongs, and funding entity) is found in the **AboutActivity** (cf., Fig. 2b). This activity also shows the current version of the application.

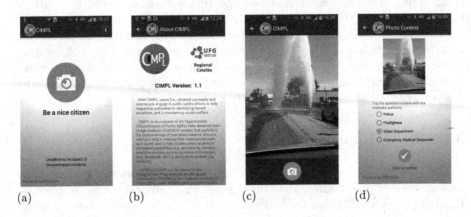

(a)   (b)   (c)   (d)

Fig. 2. Activities concerning the user interface of CIMPL.

As soon as the user wants to report on a new public safety incident (i.e., when the "Be a nice citizen" button is pressed), the **CameraActivity** is called.

This activity is used to capture a photo from such incident. When this activity is called, the camera API is initiated. When the user presses the button in the lower part of the screen (cf., Fig. 2c), the application captures the target image, saves it in a file and calls the TagActivity.

Finally, the **TagActivity** is responsible to assign a tag to the photo, which corresponds to the authority to be notified. As of now, the user has four choices, namely Police, Firefighters, Water Department and Emergency Medical Responder. Such choice list can be further extended to accommodate other authorities.

After the user selects a desired tag, a confirmation button appears in the bottom of the screen (cf., Fig. 2d). If the user confirms the chosen tag, the TagActivity gets the user location, time and selected tag, and saves this set of information to the Exif metadata of the photo. This metadata is built by means of the ExifInterface[3] that allows for setting and getting Exif tags to image files. Moreover, the user location is obtained from the Location module through the CIMPL_Service (further explained later in this section).

Then, this activity triggers a forwarding process inside the WiFiDirect through the CIMPL_service that allows this new incident to be disseminated to nearby CIMPL users upon encountering such users, and brings the application back to the main user interface, the CIMPL_MainActivity.

Concerning the background services, CIMPL comprises different modules that manage other modules, handle the database, establish Wi-Fi direct connections, and fetch user location information. These modules are CIMPL_Service, DatabaseManager, WiFiDirect, and Location.

The core of the CIMPL application is the **CIMPL_Service**. This module is always running in background and is responsible to initiate the DatabaseManager, WiFiDirect and Location modules. This module also serves as a bridge that allows user interface activities to interact with other modules (e.g., TagActivity is able to get location information from the Location module, and can trigger forwarding process in WiFiDirect module through the CIMPL_Service).

DatabaseManager is the module responsible for handling the database. It relies on the **SQLiteHelper** to create tables and respective attributes in the CIMPL database. It is worth noting that the database has no use for the current version of CIMPL. However, it will be very useful when social information is considered for forwarding purposes (i.e., the application must store the different levels of social interaction among users to help in the forwarding decisions).

The **WiFiDirect** module is based on Wi-Fi Direct technology as its name suggests. By using Wi-Fi Direct, applications are announced as services and this is used to identify other neighboring devices which run the same application. Thus, this module registers CIMPL as an available service in order to locate neighboring devices and allow the exchange of incident photos among them.

This module is responsible to send photos to and receive photos from other CIMPL devices. It is worth mentioning that the forwarding process between CIMPL devices is triggered when a request for this purpose is received, and it can be initiated (i) by the TagActivity after the user reports on a new incident

---

[3] http://developer.android.com/reference/android/media/ExifInterface.html.

(i.e., takes a new photo); or (ii) by the WiFiDirect itself when a photo reporting on a new incident is received from a neighboring CIMPL user, or when a new CIMPL user is encountered.

The WiFiDirect module also counts with **WifiDirectAutoAccept** submodule that automatically accepts Wi-Fi Direct connection requests. For security purposes, when a device receives a Wi-Fi Direct connection request, it shows a message asking whether the user wants to accept such connection. This submodule allows CIMPL to automate this process by accepting the connections without user intervention. This module is a product of Qualcomm[4]. It is worth noting that the automatic acceptance of a connection request is only possible if the application is running on the foreground. Otherwise, the user is still required to accept such connection request.

Finally, the WiFiDirect module relies on the **WiFiDirectBroadcastReceiver** submodule that keeps track of all information related to the state of the Wi-Fi Direct. It is called when the Wi-Fi Direct is enabled or disabled, a connection is made or a disconnection happens, etc.

Concerning the device (i.e., CIMPL user) location, the **Location** module uses the FusedLocationProviderAPI[5] designed by Google. This location provider API analyses GPS (if enabled), Cellular and Wi-Fi network location data in order to provide the highest location accuracy possible, considering the best and most efficient utilization of battery resource. CIMPL takes into account the most accurate location information available when building the Exif metadata of the photo in the TagActivity.

4   Evaluation of Opportunistic Dissemination Solutions

Opportunistic dissemination solutions are based on a variety of forwarding strategies, ranging from replicating data upon every contact opportunity [11], up to considering user social information (i.e., social relationships, popularity, same/different communities, shared interests) to take forwarding decisions [8].

We start this section by introducing the evaluation methodology (Sect. 4.1) and experimental settings (Sect. 4.2), which reflects a challenging networking scenario considered for the evaluation of the opportunistic dissemination solutions. Then, an analysis is done on the performance results (Sect. 4.3) as to better understand the characteristics that are valuable to the development of CIMPL. Finally, Sect. 4.2 shows a small-scale experiment to illustrate the potential of CIMPL.

4.1   Evaluation Methodology

Performance analysis is done on the Opportunistic Network Environment (ONE). Epidemic, Bubble Rap, dLife, and SCORP are the considered proposals

---

[4] https://github.com/mdabbagh88/alljoyn_java/blob/master/helper/org/alljoyn/bus/p2p/WifiDirectAutoAccept.java.

[5] https://developers.google.com/android/reference/com/google/android/gms/location/FusedLocationProviderApi.

that shall help us identifying the core features of the opportunistic data dissemination approach to be employed in CIMPL.

Despite of being completely agnostic to social similarity among users and known for its unmannerly resource consumption, Epidemic [11] is considered as it is a rather simple proposal and it serves as upper bound for delivery probability: by flooding the network with many replicas, Epidemic can reach optimum delivery rates.

Since users display different social similarities (relationships, interests, communities) among themselves [8], we are also interested on the performance of social-aware proposals. Thus, we consider the community-based Bubble Rap [3] that also relies on the local (i.e., inside communities) and global (i.e., whole network) centrality of users to decide on forwarding; dLife [7], which measures the dynamic social behavior of users and considers only socially well-connected users to perform data exchange; and SCORP [9] that is a social-aware content-based solution that considers not only how socially well-connected users are, but also the interests of these users on the content traversing the network.

To provide results with a 95 % confidence interval, each proposal is tested ten different times with different random number generator seeds. Regarding the performance evaluation metrics, we consider the average delivery probability (i.e., ratio between the number of delivered messages and the total number of messages that should have been delivered), the average cost (i.e., number of replicas per successfully delivered message), and average latency (i.e., time elapsed between message creation and delivery).

4.2   Experimental Settings

Table 1 summarizes our experimental settings and these values are based on a universal evaluation framework previously proposed [6]. The experimental scenario simulates a 10-day period in a urban environment, comprising different mobility patterns. The scenario counts with 832 nodes distributed into three groups, namely people, buses, and police patrols.

The people group is further divided into eight groups of 100 individuals. People follow the Working Day Movement mobility model, walk with speeds from 0.8 to 1.4 m/s, and may use car or bus to travel in the city. Each group has different office, home, and meeting points configurations. Each person spends 8 h at work and move in the office with pause times raging from 1 min to 4 h. Also, each person has a 50 % probability of engaging in a leisure activity, alone or with others, and can last for 2 h at most.

Regarding the bus group, it is also divided into 8 groups of two vehicles, where one vehicle follows a circular route and the other an end-to-end route as to better serve the people. The mobility model followed is the Bus Movement with waiting (i.e., pick up/drop off) times at stops varying from 10 to 30 s and speeds between 7 to 10 m/s.

Finally, the police patrol group counts with 16 vehicles following the Shortest Path Map Based Movement mobility model, in which each patrol randomly chooses a point in the city and moves there through the shortest path. Like the

Table 1. Simulation parameters.

Parameters	Values
Simulator	Opportunistic Network Environment (ONE)
Routing Proposals	Epidemic, Bubble Rap, dLife and SCORP
Simulation Time	10 days
# of nodes	800 people, 16 buses, and 16 police patrols
Mobility Models	Working Day, Bus, Shortest Path Map Based
# of Messages	40000 (39000 personal and 1000 public safety)
Message TTL	1 day
Message Size	1 – 100 kB
Node Buffer	2 MB
Node Interface	Wi-Fi (Rate: 11 Mbps / Range: 100 m)
K-Clique	k = 5 and familiarThreshold = 700 s (Bubble Rap)
Daily Samples	24 (dLife and SCORP)

buses, these patrol vehicles move with speeds between 7 to 10 m/s; however, their waiting (i.e., watch) time ranges from 100 to 300 s.

Independently of the group, each node is equipped with a Wi-Fi interface (11 Mbps and 100 m communication range), which is commonly found in today's devices. Despite the increased capabilities of devices, the storage space considered for message relaying in each node is of 2 MB since their owners may not be willing to share all their storage to carry content to others.

The generated load is of 39000 messages between specific source/destination pairs (hereafter referred to as personal messages), plus 1000 that are public safety messages directed to the police group. Since we are interested in the dissemination of public safety messages in urban setting, the load accounts for personal messages as well in order to allow the assessment of the considered benchmark with the burden of dealing with high traffic load. Messages are generated since the beginning of simulation until the end of the 9th simulation day, allowing a full day for remaining deliveries to take place prior to simulation end.

For Epidemic, Bubble Rap and dLife, the daily rates in which messages are generated are 3900 personal and 100 public safety messages. In order to achieve the same load with SCORP (as it is an interest-driven approach), each people group holds 10 different interests, which may overlap among groups. The different interests combined with the creation of 170 personal and 63 public safety messages, allows for a daily rate of 3000–8000 personal and 96–144 public messages, coming to a total of 40000 messages altogether.

Message TTL is of 24 h and size varies between 1 and 100 kB to simulate data produced by different user applications (e.g., email, chats).

Regarding the specific proposal-related parameters, dLife and SCORP consider 24 daily samples (i.e., each of one hour) [7,9], and Bubble Rap uses K-Clique for community formation and and cumulative window approach for node centrality computation as proposed by Hui et al. [3]. The parameter k $(= 5)$ is chosen based on simulations in which Bubble Rap has the best overall performance in terms of the considered evaluation metrics.

4.3   Result Analysis

The ODISSEIA project targets a urban, dense scenario with nodes following different mobility patterns, having many different contact opportunities throughout their daily routines, and producing/consuming a large amount content.

The designed scenario (cf., Sect. 4.2) attempts to capture that by reproducing a city setting with people using different modes of transportations and communicating with one another. Since ODISSEIA aims at the dissemination of public safety information within such setting, the exchange of personal and safety-related data is considered as to observe how the different benchmark solutions deal with a demanding traffic load.

Figure 3(a) shows the delivery performance. It is worth remembering that the scenario counts with a daily load of 3900 personal messages between specific source/destination pairs, and 100 public safety messages directed to the respective authority, i.e., the police. Moreover, message TTL is set at 24 h, and nodes count with 2 MB storage. Under these conditions, Epidemic and Bubble Rap are able to deliver only 18.01 % and 15.42 %, respectively, of the total number of messages expected to be received throughout the simulation.

Since nodes have many encounters during their daily routines, Epidemic takes this opportunity to replicate messages at every encounter, and this plays against it as storage is limited (despite the fact that devices have large storage space, users may not be willing to share this resource in its entirety). Also, message TTL further degrades Epidemic's performance as the longer messages stay in the system, the more they are replicated. This may exhaust storage space and take the opportunity of newly generated messages to be received/propagated.

Bubble Rap works with the notion of communities and node centrality to perform data exchange. Since the considered scenario includes many nodes who often encounter one another, the number of formed communities is high (approximately 200 communities including up to 98 % (816) of the total number of nodes in each of them). Thus, the aspect (number of different clusters) considered by Bubble Rap to differentiate groups of nodes, and to reduce replication with increased delivery probability ceases to exist (i.e., roughly almost all nodes belong to the same, many communities).

Moreover, the nodes is such dynamic scenario count with an average centrality[6] of 25.14, with around 36 % of such nodes with centrality (ranging from 26 to 119) higher than this average. Given the daily load in the scenario, this becomes an issue as Bubble Rap relies on high centrality nodes to reach destinations. Since such nodes are few compared to the total number, they end up becoming

---

[6] Centrality is given by the number of times a node acted as relay in the communication path of two other nodes.

bottlenecks suffering with exhaustion of storage space and discarding messages that could have been delivered if other relay nodes had been chosen.

This results in a poor delivery performance as nodes belong to many overlapping communities and count with few important (i.e., high centrality) nodes, which makes it difficult to Bubble Rap find the best next hop and waste contact opportunities for the exchange of newly generated messages (which explains its 2.59 % performance disadvantage against Epidemic).

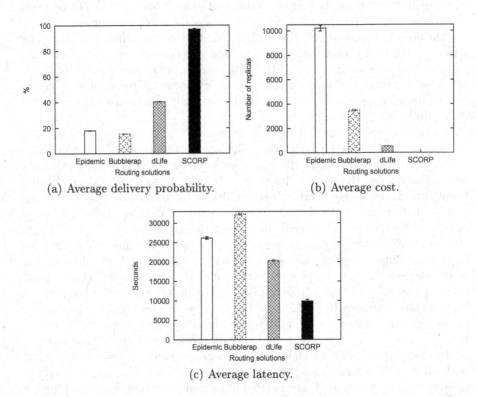

(a) Average delivery probability.   (b) Average cost.

(c) Average latency.

Fig. 3. Performance evaluation of opportunistic dissemination solutions.

Despite their poor performance, Epidemic and Bubble Rap still manage to deliver 47.67 % and 35.8 %, respectively, of the generated public safety messages.

As for dLife, it relies on the different levels of social engagement as well as the node importance to perform data exchange. Since the number of encounters among nodes in the simulation scenario is very high and frequent, dLife takes longer to have a stable knowledge concerning the levels of social engagement among nodes and their respective importance. This results in a delivery of 40.44 % of the total generated messages, as dLife chooses to hold messages longer until finding a socially well connected node to the messages' recipients or a more important node to reach such recipients. Despite the heavy load and regular performance, dLife still delivers 91.56 % of the generated public safety messages.

Considering the particularities of the scenario (i.e., high numbers of frequent encounters, heavy load), SCORP is the only solution that manages to overcome the other benchmark solutions, reaching a 97.07 % delivery of messages.

As dLife, SCORP focuses on the level of social engagement. However, it also considers the interests that nodes have in the content traversing the network. This combination allows the proposal to focus on data exchange only among socially well-connected nodes that share interests, or among nodes that have a high social engagement with others interested in the content to be replicated.

Nodes representing police patrols and buses are a good example of this latter case (their centrality lie between 41 and 119, meaning they are in contact with many different nodes) which make them good information dissemination points.

Regarding the public safety information, SCORP manages to successfully deliver 99.95 % of such messages.

When it comes to average cost (cf., Fig. 3(b)), the performance of the benchmark solutions reflects the forwarding mechanisms behind them. As one could expect, Epidemic is the most costly solution as it will take advantage of every contact to replicate every message carried by nodes. Given the high numbers of contacts in the scenario, this produces a very high number of replicas (average of 10218 replications) to achieve a successful delivery.

As nodes belong to same communities, this differentiating aspect loses its usefulness to Bubble Rap, which will rely its forwarding decisions on centrality. Since only few of these nodes are important (i.e., display high centrality), not many replications take place, which explains its lower cost (3495 replicas to perform a successful delivery). Still, we believe that the cost for Bubble Rap could have been higher. However, as storage space at these 'hub' nodes becomes a constraint, this benchmark solution is unable to perform further replication.

dLife was conceived for rather dynamic scenarios. This allows this benchmark solution to capture the different levels of social engagement between nodes and their respective importance within the network to perform forwarding decisions. So, based on the users' social life, dLife is much more careful while creating messages replicas. This solution only chooses next hop nodes if they are well socially connected with the intended destinations or if they are very important in the system at the time of message replication. This results in a cost (550) 94.61 % and 84.24 % less when compared to Epidemic and Bubble Rap, respectively.

SCORP outperforms the other benchmark solutions by producing 99 % (2.25) less replicas. By considering how socially well-connected nodes are to destinations interested in the content, replications happen at a very low frequency as this solution chooses to wait (very long) for the best next hop (average buffer time for SCORP ranges between 94 % and 99 % more than the other solutions). The longer wait is indeed worth, having a positive impact on the associated cost of this benchmark solution.

Finally, the latency (cf., Fig. 3(c)) experienced by the messages successfully delivered by these benchmark solutions reflects their choices for the next hop nodes while attempting to reach the intended destinations.

Epidemic replicates messages at every encounter. As some of these replications can find their way quickly to their respective destinations, some may require a long journey: in some cases, we observed messages traversing up to 165 different hops (experiencing a total latency of above 86000 s). This clearly impacts the average latency of Epidemic (26174 s).

Despite of not experiencing similar hop count as Epidemic (with an average count of 4.41 hops), Bubble Rap has the worst latency performance among the analyzed benchmark proposals, 32282 s. Since the choice for next hop nodes is almost solely based on node centrality, some of the chosen nodes are not the best for reaching specific destinations. That is, messages reach such nodes because they are indeed included in the communication path of different others, but this does not mean that they will quickly reach certain destinations. Thus, messages end up taking too long to be delivered.

dLife presents latency (20313 s) performance 22.39 % and 37.08 % lower than Epidemic and Bubble Rap, respectively. Its messages experience an average count of 4.85 hops, which is even a bit higher than the one of Bubble Rap.

However, dLife's advantage lies on the fact that it can still capture the dynamic social behavior of users. Although it takes a while to have a stable view of the network in terms of social engagement among users and their importance, dLife still makes routing decisions that are favorable in what concerns the latency experienced by messages: high levels of social engagement indicate that nodes tend to spend more time together, allowing for a faster delivery of messages.

By exploit the social engagement among users sharing similar interests or users that are socially well connected to others with target interests, SCORP outperforms the other solutions in what concerns latency (9865 s).

Despite the fact that messages may experience an average count of approximately 7 hops to reach their destinations, the chosen nodes in this communication path are those who really matter: either they display very high social engagement or they often encounter nodes that are interested in the content to be shared. This is enough to reduce the latency of messages exchanged through the SCORP mechanism by 62.31 %, 69.44 %, and 51.43 % when compared to Epidemic, Bubble Rap, and dLife, respectively.

4.4   Small-Scale Experiment

From the performance evaluation, we can see that even the simplest dissemination solution (Epidemic), just by exploiting the *opportunistic contacts* among nodes, is able to deliver a fair amount of public safety information (47.67 %). Delivery is further improved when the solution consider the *dynamic social behavior of users*, i.e., dLife, delivering up to 91.56 % of the produced public safety data. If the dynamic social behavior is complemented by the *interest of users* as in SCORP, the delivery of public safety information is close to optimum, 99.95 %.

Thus, clearly the current epidemic version of CIMPL is enough to allow the exchange of incident photos among users throughout the users' daily routines.

To validate the feasibility of CIMPL, we perform a small-scale experiment to illustrate how users can exchange data through opportunistic contacts, without resorting to internet connectivity. As show in Fig. 4, the experiment comprises three Android devices, representing two CIMPL users (U1 and U2) and the city maintenance authority (CMA). At 2:00 p.m., U1 creates an incident photo about a hole on the road, which could lead to accidents. As U1 does not have connectivity, the incident is saved locally. Upon encountering U2 at 2:03 p.m., U1 replicates the incident photo as to allow it to faster reach one the CMA representatives. Finally, at 2:07 p.m., U2 replicates the incident photo to the CMA worker, who can intervene to avoid accidents in the affected road.

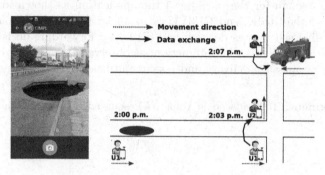

Fig. 4. Small-scale real-world experiment.

By taking advantage of the opportunistic contacts, CIMPL users are able to exchange content in faster way, allowing the CMA worker to be notified upon an encounter with any user who holds the incident photo. In terms of latency (i.e., the time since U1 created the content up to its arrival at the CMA worker's device), most of it accounts for the time the content remains at the device storage (in this experiment, 7 min). The transmission time is almost irrelevant for the experienced latency: the incident photo (including location and authority information) has approx. 132 KB, taking 0.055 s to be transmitted. As for the associated cost, in this experiment 2 replicas are needed to achieve a successfully delivery given the epidemic nature of CIMPL. While this is not a problem for this experiment, replications may become a burden if many users start to report various incidents, and storage is limited as observed in Sect. 4.3. However, we are still able to show the potential of CIMPL when it comes to exploiting opportunistic contact to allow the exchange of public safety data, thus improving social welfare.

5 Conclusions and Future Work

Concerning the future version of the CIMPL application, we must remove **TagActivity** (cf., Fig. 1) as photo tagging is the key aspect of the application and must be done automatically, by means of image analysis approaches.

In the current version of CIMPL this is done by the users themselves. The operation as described in Sect. 3 remains the same, except for the fact that no user intervention shall be required. This activity will be adapted to suitably compose the **Tagging** module, which will do the photo tagging. For that, the OpenCV computer vision library shall be considered along with the adapted version of the TagActivity to allow the recognition of the authority to be notified upon the reporting of incident photo through the CIMPL application.

Another potential future work concerns the employed opportunistic dissemination approach as the one employed by Oi! application [1]. The current version of CIMPL relies on an epidemic solution implemented at the WiFiDirect module, which is enough for the purpose of the application as shown in Sect. 4.4. However, we intend to extend CIMPL as to allow: (i) for the employment of other forwarding schemes as desired; and (ii) better forwarding decisions based on the different levels of social interaction, as observed in Sect. 4.3, in order to further improve message delivery and resource utilization (i.e., storage).

Acknowledgment. To Fundação de Amparo à Pesquisa do Estado de Goiás (FAPEG) and PNPD/CAPES.

References

1. Amaral, L., Sofia, R.C., Mendes, P., Moreira, W.: Oi! - opportunistic data transmission based on Wi-Fi direct. In: IEEE Infocom 2016 Live/Video Demonstration (Infocom 2016 Demo), San Francisco, USA, April 2016
2. Braga, A.A., Welsh, B.C., Schnell, C.: Can policing disorder reduce crime? a systematic review and meta-analysis. J. Res. Crime Delinquency **52**(4), 567–588 (2015)
3. Hui, P., Crowcroft, J., Yoneki, E.: Bubble rap: social-based forwarding in delay-tolerant networks. IEEE Trans. Mob. Comput. **10**(11), 1576–1589 (2011)
4. Longstaff, A., Willer, J., Chapman, J., Czarnomski, S., Graham, J.: Neighbourhood policing: Past, present and future
5. Moreira, W., Mendes, P.: Pervasive data sharing as an enabler for mobile citizen sensing systems. IEEE Commun. Mag. **53**(10), 164–170 (2015)
6. Moreira, W., Mendes, P., Sargento, S.: Assessment model for opportunistic routing. In: 2011 IEEE Latin-American Conference on Communications (LATINCOM), pp. 1–6, October 2011
7. Moreira, W., Mendes, P., Sargento, S.: Opportunistic routing based on daily routines. In: 2012 IEEE International Symposium on World of Wireless, Mobile and Multimedia Networks (WoWMoM), pp. 1–6, June 2012
8. Moreira, W., Mendes, P.: Social-aware opportunistic routing: the new trend. In: Woungang, I., Dhurandher, S.K., Anpalagan, A., Vasilakos, A.V. (eds.) Routing in Opportunistic Networks, pp. 27–68. Springer, New York (2013). http://dx.doi.org/10.1007/978-1-4614-3514-3_2

9. Moreira, W., Mendes, P., Sargento, S.: Social-aware opportunistic routing protocol based on user's interactions and interests. In: Mellouk, A., Sherif, M.H., Bellavista, P., Li, J. (eds.) ADHOCNETS 2013. LNICST, vol. 129, pp. 92–107. Springer, Heidelberg (2013)

10. Tonry, M., Farrington, D.P.: Strategic approaches to crime prevention. Crime Justice **19**, 1–20 (1995)

11. Vahdat, A., Becker, D.: Epidemic routing for partially connected ad hoc networks (2000). http://citeseerx.ist.psu.edu/viewdoc/summary?doi=10.1.1.34.6151

Sensors/IoT

A Service Based Architecture
for Multidisciplinary IoT Experiments
with Crowdsourced Resources

Panagiotis Alexandrou[1,2], Constantinos Marios Angelopoulos[3,7],
Orestis Evangelatos[3(✉)], João Fernandes[5], Gabriel Filios[1,2],
Marios Karagiannis[3], Nikolaos Loumis[6], Sotiris Nikoletseas[1,2],
Aleksandra Rankov[4], Theofanis P. Raptis[1,2],
José Rolim[3], and Alexandros Souroulagkas[1,2]

[1] Computer Engineering and Informatics Department,
University of Patras, Patras, Greece
aleksandro@ceid.upatras.gr
[2] Computer Technology Institute and Press "Diophantus", Patras, Greece
[3] University of Geneva, Geneva, Switzerland
orestis.evangelatos@unige.ch
[4] DunavNET, Novi Sad, Serbia
[5] Alexandra Institute, Aarhus, Denmark
[6] University of Surrey, Guildford, UK
[7] Bournemouth University, Poole, UK

Abstract. Research on emerging networking paradigms, such as Mobile
Crowdsensing Systems, requires new types of experiments to be conducted
and an increasing spectrum of devices to be supported by experiment-
ing facilities. In this work, we present a service based architecture for IoT
testbeds which (a) exposes the operations of a testbed as services by following the Testbed as a Service (TBaaS) paradigm; (b) enables diverse facilities to be federated in a scalable and standardized way and (c) enables
the seamless integration of crowdsourced resources (e.g. smartphones and
wearables) and their abstraction as regular IoT resources. The architecture
enables an experimenter to access a diverse set of resources and orchestrate
experiments via a common interface by hiding the underlying heterogeneity and complexity. This way, the field of IoT experimentation with real
resources is further promoted and broadened to also address researchers
from other fields and disciplines.

Keywords: Internet of Things · Testbeds · Architectures · Platforms ·
Crowd

1 Introduction

Experimental facilities also known as testbeds, provide the controlled environment needed for implementing and testing novel technologies and architectures.

© Springer International Publishing Switzerland 2016
N. Mitton et al. (Eds.): ADHOC-NOW 2016, LNCS 9724, pp. 187–201, 2016.
DOI: 10.1007/978-3-319-40509-4_13

Among others things, they follow agile architectures that are easy to re-configure in the context of experiments and provide additional services and tools for collecting meta-information on the experiment execution (e.g. monitoring several performance metrics or providing execution logs for post-experiment processing). Also, their controlled environment constitutes a base reference that helps in evaluating and comparing different architectures and protocols. However, despite the services and the advantages provided, testbeds also pose some limitations. By nature, each testbed facility focuses on a specific area of interest (e.g. in IoT applications or M2M communication protocols) and therefore it's architecture and the services provided define the experiments supported. Other indicative limiting factors include the number and type of available resources and the number of simultaneous experiments the facility can support.

In order to overcome such limitations, experimenters have been working towards federating different testbeds. Such federated meta-testbeds enable different research groups to join forces towards diversifying and extending the existing experimental facilities. From these efforts emerged the prevailing paradigm of Testbed as a Service (TBaaS). According to this paradigm, the resources and the services of an individual facility are exposed to third parties via some RESTful APIs, over the Internet. This virtualization while obfuscating the underlying implementation details, enables an experimenter to utilize the facility while being agnostic of its complexity. Facilities that are virtualized by complying with commonly understood APIs can then be federated under the umbrella of a webservice. Hence, experimenters are provided with a single point of entry towards several facilities. Depending on the federation architecture, experimenters are able to provision resources for their experiment's which are provided by different individual facilities.

Our Contribution. The existing experimental infrastructures mainly focus on resources that are statically deployed or characterized by low dynamics. For instance, they either refer to regular computer networks or to IoT infrastructure deployed within a static context, such as a smart building. In this work we present a holistic architecture for TBaaS, implemented in the context of IoT Lab European research project. The architecture has been adopted by the IoT Lab platform that can be accessed at http://www.iotlab.eu/. Through the IoT Lab one can access the resources of sensor testbeds from the Universities of Patras, Geneva, Surrey, from Mandat International and from the Ekonet mobile testbed. In addition resources of smartphone sensors can be accessed. The IoT Lab architecture enables us in particular to:

- Extend the range of resources by considering crowdsourced resources provided by the general public.
- Use a novel, generic yet specialized experiment mechanism. It allows a smart combination, composition and execution of diverse experiments to the users of the architecture.
- Adopt abstraction mechanisms that leverage devices, such as smartphones and smart wearables, as a distributed experimental infrastructure.

Related Work. Internet of Things facilities cover a large number of topics from purely technical issues (e.g. routing protocols, semantic queries), to a mix of technical and societal issues (security, privacy, usability), as well as social and business themes [14,15]. Federation of such facilities can be feasible with tools such as those introduced in [9]. The OneLab experimental facility, presented in [13], is a leading prototype for a flexible federation of testbeds that is open to the current Internet. GENI, the Global Environment for Networking Innovation [10], is a distributed virtual laboratory for transformative, at scale experiments in network science, services, and security. The Fed4FIRE federation framework [2] is gradually enabling experiments that combine facilities from the different FIRE research communities. Last but not least, the GEANT World Testbed Facility [12] focuses on regular computer networks.

Related crowdsourcing and crowdsensing platforms such as [1,3,7] solely focus on mobile resources as a source of sensing data. Therefore they do not include any data annotation, or any other data coming from the knowledge of the crowd. Other platforms like EpiCollect [5] and PhoneLab [17] introduce the crowdsourcing concept. However they do not integrate other types of resources, such as testbed resources nor do they include any user profiling through which they can filter the crowd, or provide support mechanisms for incentives.

2   Resource Handling

Due to the increasing amount of electronic devices and sensors that are available, either by portable devices or through testbeds and experimental facilities, there has emerged a need to migrate all the available resources under the same umbrella. This migration allows an easy interaction between end-users and experimenters, on the one hand and available resources on the other. Key requirements in our architecture design were the federation of heterogeneous resources (e.g. static, mobile, portable, and crowd-sourced resources), the scalability of architecture in terms of mobile users and IoT integration and the simultaneous handling of a large number of resources and data during the experiment execution

2.1   Resource Description

Our architecture is gathering heterogeneous resources provided either by testbeds facilities, or by crowdsourcing (e.g. by the end-users of a facility or even the general public). In order to overcome complex migration problems with the heterogeneity of the resources, we adopted the RSpec (resource specifications) scheme [4]. RSpec was used for network related resources and had to be adapted for use by IoT ones. The RSpec, is an XML schema used by the resource providers in order to describe all the available resources in the architecture. This schema is simple, yet powerful, and includes all the necessary information to describe the resources adequately. The RSpec was mainly used for network resources and for this reason we had to expand its capabilities so that they would fit in our architecture requirements.

RSpec provides tags that describe several properties of each resource such as an IP address, a protocol for communication, an access port, or a location. In particular, the aforementioned tags are aligned with the types and function sets defined by the IPSO Application Framework. For instance, a luminance sensor following the IPSO Application Framework is categorized as "ipso.sen.lum". All resource providers generate an RSpec compliant XML file that is aggregated to a single architecture-wide description of available resources in the Resource Directory.

The schema provides tags that describe nodes (<node> </node>) which include properties allowing the system to directly access the resources of each node. These properties include the IP address (ip), the protocol the node understands (protocol) and the port (port).

Inside the <node> tag, the schema provides tags for individual resources (<resource> </resource>) that describe in detail the relative path that must be used by the architecture in order to request values from each resource, as well as the type of the resource (e.g. sensor or actuator). Inside the <resource> tag, the schema describes the resource using tags that follow the types and function sets defined by the IPSO Application Framework.

Other information that is contained inside the <node> tag includes an <interface> tag that provides more information about the component ID and a <location> tag that provides information about the physical location of the node. In the link below, we provide a snippet from an indicative RSpec XML file that describes some nodes in the Geneva's testbed.[1]

Through RSpec we can also describe RESTful web services that add functionalities to testbed resources. For example, we can convert a pulse meter to an energy one, make a temperature IoT resource from a weather API, or even provide alarm and notification services. For example, the resources available in the architecture are described by each resource provider using RSpec and can include all the necessary information to describe the resources needed to compose an experiment.

2.2 Diverse Resource Types

Static IoT Resources. Each testbed which may be comprised of actuator motes, and either wireless, fixed or mobile sensors, provides them as resources to the platform. Each resource has a specific URL which invokes an API call. The resources of our testbeds follow a RESTful implementation via which GET, PUT or POST methods can be used to access them. Typical examples of such resources are the TelosB [11], Z1 [19] and Arduino IoT [8] devices.

Mobile/Portable IoT Resources. In addition to static IoT resources, mobile and portable testbeds can also be integrated. By doing so, we create networks of moving resources with multiple sensors that are capable of providing data and properties of temporal, technological, and spatial diversity. Existence of this

---

[1] http://129.194.70.52:8111/ero2proxy/service/type/xml_rspec.

type of testbeds provides the users of the architecture with more control over the choice of environment within which they want to deploy their experiments.

Virtual/Modelled Resources. Virtual resources are simulated nodes that act identically to the IoT resources and are running the same executable. The difference to the physical ones comes from the way the reported values are generated. Those values are estimated by either taking into account other resources in the virtual environment only, or they can be interpolated by the physical resources of the same provider. The number of virtual resources that are deployed in each testbed side is fixed and set by the owner of the testbed. The functionality to add and integrate additional (virtual) modelled resources is also provided.

Crowdsourced Resources - Opportunistic/Participatory Sensing. As presented in [16], each embedded sensor of a smart electronic device (e.g. smartphones), can be categorized as inertial, positioning, or ambient. Combining these categories, we can identify and measure the acceleration and rotational forces of a solid object, as well as measure the physical position of a device and various environmental parameters. The collection of measurement can be opportunistic or participatory. Opportunistic sensing takes place in the background, without needing the users to interact and have an active participation. Alternatively, participatory sensing urges users to be involved and provide the needed information, or data (such as scanning a QR code for localization purposes or answering a questionaire).

3   Experiment Composition and Execution

The experimenter is provided with a list of available resources that can view and reserve for their experiment. After the experimenter chooses and reserves the desired resources, he/she is prompted to the experiment composition module. In the background, the same RSpec XML schema is used to transfer the information regarding the resources reserved between the reservation module and the experiment composition module along with some meta-information on the experiment itself; e.g. duration and period of execution, human readable description of the experiment, etc. This information is incorporated in the RSpec document via tags such as the <research_id> tag, that provides the id of the parent research of the experiment to be composed, the <experiment_title> tag which provides the title the experimenter has given to the experiment to be composed and the <experiment_desc> which provides a short description of the experiment.

3.1   Experiment Composition

The experiment composition module receives this information and provides a simple but powerful mechanism with which the experimenter can define the details of how resources will be used in the context of "If This Then That" (IFTTT) scenarios. The final experiment consists of a set of these scenarios.

The experiment composition module allows the experimenter to set the following actions:

Get a Value from Specified Resources. The frequency of the reading request is set in minutes or hours and include one or more resources. The resources must be of type "sensor" and must be included in the experiment before the experimenter enters the main composition module. This action is called "reading". As an example, a reading can be "Get a value from sensor 1 every 5 min between these 2 dates and times".

Set a Condition. A condition can be the average, absolute, minimum or maximum value of one or more resources being greater, equal or lesser than a set value. In the case of multiple resources a logical operator can be set. An example of a condition can be "The maximum value of sensor 1 OR the maximum value of sensor 2 to be greater than 5".

Set an Outcome. An outcome is an action that can be taken. This action is either to take more measurements from sensors or to actuate an actuator. Outcomes also include a logical operator in case there is more than one conditions. An example of an outcome could be "Actuate actuator 1, if all conditions are met (with logical AND)".

Define an Action. Actions are combinations of conditions and outcomes. Actions are set in an "IF-THEN" form in order to clarify their meaning. An example of an action can be "IF condition 1 AND condition 2 are true THEN perform outcome 1". The logical operator AND is actually defined in the outcome and not in the conditions, as specified above.

After the experiment scenario has been defined, it is dispatched to the execution module. The scenario is described in an XML schema called Experiment Description XML schema (ED XML). The Experiment Description XML defines a parent tag <experiment> </experiment> that encloses all other elements. The <measurements> tag defines the measurements database server information along with the <ip> and <port> sub-tags inside it. The next tag is a random identifier tag. This is generated during the ED creation randomly and is used to uniquely identify the experiment description. The tag that provides this identifier is the <identifier> tag.

Readings are included in the <reading> tag. Inside this tag, a <frequency> tag with a "unit" property defines the frequency of the reading while <start> and <end> tags define the start and end of the readings period for the specified reading. The <resources> tag then defines which resources have to be probed for a reading every time it's needed. These are defined using <id> tags that include properties "component", "resource_id", "port", "ip", "protocol" and "path". The combination of these properties allow the execution engine to identify and reach the resources directly.

Actions are defined using the <action> tag. These include <conditions> and <outcome> tags. The <conditions> tag include the aggregation and logical operations as a tag and property respectively (e.g. <average logic = "and">).

Inside this tag, the resources are defined using an <id> tag and also the threshold is defined using a <threshold> tag. The <outcome> tag includes a property for the logical operator and inside the tag, resources are defined (either sensors or actuators) using <id> tags as above. An example of an ED XML is shown in Listing 1.1 in the Appendix.

3.2 Experiment Execution

When an experimenter finalizes the definition of an experiment at the Experiment Composition module, an Experiment Description XML document is created which is transferred to the Experiment Execution module which proceeds in parsing it and finding all necessary information in order to start running the experiment.

At first, the research ID, the experiment title and the experiment description are identified and posted as a new 'research' entity in the Resource Directory database. As already described, the Experiment Description XML document contains a number of readings and action tags. Each of these tags will spawn a thread to handle their tasks. A queue and two objects are used to handle communication between the readings and the actions. The *readingObject* notifies that a new measurement was taken. The *finishedObject* denotes that a reading thread was terminated.

Each reading tag has several resources with their contact information and a frequency with which they are to be read. Every one of those readings, spawns a new *getMeasurements()* thread tasked with obtaining the measurements from the resources in the time and with the frequency specified by the experimenter. The thread sleeps until it is time to take a measurement. When the measurement comes, the thread will wake up and call each resource associated with it for a measurement. After the measurements are taken the thread puts a *readingObject* to the queue and proceeds to sleep until the time comes to take a new measurement. When the time to finish the readings comes, the thread puts a *finishedObject* in the queue and then terminates.

Inside the actions tag there are a number of tied conditions and outcomes. Their information is parsed and summarized in two lists: one for the conditions called *conditionsList* and one for the outcomes called *outcomeList*. Then a thread for a function called conditionChecker(), with the two aforementioned lists as parameters is spawned. This thread reads the queue responsible for the communication between readings and actions. When a *readingObject* is read, it will evaluate the logic of *conditionsList* as specified in the Experiment Description XML. If it is evaluated to 'True', then the outcomes from *outcomeList* will be executed. When a *finishedObject* is seen, the number of the aforementioned threads will decrease by one. The thread will run as long as there are any *getMeasurements()* threads active.

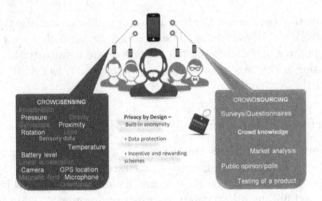

Fig. 1. Crowd participation in TBaaS architecture.

3.3   Crowd Interactions

Crowd interactions require inputs from the smartphone users through surveys and questionnaires (Fig. 1). The process of filtering and selecting the user in order to engage him/her in the specific research includes the following mechanisms available through the architecture: survey queries, survey lists and geofencing.

Survey Queries: A query is a mechanism that allows the experimenter to filter crowd users in a meaningful way in order to select the users needed for the post of a mobile query. The filtering function is based on the socio-economic profile of the user which they voluntarily include during anonymous registration through the mobile app. The query is defined and then saved in the experimenter's profile so that it can be easily reused in the future, which makes it a very powerful tool as the crowd users constantly change in number throughout the architecture's lifetime. Queries, although static themselves, provide dynamic results in the form of sets of users that fit the set criteria.

Survey Lists: Every time a query is used, an up-to-date list of crowd users that meet the query's criteria is presented. The experimenter then has the opportunity to select individual recipients to form a survey list. A survey list is a static list of survey recipients that is used to send a survey to the mobile devices recipients. The content of the user list is anonymous and only social and economic data are associated with each entry. When the final survey list is compiled, it is saved under the experimenter's profile and can be used as the destination list in which to post a survey. A special case of a survey recipient list is the "all users" static list which includes all available users of the architecture.

Geofencing: Geofencing refers to the experimentation activity in which it is possible to setup a virtual perimeter on a real world geographic area and utilize this perimeter for determining if a mobile resource enters the area defined by the perimeter, exits such an area or is located inside or outside this area. This could be achieved, for example through the use of the GPS sensors, which are usually available on modern smartphones.

4   Engaging the Crowd

Contrary to traditional sensing systems that are designed specifically to monitor and collect data from fixed positions in their immediate environment and whose behavior can thus be engineered, in mobile crowdsensing systems each sensing point is controlled by a person that needs to give his/her consent in order for its device to participate in the system. This consequently introduces a high degree of unpredictability and unreliability to the system and thus raises demands for incentive and reputation mechanisms to engage the device owners by taking into account their individual preferences and behavior.

4.1   Incentives Framework

To support the envisioned incentive models, apart from the experimenters and the regular users, new types of users were introduced:

- *Sponsors* can be individuals, companies or institutions following experiments and backing up the ones they find interesting.
- *Charities* are organizations or causes that the crowd can support through the allocation of their incentives.

A list of functionalities for the incentives and reputation framework has been specified to support the model as the most applicable to the architecture. Indicatively, the functionalities of this framework need to enable:

- Sponsors to back a specific research, specifying the amount of their contribution, which is transferred to the architecture and allocated to the research.
- Triggering of payments either periodically or when the research is completed, notifying users of the credits they have accumulated during their participation period.
- The user to have up-to-date information regarding their contribution(s), and the overall accumulated information about the credits for each research participation.

4.2   Reputation Mechanisms

Another mechanism to motivate the crowd to participate in the research process is a reputation scoring scheme that provide users with information and statistics about the whole architecture as well as their part in it. The reputation mechanisms monitor user activities and calculate the their rating in a semi-automatic way.

The core module of the reputation mechanisms is a set of ranking functions that calculate the rank and the reputation of the experimenters, the experiments, the users, the devices and the platform itself. There are different types of ranking functions that adjust the rates with negative/positive contributions, i.e. with a five star rating scale or with a flag functionality that characterizes a user or an experiment (e.g. blocking a user). These functions run automatically in the back-end of the architecture and calculate the rates taking into account some statistics about the usage of the architecture, on one hand, and the rate that the users give through the mobile crowdsourcing tool and the experimenters through the website portal, on the other.

The ranking functions that calculate automatically the rates for experimenters take into account the statistics of each experimenter (e.g. the number of his completed experiments, if he provides reports with results from experiments, etc.), the evaluation of users for the experimenter's experiments and the rank of his ideas for proposed experiments. The rate of users that participate in experiments via the mobile crowdsourcing tool is calculated by functions based on the participation of each user (e.g. since when he is using the architecture, his response rate to experiments, the resources he provides for experiments, etc.) and the rate of their devices.

5   Architecture Scalability

In order to evaluate the performance, an extensive analysis of different non-functional properties of the architecture has been conducted (scalability, reliability and availability). This section describes an extensive scalability study, where we have identified three scenarios with different demands in terms of network bandwidth and analyzed the overall performance of the system. Figure 2 depicts the architecture's network architecture with all of its components (application, testbeds and TBaaS server).

We evaluate our architecture on the IoT Lab platform that is running on a server hosted in a Swiss data center providing a bandwidth of 100 Mbps (sym-

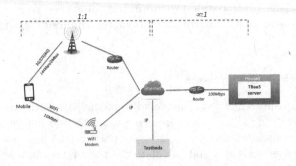

Fig. 2. Architecture's network setting.

metrical, no SLA). For this study three scenarios with different demands in terms of both sensing and sourcing data were identified, as follows:

- *Use Case 1:* High-end scenario: sensing every 10 s and sourcing every minute.
- *Use Case 2:* Average scenario: sensing every 1 min and sourcing every 30 min
- *Use Case 3:* Low-end scenario: sensing every 5 min and sourcing once daily

In regards to bandwidth requirements, use cases one and two can be characterized as high-end and average scenarios since they require constant environmental monitoring as to not disrupt the buildings usage whereas the third use case can be characterized as a low scenario since it doesn't require online responsiveness. For each of the scenarios, the following packages and sizes are considered:

- Sensing package: The sensing package is related to the message sent by the smartphone containing sensing observations (accelerometer, GPS, luminosity, humidity, temperature, etc.) or sensed values coming from a testbed resource (thermometer, humidity sensor, light sensor, noise sensor, etc.). In the mobiles case, we use 5 Kb as packet size having measured on the phone messages averaging 1,15 Kbs of size, whereas for the testbed part, we consider the message size of 1 Kb, having measured an average size of 400 bytes for this type of messages;
- Sourcing package: The sourcing package is related typically to answering a questionnaire. Each question's package size is measured around 20 Kbs and each questionnaire consists of around 5 questions or 100 Kbs in total.

Table 1 shows the required bandwidth for each of the scenarios and the number of connections with the server at 100 % and 50 % of its capacity. For the mobile part, two bandwidths are calculated: the first using 3G/LTE for communication (1 to 1 communication) and the second using Wi-Fi with 20 connections to the same hotspot as a plausible and "safe" number. The 3G and 4G/LTE average bandwidths are around 0,5 Mbps and 2 Mbps to 12 Mbps.

Table 2 shows that for the high-end scenario the server can handle 8M or 4M connections at 100 % or 50 % of its capacity respectively. For the average scenario 48M or 24M connections and finally 240M or 120M connections for the low-end scenario. Considering the average scenario with the server with capacity at 50 %, we can have up to 24M testbed resources connected and communicating their values to the architecture.

Table 1. Mobile side - bandwidth requirements, number of connections with server,

Mobile side	BW (3G/LTE Kbps)	BW (Wi-Fi)	Capacity 100 %	Capacity 50 %
High-end scenario	0,271	5,417	369.231	184.615
Average scenario	0,017	0,347	5.760.000	2.880.000
Low-end scenario	0,002	0,045	44.883.117	22.441.558

Table 2. Testbed side - bandwidth requirements, number of connections with server

Testbed side	Bandwidth (Kbps)	Capacity at 100 %	Capacity at 50 %
High-end scenario	0,013	8.000.000	4.000.000
Average scenario	0,002	48.000.000	24.000.000
Low-end scenario	0,0004	240.000.000	120.000.000

6   Practical Applications

Testbed as a Service offers the capability to conduct researches involving both crowd and IoT interactions. The following three scenarios showcase the capabilities and range of scenarios feasible by the architecture.

Light Control Scenario for a Building Management System [6]. The end goals are to monitor the energy consumption, to automate the lighting and to save energy. It uses static and crowd lent IoT devices together with surveys as a way to learn the crowd's opinion. The first step is to monitor the energy consumption. Then a group of crowd users based on their geolocation is created and a message is sent, informing them about the experiment and their role in it. The research requires passive light measurements from their sensors as well as opportunistic ones for their location within the building. These values determine whether or not the lights will be turned on. Questionnaires forwarded via the architecture determine the user's satisfaction and the need to readjust the parameters of the experiment.

Environmental Monitoring Scenario [18]. It's a cross disciplinary scenario which involves monitoring indoor and outdoor environmental data and correlating them with the crowd happiness. It uses both the crowd's opinion as well as IoT resources. Interaction with the crowd is realized through surveys. A collection of the crowd geolocation data at the time of posting the survey is necessary, in order to relate the survey responses with the geolocalized environmental data obtained from testbeds like ekoNET [18]. ekoNET is a network of mobile IoT sensor devices capable of monitoring temperature, humidity, pressure and air quality. These sensing data are also tied with a location measurement. Over a period of time some correlation between the gathered sensor data and the crowd's opinion may appear. To keep user participation high, proper incentivization is essential.

Virtual and Modelled Resources Scenario. The objective as with first research is to run an energy efficiency scenario. The traditional way of doing so is by deploying static IoT devices tasked with measuring the luminance level and based upon their readings actuating the lights. With the help of virtual resources these measurement points can be augmented with virtual sensors. To do so the values produced by the sensors in the outer boundary of the building are used to create a dataset of external light data along with timestamps in order to identify patterns of external light coming into the rooms. In this fashion the number of physical resources is decreased. To run this scenario prior knowledge of the position of the static sensors in the building is required.

7   Conclusions

In this paper we presented a service based architecture for IoT testbeds which exposes the operations of a testbed as services, enables diverse facilities to be federated in a scalable and standardized way and enables the seamless integration of crowdsourced resources. The architecture enables an experimenter to access a diverse set of resources and orchestrate experiments via a common interface. Moving forward we plan to increase the control and capabilities of the experimenters and integrate additional devices and functionalities to the experiment composition module and present real world applications of the platform.

Acknowledgments. This work was supported by the EU/FIRE IoT Lab project-STREP ICT-610477.

A   Appendix

In the following Listing 1.1 it is shown an Experiment Description XML example. In this example, a reading is requested between two specified date-times to be taken every 1 min from a resource. The experiment also defines that if the average value of one of these resources is less than 1 the light control defined must be actuated. All measurements recorded through experiments are stored in the MongoDB measurements database. This means that experiments can also be conducted without even defining conditions and actions, if what is needed is only data from specific sensors to be taken.

Listing 1.1. Experiment Description XML

```xml
<?xml version='1.0' encoding='utf-8'?>
<experiment>
  <measurements><ip>129.194.70.52</ip><port>9000</port></measurements>
  <identifier>IemNuXCQTGasLMo5mMjkqxPYKewJYhkh</identifier>
  <reading>
    <frequency unit='minutes'>1</frequency>
    <start>2015-06-19 14:54</start>
    <end>2015-06-19 14:54</end>
    <resources>
      <id component=urn:publicid:IDN+iotlab:mitestbed:mitestbed+node+
      node7.mitestbed' resource_id='undefined' port='61616'
      ip='2001:620:607:5800:0:0:0:1c' protocol='coap'
      type='sensor' path='/co2' unit='ppm'></id>
    </resources>
  </reading>
  <action>
    <conditions>
      <average logic='and'>
        <id component='urn:publicid:IDN+iotlab:unigetestbed:unigetestbed+node+
        C3S7A1-LightLevel' resource_id='undefined' port='8111' ip='129.194.70.52'
        protocol='http' type='sensor' path='/lum' unit='lx'></id>
        <threshold type='less' value='100'></threshold>
```

```
      </average>
    </conditions>
    <outcome logic='and'>
      <id component='urn:publicid:IDN+iotlab:ctitestbed:ctitestbed+node+
      node_light_control' port='568' unit='none' resource_id='undefined'
      ip='2001:620:607:5f00::15' protocol='coap' type='actuator' path='PUT-dev0-1'></id>
    </outcome>
  </action>
</experiment>
```

References

1. APISENSE - Crowd-sensing made easy! www.apisense.com/. Accessed April 2016
2. Fed4FIRE project. http://www.fed4fire.eu/. Accessed April 2016
3. Funf - Open sensing framework. http://funf.org/. Accessed April 2016
4. Rspec, fed4fire project. http://fed4fire-testbeds.ilabt.iminds.be/asciidoc/rspec. html. Accessed April 2016
5. Aanensen, D.M., Huntley, D.M., Feil, E.J., al Own, F., Spratt, B.G.: EpiCollect: linking smartphones to web applications for epidemiology, ecology and community data collection. PLoS ONE 4(9), e6968 (2009)
6. Angelopoulos, C., Evangelatos, O., Nikoletseas, S., Raptis, T., Rolim, J., Veroutis, K.: A user-enabled testbed architecture with mobile crowdsensing support for smart, green buildings. In: 2015 IEEE International Conference on Communications (ICC), pp. 573–578, June 2015
7. Angelopoulos, C.M., Nikoletseas, S., Raptis, T.P., Rolim, J.: Design and evaluation of characteristic incentive mechanisms in mobile crowdsensing systems. Simul. Model. Pract. Theor. 55, 95–106 (2015)
8. Arduino: Arduino motes. https://www.arduino.cc/. Accessed April 2016
9. Aug, J., Parmentelat, T., Turro, N., Avakian, S., Baron, L., Larabi, M.A., Rahman, M.Y., Friedman, T., Fdida, S.: Tools to foster a global federation of testbeds. Comput. Netw. 63, 205–220 (2014). Special issue on Future Internet Testbeds
10. Berman, M., Chase, J.S., Landweber, L., Nakao, A., Ott, M., Raychaudhuri, D., Ricci, R., Seskar, I.: GENI: a federated testbed for innovative network experiments. Comput. Netw. 61, 5–23 (2014). Special issue on Future Internet Testbeds Part I
11. Crossbow, T.: Telosb. www.willow.co.uk/TelosB_Datasheet.pdf. Accessed April 2016
12. Farina, F., Szegedi, P., Sobieski, J.: GEANT world testbed facility: federated and distributed testbeds as a service facility of GEANT. In: 2014 26th International Teletraffic Congress (ITC), pp. 1–6, September 2014
13. Fdida, S., Friedman, T., Parmentelat, T.: OneLab: an open federated facility for experimentally driven future internet research. In: Tronco, T. (ed.) New Network Architectures. SCI, vol. 297, pp. 141–152. Springer, Heidelberg (2010)
14. Gluhak, A., Krco, S., Nati, M., Pfisterer, D., Mitton, N., Razafindralambo, T.: A survey on facilities for experimental internet of things research. IEEE Commun. Mag. 49(11), 58–67 (2011)
15. Horneber, J., Hergenroder, A.: A survey on testbeds and experimentation environments for wireless sensor networks. IEEE Commun. Surv. Tutor. 16(4), 1820–1838 (2014). Fourthquarter
16. Hoseini-Tabatabaei, S.A., Gluhak, A., Tafazolli, R.: A survey on smartphone-based systems for opportunistic user context recognition. ACM Comput. Surv. 45(3), 27:1–27:51 (2013)

17. Nandugudi, A., Maiti, A., Ki, T., Bulut, F., Demirbas, M., Kosar, T., Qiao, C., Ko, S.Y., Challen, G.: PhoneLab: a large programmable smartphone testbed. In: Proceedings of First International Workshop on Sensing and Big Data Mining, SENSEMINE 2013, pp. 4:1–4:6. ACM, New York (2013)
18. Pokric, B., Krco, S., Drajic, D., Pokric, M., Jokic, I., Stojanovic, M.: ekoNET - environmental monitoring using low-cost sensors for detecting gases, particulate matter, and meteorological parameters. In: 2014 Eighth International Conference on Innovative Mobile and Internet Services in Ubiquitous Computing (IMIS) (2014)
19. Zolertia: Zolertia motes. http://zolertia.io/. Accessed April 2016

RPL Border Router Redundancy
in the Internet of Things

Quang-Duy Nguyen[1], Julien Montavont[1(✉)],
Nicolas Montavont[2], and Thomas Noël[1]

[1] ICube laboratory (CNRS), University of Strasbourg, Strasbourg, France
{qdnguyen,montavont,noel}@unistra.fr
[2] Institut Mines-Telecom/Telecom Bretagne, Rennes, France
nicolas.montavont@telecom-bretagne.eu

Abstract. The Internet of Things (IoT) refers to a broad variety of objects with communication capabilities that are integrated into Internet. The interconnection between those objects and the Internet is enabled thanks to border routers. In this article, we investigate the aftermath of the failure of border routers on ongoing communications. Next, we propose to overcome the exposed problems by providing objects with multiple border routers. The corresponding subnet is therefore multi-homed, i.e. all objects in this subnet are reachable via multiple paths, one per active border router. Whenever a border router fails, we dynamically re-route traffic to an active border router. Such flows redirection remains transparent to remote peers. Our solution, referred to as Syn-RPL, is based on the well-known IPv6 Routing Protocol for Low-Power and Lossy Networks (RPL). Syn-RPL is evaluated through experimentations on a real testbed.

Keywords: Internet of Things · 6LoWPAN · RPL · Multihoming · Failover

1 Introduction

In the recent years, the rapid development of low-power wireless technologies together with the miniaturization of electronic components gave birth to what we commonly call the Internet of Things (IoT). The IoT refers to a set of physical objects (ranging from sensors to common household electrical goods) with communication capabilities that are able to collect, exchange and receive information throughout the Internet. The IoT enables a large variety of new applications, ranging from scientific observations [16] to personal home automation [14].

In the IoT, objects in a given neighborhood use their wireless communication capabilities to form a multihop wireless network known as Low-power and Lossy wireless Network (LLN). Such networks are characterized by a variety of lossy links (low speed, low energy consumption and unstable connectivity) and constrained devices (limited computational power, memory and energy). Interconnecting LLNs with the Internet is made possible by the IPv6 over Low power

© Springer International Publishing Switzerland 2016
N. Mitton et al. (Eds.): ADHOC-NOW 2016, LNCS 9724, pp. 202–214, 2016.
DOI: 10.1007/978-3-319-40509-4_14

Wireless Personal Area Network (6LoWPAN) IETF standard [10]. 6LoWPAN introduces IPv6 header compression and provides a fragmentation and reassembly adaptation layer below IP, enabling the transport of IPv6 packets over LLNs. IPv6 packets originated from or destined to a LLN are processed by the 6LoW-PAN Border Router (BR) [17]. This entity is located at the junction between the LLN and the IPv6 Internet and is responsible to compress/decompress or fragment/defragment IPv6 packets regarding the 6LoWPAN standard before forwarding them towards the destination. Inside a LLN, packets are routed with the IPv6 Routing Protocol for Low-Power and Lossy Networks (RPL) [15]. RPL builds a Destination Oriented Directed Acyclic Graph (DODAG) rooted at the BR. As a result, all the traffic between the LLN and the IPv6 Internet goes through the BR.

Similarly to wired IPv6 networks with access routers, the IPv6 connectivity of each smart object is directly dependent of the BR status. Whenever the BR becomes unreachable (as a result of system failure, congestion due to funneling effect [6], lack of connectivity due to power outage on neighbor nodes, etc.) the whole LLN is disconnected from the Internet, terminating all ongoing communications with no possibilities to start new ones. In this article, we address such issue by providing LLNs with multiple BRs. In addition to increasing the overall network bandwidth and coverage (the overall throughput increases linearly with the number of egress points), the cooperation of multiple BRs enables a failover mechanism to prevent network disconnection. Our proposal, referred to as Syn-RPL, extends RPL and introduces a virtual BR that federates each graph rooted at a single BR into a unique DODAG. In addition, Syn-RPL only extends the BR and does not require additional software on leaf or intermediate nodes located inside the LLN. Syn-RPL was implemented in Contiki OS and evaluated throughout an extensive experimentation campaign. The obtained results show that Syn-RPL allows smart objects to remain connected to the Internet even after the failure of BR. We show that the traffic redirection from one BR to another is almost transparent for the remote hosts.

The rest of the paper is organized as follows. Next, we present the motivations and advantages of a LLN served by multiple BRs. Section 3 presents solutions currently available in the literature that consider multiple BRs. Then, our proposal referred to as Syn-RPL is introduced in Sect. 4, followed by an overview of the experimentation campaign and performance analysis. Finally, conclusions and future work are presented in Sect. 6.

2   Problem Statement

In constrained environments, routing is usually provided with the IPv6 Routing Protocol for Low-Power and Lossy Networks (RPL) [15]. RPL builds a Destination Oriented Directed Acyclic Graph (DODAG) rooted at the border router (BR) of the network. The DODAG is shaped according to link metric(s) and an objective function which defines how to compute the paths. Each node periodically broadcasts a DODAG Information Object (DIO) message to announce a potential attachment to the DODAG. When a node receives a DIO, it updates a list of potential next hops to the BR, also known as the parent set, and selects a preferred parent from this set based on the objective function and the link metric. A rank is also computed, giving the relative position of node in the DODAG. The preferred parent will serve as the next hop in the default IPv6 route towards the BR. Nodes can also solicit the transmission of DIO by sending DODAG Information Solicitation (DIS) messages. RPL supports point-to-point and point-to-multipoint communications using DODAG Destination Advertisement Object (DAO) messages. After computing its rank, a node can send DAO to its preferred parent in order to advertise a new downward destination. In non-storing mode, where source routing is used, DAO are simply propagated towards the BR. In storing mode, nodes store routing table entries for destinations learned from DAO. DAO are therefore forwarded upward until reaching a node for which the advertised destination is already known.

Due to the broadcasting nature of the wireless communications and because the node density can be important, there is generally a multitude of paths towards the BR. RPL can take advantage of such a situation by using alternatively or simultaneously multiple paths [3]. Whenever a node fails, RPL could be able to detect this failure using unreachability detection mechanisms [5] and compute alternative routes that bypass this node. However, there is currently no solutions to recover from a situation where the BR itself becomes unreachable. Such a situation could be the result of a system failure of the BR itself, a serious congestion occurring at the BR due to funneling effect [6] (all upward traffic and point-to-point traffic in non-storing mode are routed towards the BR) or a lack of connectivity (all BR's neighbors experience a power outage). For readability reasons, we will refer to one of these causes by the terms *BR failure* in the rest of the article. In RPL, and more generally in the IoT, the BR represents a single point of failure for the LLN located behind. When the BR fails, all nodes in the LLN are affected as all ongoing communications with remote pairs are instantly broken and no new communication could be initiated. A failure of the BR also results in breaking local communications, especially in non-storing mode of RPL. One solution to resolve those problems is to provide a LLN with multiple BRs. Deploying multiple BRs not only allow to have alternative paths to the Internet when one of the BR fails, but it also provides load sharing through multiple egress interfaces towards the Internet. However, how RPL can be extended to efficiently support multiple BRs remains an open problem. In the next section, we investigate the solutions currently available in the literature before introducing our own contribution called Syn-RPL.

3   Related Work

In IP networks, routers are in charge of forwarding data packets along networks towards their final destinations. Furthermore, access routers enable the interconnection of local networks to the Internet. For this reason, the failure of an access router leads to the disconnection of the hosts located behind this router. In static networks, one of the standard solutions for solving this problem is the Virtual Router Redundancy Protocol (VRRP) [9]. VRRP allows the deployment of one master router and several backup routers. Whenever the master router fails, one of the backup routers dynamically takes in charge the forwarding responsibility. However, VRRP makes a heavy use of multicast communications which are not desirable in LLNs due to energy conservation.

Another common solution is to use multiple routers simultaneously [13]. The corresponding subnet is said to be multihomed, i.e. all hosts in this subnet are reachable via multiple paths, one per active router. Such a situation allows redundancy (if a router fails a host can use one of the other active routers) and load sharing (traffic can be distributed among all active routers). Again, such solutions are not adapted to the characteristics of LLNs. The subnet concept does not apply in LLNs that are composed of a large number of overlapping radio ranges, forming a complex Non-Broadcast Multiple Access (NBMA).

In LLN, there are many proposals that take advantage from multiple BRs (also referred to as sinks in the literature). Increasing the number of sinks allows increasing the lifetime of the network together with the reduction of the number of hops towards a sink and load sharing [2,8,11]. However, LLNs with multiple sinks require complex signaling protocols to operate [4]. By contrast, RPL [15] is designed to operate either as a single DODAG with a single root, as multiple uncoordinated DODAGs with independent roots or as a single DODAG with a virtual root that coordinates multiple BRs. However, the coordination between multiple BRs is not yet defined by the IETF. Nevertheless, the authors of [7] study the usage of a RPL virtual root together with multiple BRs. They show that using such an architecture allows reducing the energy consumption (by a factor of 30 %) and reducing packet loss (by a factor of 39 %). However, this solution does not address the failure of one BR and how incoming and outgoing packets can be re-routed to another active BR.

The present article focuses on a failover solution to prevent nodes disconnection whenever a BR fails in the context of LLNs. We will see that our solution also proposes load balancing between the BRs. By contrast to [7], our solution considers multiple BRs that can be connected to the Internet independently via the same or different access networks.

4   Contribution

This section presents our contribution, referred to as Syn-RPL, which consists in extending RPL with multiple BR support. In stock RPL, deploying multiple BRs in the same area will result in multiple DODAGs that are independent from each others. Nodes may attach to each DODAG but switching from one BR to another (in case of a BR failure) will not be seamless as BRs do not necessarily share the same IPv6 prefix (BRs can interconnect the LLN via different access networks). Once a node changes its default BR, it is also required to change its IPv6 address to the one operated by the new BR (in order to avoid ingress filtering). In addition, all incoming packets destined to its previous IPv6 address are still routed to the old BR, resulting in packet loss.

To overcome those problems, Syn-RPL uses the virtual node (VR) introduced by RPL [15]. The VR acts as the unique root of all branches anchored at each BR. In the following, we will call these branches sub-DODAG. All cooperating BRs will therefore construct a single DODAG rooted at the VR instead of creating their own DODAG. Each node will select the best BR using legacy RPL operations, but whatever its choice, each node belongs to the same (and unique) DODAG. As a result, the traffic load is automatically shared as each packet will be forwarded towards the closest (regarding the objective function and metrics used by RPL) BR. Communications between nodes belonging to different sub-DODAGs are considered as off-link traffic and nodes use their default BR to forward such traffic over the Internet. With this set up, a BR failure is managed almost just as a node failure in a more standard RPL DODAG. Orphan nodes simply re-attach to another sub-DODAG and their traffic is redirected via the new selected BR.

Obviously, the cooperating BRs should share some information to build a unique DODAG. Syn-RPL introduces a new entity known as the Anchor Agent (AA). The AA has all parameters required to build the DODAG and sends them to each cooperating BR whenever necessary. To maintain the IPv6 connectivity when one or more BRs become unavailable, the AA also acts as a relay station to forward the traffic destined to or originated from the LLN. By default, all IPv6 prefixes used by the BRs inside the LLN are routed towards the AA. As a result, connectivity is maintained between the AA and each BR through a bi-directional IPv6-in-IPv6 tunnel created during the bootstrap. Upon the failure of a specific BR, the AA will update the endpoint address of the corresponding tunnel with one of the BR still operating this DODAG. The AA is located in a more standard IPv6 link and is therefore not as prone to failure as BR (no energy constraints on neighbors, no contention thanks to wire links, etc.). Nevertheless, the AA can be a single point of failure but we can adapt well-known failover mechanisms (such as VRRP [9]) to allow a backup AA to take over and continue providing service to LLN. This is a part of our future work. All Syn-RPL operations are carried out using two new messages referred to as *register* and *register acknowledgment*. Figure 1 illustrates the Syn-RPL framework.

4.1   Bootstrap Operations

At bootstrap, the AA is pre-configured with all the necessary parameters to shape DODAGs. These parameters include for each DODAG, the RPL Instance ID (a unique identifier), the DODAG ID (the identifier of the DODAG root, i.e. the VR in Syn-RPL), the DODAG Version Number (the current version of the DODAG), the objective function (i.e. how RPL nodes select and optimize routes within a RPL Instance), the routing metric, the lifetime (the maximal duration during which a BR is considered as reachable), a list of BR identifiers (e.g. the EUI-64 of each BR that belongs to this DODAG), one or more IPv6 prefix(es) to delegate and a shared secret (to authenticate an authorized BR to be part of this DODAG). Note that the (RPL Instance ID, DODAG ID, DODAG Version Number) tuple uniquely identifies a DODAG version [15]. This is why all

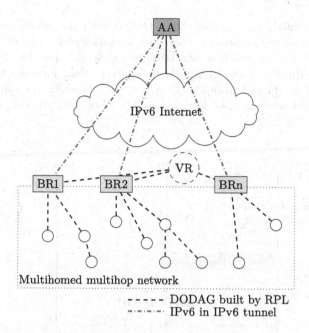

Fig. 1. Framework of Syn-RPL (Color Figure online)

cooperating BRs must share those parameters. An AA that manages multiple LLNs should be pre-configured with several 9-tuples, each corresponding to a specific LLN. The AA also maintains a registration cache in which it records, for each managed LLN, the BRs currently operating the network.

On the BR side, each BR is pre-configured with its identifier (e.g. the EUI-64), the IPv6 address of the AA and the shared secret corresponding to the DODAG the BR belongs to. At bootstrap, a BR sends a *register* message to the AA in order to retrieve the necessary information to start building the DODAG. Register message includes the identifier of the BR, the current IPv6 address of the BR and the shared secret. Upon reception, the AA looks up in its database to retrieve the DODAG information corresponding to this BR. If a corresponding entry is found and the provided shared secret is valid, the AA proceeds to the registration of this BR. First, it sends back a *register acknowledgment* that includes the registration status (successful, rejected). In the case of a successful registration, the register acknowledgment also includes all the parameters to build/expand the DODAG (the RPL Instance ID, the DODAG ID, the DODAG Version Number and the lifetime) together with the IPv6 prefix to use for IPv6 auto-configuration inside the LLN. The AA also adds a new entry in the registration cache containing the identifier and IPv6 address of the BR together with the delegated IPv6 prefix. Each registration cache entry is only valid for a period of time. As a result, BRs periodically send new register messages before the expiration of the lifetime period, otherwise the entry is removed. Finally the BR and the AA set up a bi-directional IPv6-in-IPv6 tunnel. In the same time, the BR starts building/expanding the DODAG with the information provided by the AA. Once the bootstrap is complete, packets sent to the delegated prefix are routed to the AA which forwards them to the BR via the tunnel. When packets reach the BR, they are routed normally to their final destination with RPL. Packets originated from the LLN are also tunneled via the AA before being forwarded to their final destination. The bootstrap phase of Syn-RPL is illustrated on Fig. 2.

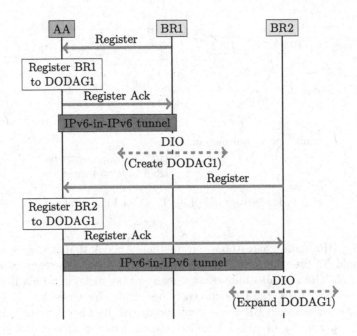

Fig. 2. Syn-RPL operations at bootstrap

4.2   Failover Operations

Once a BR fails, the nodes located in its sub-DODAG should first detect that their default BR is no longer reachable. The nodes directly connected to this BR will likely be the first to detect its failure with the use of external unreachability detection mechanisms [5]. Upon unreachability confirmation, these nodes will send new DIO messages advertising an infinite rank in order to poison routes towards the failed BR. All nodes in the failed sub-DODAG can now accept new DIO from nodes that belongs to another sub-DODAG. Upon reception of such DIO, a node re-attach to the DODAG and sends a DAO to advertise a new node destination information to the BR in charge of the related sub-DODAG. It is important to note that a node will keep its IPv6 address when changing BR, even if its prefix does not match the one used in its new sub-DODAG. The new BR will inform the AA to delegate the corresponding IPv6 prefix, and use the existing tunnel between the AA and the BR to forward this traffic. For this, the BR sends a new register message including the new IPv6 prefix(es) to delegate. Upon reception, the AA will update all tunnel endpoints related to the failed BR to the requesting BR to reflect the new organization of the DODAG. Next, the AA sends back a register acknowledgment to the requesting BR, which in turn updates its own tunnel to the AA. From now on, all traffic destined to or originated from the IPv6 prefix used by the failed BR will be routed to the new BR in charge of this prefix. Figure 3 illustrates Syn-RPL operations when a BR fails. Note that if nodes that were using the same IPv6 prefix attach to different BR, /128 prefixes might be used to add routes for each of these nodes.

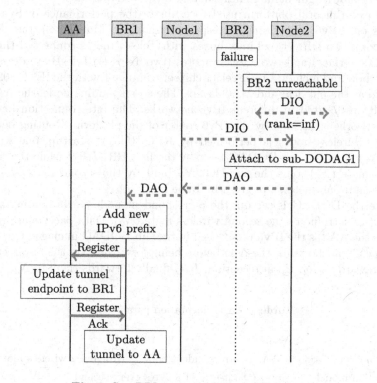

Fig. 3. Syn-RPL operations upon BR failure

5   Experimentation Campaign and Results

5.1   Implementation and Platform Specifications

The Syn-RPL framework involves an Anchor Agent (AA) and 6LoWPAN Border Routers (BR) that interconnect Low-power and Lossy wireless Networks (LLN) with the Internet. On the BR side, we implemented Syn-RPL in Contiki (version 3.x). Contiki is an open source operating system designed for embedded systems and wireless sensor networks. This operating system includes an IP network stack in addition to standards dedicated to LLN such as 6LowPAN [10] and RPL [15]. Moreover, a communication serial interface between a Linux box and a Contiki device can be established by using the extra tools provided by Contiki such as tunslip6. For this reason, we can turn a Linux box with a Contiki device into a BR that interconnects a LLN with legacy IPv6 networks. On the AA side, we implemented Syn-RPL for the Linux operating system as a userland application. The application opens a UDP socket and waits for messages from BR. The input parameters (IPv6 prefixes, shared secret, etc.) are retrieved from an XML file. The messages exchanged between the BR and the AA (register and register acknowledgment) are encapsulated in UDP datagrams.

The experimental platform used to evaluate the performance of Syn-RPL includes one IPv6 router, one corresponding node (a Linux box), one AA (a Linux box), two BRs (two Linux boxes with TelosB motes connected through the USB interface) and two wireless motes (two TelosB). TelosB are developed by Crossbow and include a transceiver chipset compliant with the IEEE 802.15.4 standard at the physical and MAC layers. The corresponding node, the AA and each BR are located in different IPv6 networks. The interconnection between those networks is enabled by the IPv6 router of the platform. Routing between BRs and wireless nodes is performed by RPL [15]. At startup, one wireless mote (N_1) joins the sub-DODAG built by the first BR (BR_1) while the second wireless mote (N_2) joins the sub-DODAG built by the second BR (BR_2). All experimental parameters are given in Table 1.

Once the DODAG is set up, the correspondent node starts sending a constant bit-rate traffic to the second wireless mote (N_2). Data packets are routed towards the AA (as the IPv6 prefix used in the sub-DODAG managed by BR_2 is topologically anchored at the AA) before being forwarded via the bi-directional tunnel towards BR_2. Upon reception, BR_2 finally forwards them to N_2.

Table 1. Experimentation parameters

Parameters	Values
Platform organization	1 AA, 1 correspondent node, 2 BRs and 2 wireless motes
Application model	Constant bit-rate of 8 bytes every second
Syn-RPL	Lifetime 10 s
RPL	DIO sending rate fixed by [12] Objective function zero and MinHop Storing mode
Phy and Mac	802.15.4@2.4 GHz at -25 dBm, contention-based

5.2   Experimentation Results

The results presented in this section are an average of the overall data collected over 10 experiment trials. We also calculated the 95 % confidence interval for each values to measure the reliability of our measurements. For readability reasons, we do not show the obtained confidence intervals since they were very small.

First, we evaluated the duration of the bootstrapping phase as presented in Fig. 4. In stock RPL, a BR starts sending its first DIO 5.5 s after its startup. When N_1 receives this DIO, it can attach to the DODAG and sends back a DAO to BR_1, which in turn updates its routing table accordingly. Then, the first incoming data packet is forwarded to N_1 at $t = 9.16$ s after the startup of BR_1. With Syn-RPL, a BR should first retrieve the DODAG parameters from the AA by the mean of register and register acknowledgment messages. As we can see on Fig. 4, BR_1 sends the register message right after its startup (at $t = 0.68$ s) and receives the register acknowledgment only 740 ms later. Then, the first incoming data packet is forwarded to N_1 at $t = 10.65$ s after the start up of BR_1. As a result, Syn-RPL only adds 1.5 s on average as extra bootstrap delay compared to standard RPL. This extra delay is mainly due to the exchange of register and register acknowledgment messages. In our testbed, the RTT between the BRs and the AA is lower than 10 ms which explains the short registration delay. It is obvious that a larger delay to reach the AA will increase the duration of the Syn-RPL bootstrap phase. However, this delay remains low (within a few seconds) and only occurs at bootstrap when an extra delay does not usually affect IoT applications.

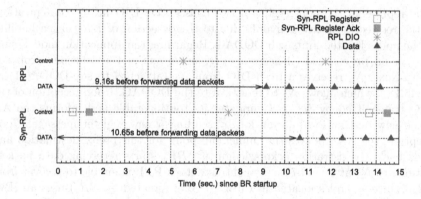

Fig. 4. Bootstrap delay (Color Figure online)

Next, we evaluated the delay needed to redirect flows from a failed BR to an active one. Obtained results are shown on Fig. 5. Each dot represents the transmission or reception of a packet at the time indicated on the Y-axis. At $t = 10$ s we shutdown BR_2 in order to emulate its failure. As we can see, the flow redirection from BR_2 to BR_1 takes approximately 36.7 s on average. During this period of time, 24 data packets are lost. At approximately $t = 35$ s N_2 detects that BR_2 is unreachable and starts sending DIO with an infinite rank in order to poison routes towards BR_2. For implementation ease, we chose to use a fixed timeout of 25 s upon reception of DIO to detect that a BR is unreachable. This

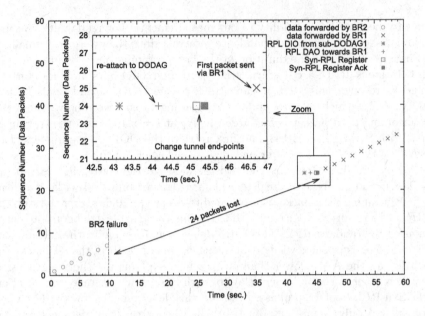

Fig. 5. Flow redirection upon failure of BR (Color Figure online)

value represents the non-receipt of two consecutive DIO in our configuration. So the redirection delay is mainly due to the detection of BR_2 unreachability by the nodes located in its sub-DODAG. Regarding the application model, such delay could be reduced (if necessary) by using one of the solutions proposed in [5]. Next, N_2 receives a fresh DIO from N_1 (attached to sub-DODAG1) at $t = 43.18$ s, which allows N_2 to re-attach to the DODAG. Upon reception of the DAO transmitted by N_2, BR_1 updates its routing table and informs the AA of the new destination information by sending a new register message. Upon reception, the AA updates the tunnel end-point for this prefix/destination and sends back a registration acknowledgment. BR_1 starts receiving data packets destined to N_2 at $t = 46.74$ s. By contrast, stock RPL is unable to recover from BR_2 failure as nodes located in sub-DODAG2 (created by BR_2) uses an IPv6 prefix different from the one used in sub-DODAG1. Although a node can re-attach to the DODAG once it has detected the failure of its BR, it still needs to change its IPv6 address, which will break all ongoing communications. Remote hosts should be informed of such change in order to re-start their communication towards such nodes. As we have shown, Syn-RPL enables transparent flow redirection from or towards remote hosts in the case of a BR failure.

6   Conclusions and Future Work

The interconnection between a network composed of objects (referred to as Low-power and Lossy Network - LLN) with the Internet is usually enabled by Border Routers (BR). On their egress interface, BRs act as legacy IP access routers.

On their ingress interface, they support the communication stack designed for LLN. As a result, they act as gateways between the IP world and the LLN world. However, BRs introduce single point of failure for the IoT. Whenever the BR becomes unreachable (as a result of system failure, network congestion, lack of connectivity, etc.) all objects located in the corresponding LLN become disconnected from the Internet, breaking all ongoing communications. In this article, we presented Syn-RPL, an extension to RPL [15] that provides a LLN with multiple BRs. Syn-RPL allows transparent load sharing (each mote will automatically attach to a specific sub-graph anchored to a specific BR regarding the objective function and metrics used by RPL) and failover upon BR unreachability confirmation. The failover is provided by an Anchor Agent (AA) which is able to redirect IPv6 prefixes towards different BRs according to their availability. Experimentation results showed that Syn-RPL adds a short extra delay at startup (approximately 1.5 s) while it enables fast flow redirection upon BR failure (the disconnection approximately lasts for 37 s). The delay introduced by flow redirection is composed of the delay required to detect the failure of the BR plus the delay required to update the AA in order to update tunnel endpoints. Since it is the BR failure detection that represents most of the delay, we could use more reactive trigger if the application requires it [5]. As a result, Syn-RPL allows fast recovery from a BR failure without involving any actions from correspondent nodes.

Encouraged by the results presented here, we plan to further analyze Syn-RPL via large-scale experiments including several practical scenarios. For this, we will use the FIT IoT-Lab experimental platform [1] which will allow us to scale the number of nodes up to a thousand of motes. With this amount of nodes, we will also be able to further study the load balancing property of Syn-RPL. We also plan to study AA redundancy, which is a less sensible router compared to RPL BR. AAs are located in a more standard IPv6 link, and we will develop methods to provide failover mechanisms using backup AAs. Finally, we plan to extend Syn-RPL to support mobile LLNs.

References

1. Future Internet (FIT) - Internet of Things testbed. http://www.iot-lab.info
2. Bogdanov, A., Maneva, E., Riesenfel, S.: Power-aware base station positioning for sensor networks. In: Proceedings of the Annual Joint Conference of the IEEE Computer and Communications Societies (INFOCOM), March 2004
3. Pavkovic, B., Teholeyre, F., Duda, A.: Multipath Opportunistic RPL Routing over IEEE 802.15.4. In: Proceedings of ACM International Conference on Modeling, Analysis and Simulation of Woreless and Mobile Systems (MSWIM), July 2011
4. Buratti, C., Conti, A., Dardari, D., Verdone, R.: An overview on wireless sensor networks technology and evolution. Sensors 9(9), 6869–6896 (2009)
5. Cobârzan, C., Montavont, J., Noel, T.: Integrating mobility in RPL. In: Abdelzaher, T., Pereira, N., Tovar, E. (eds.) EWSN 2015. LNCS, vol. 8965, pp. 135–150. Springer, Heidelberg (2015)

6. Wan, C.-Y., Eisenman, S.B., Campbell, A.T., Crowcroft, J.: Siphon: overload traffic management using multi-radio virtual sinks in sensor networks. In: Proceedings of the 3rd ACM International Conference on Embedded Networked Sensor Systems (SenSys 2005), November 2005

7. Carels, D., Derdaele, N., Poorter, E.D., Vandenberghe, W., Moerman, I., Demeester, P.: Support of multiple sinks via a virtual root for the RPL routing protocol. EURASIP J. Wirel. Commun. Networking **2014**(1), 1–23 (2014)

8. Oyman, E.I., Ersoy, C.: Multiple sink network design problem in large scale wireless sensor networks. In: Proceedings of the IEEE International Conference on Communications (ICC), June 2004

9. Nadas, S., (ed.): Virtual Router Redundancy Protocol (VRRP) Version 3 for IPv4 and IPv6, IETF Request for Comments (RFC) 5798

10. Hui, J., Thubert, P.: Compression Format for IPv6 Datagrams over IEEE 802.15.4-based Networks, IETF Request for Comments (RFC) 6282, September 2011

11. Li, J., Ji, S., Jin, H., Ren, Q.: Routing in multi-sink sensor networks based on gravitational field. In: Proceedings of International Conference on Embedded Software and System (ICESS), July 2008

12. Levis, P., Clausen, T., Hui, J., Gnawali, O., Ko, J.: The Trickle Algorithm, IETF Request for Comments (RFC) 6206, March 2011

13. Kuntz, R., Montavont, J., Noel, T.: Multihoming in IPv6 mobile networks: progress, challenge and solutions. IEEE Commun. Magazine **51**(1), 128–135 (2013)

14. Hussain, S., et al.: Applications of wireless sensor networkds and RFID in a smart home environment. In: Proceedings of the 7th Annual Conference CNSR, May 2009

15. Winter, T., Thubert, P., Brandt, A., Hui, J., Kelsey, R., Levis, P., Pister, K., Struik, R., Vasseur, J.P., Alexander, R.: IPv6 Routing Protocol for Low-Power and Lossy Networks, IETF Request for Comments (RFC) 6550, March 2012

16. Dyo, V., et al.: Evolution and sustainability of a wildlife monitoring sensor network. In: Proceedings of the ACM Conference on Embedded Networked Sensor Systems (SenSys), November 2010

17. Shelby, Z., Chakrabarti, S., Nordmark, E., Bormann, C.: Neighbor Discovery Optimization for IPv6 over Low-Power Wireless Personal Area Networks (6LoWPANs), IETF Request for Comments (RFC) 6775, November 2012

Linking the Environment, the Battery, and the Application in Energy Harvesting Wireless Sensor Networks

Jad Oueis[✉], Razvan Stanica, and Fabrice Valois

Univ Lyon, INSA Lyon, Inria, CITI, 69621 Villeurbanne, France
{jad.oueis,razvan.stanica,fabrice.valois}@insa-lyon.fr

Abstract. In this paper, we study photovoltaic energy harvesting in wireless sensor networks. We build a harvesting analytical model for a single node, linking three components: the environment, the battery, and the application. Given information on two of the components, limits on the third one can be determined. To test this model, we adopt several use cases with various indoor and outdoor locations, battery types, and application requirements. Results show that, for pre-defined application parameters, we are able to determine the acceptable node duty cycle given a specific battery, and vice versa. Moreover, the suitability of the deployment environment (outdoor, well lighted indoor, poorly lighted indoor) for different application characteristics and battery types is discussed.

1 Introduction

Wireless sensor networks (WSNs) are formed of multiple small-sized, low-cost, low-power embedded devices with sensing, computation and wireless communication capabilities [1]. The nodes self-organizing abilities, and the collaboration among them allows a wide range of applications such as military operations [2], disaster relief [3], and environmental monitoring [4], to name a few. Energy efficiency gained particular attention in WSNs design, since networks are required to run for long durations. However, node lifetime is limited by the finite capacity battery powering it. Regularly charging or replacing depleted batteries in a WSN is a complex and costly procedure, especially in large-scale networks, or in hard-to-reach deployment locations [5]. As a solution to this problem, Energy Harvesting WSNs (EH-WSNs) emerged.

EH-WSN Background. Energy harvesting consists of collecting energy from the surrounding environment and converting it to electrical energy [6]. By implementing energy harvesting, the sensor nodes become self-powered, using the renewable environmental energy as their own power source. Harvestable energy can be provided by several sources in the node ambient environment, hence the numerous energy harvesting techniques such as photovoltaic [7], piezoelectric [8], and wind energy harvesting [9]. The node energy intake depends on the output of

© Springer International Publishing Switzerland 2016
N. Mitton et al. (Eds.): ADHOC-NOW 2016, LNCS 9724, pp. 215–228, 2016.
DOI: 10.1007/978-3-319-40509-4_15

the harvestable power source which may vary as a function of time. Thus, adapting power usage to the harvestable power generation pattern has been largely studied, and many harvesting-aware communication protocols were proposed [5]. In EH-WSN, the main focus shifts from energy conservation schemes to energy management, in order to optimally use harvested energy to enhance network performance [6].

Paper Contribution. In this work, we focus on photovoltaic (PV) energy harvesting. We build an analytical PV energy harvesting model for a single node in a WSN based on three components: the environment, the battery, and the application. By linking energy collection, energy management, energy consumption and the interactions between them, the operational limits of any of the three components can be specified, given information on the two other components. To test our model, we adopt several use cases covering a wide range of environments, batteries and applications. We study two application classes: outage intolerant applications, where once the node suffers from a power outage, it stops its activity permanently; outage tolerant applications, where if a power outage takes place, the node is allowed to cease its activity, recharge the battery, and then return to the network. Using our model, we are able to determine the suitability of a battery, given the node duty cycle, and vice versa. Moreover, the feasibility of an application in a particular environment can be studied. Results show that even a small power outage tolerance leads to significantly higher duty cycles.

This paper is organized as follows. Section 2 gives an overview on the energy harvesting model. We detail the characterization, the modeling and the adopted use cases for each of the three components: environment, battery and application in Sects. 3, 4, and 5, respectively. Results on outage intolerant applications are discussed in Sect. 6, while the outage tolerant applications are studied in Sects. 7 and 8, respectively. Finally, Sect. 9 concludes the paper, with a perspective on future work.

2   General Objective

Three major factors contribute to the node activity in a photovoltaic EH-WSN: the environment, the battery, and the application.

The environment, where the network is deployed, determines light intensity. Each location is distinguished by specific lighting conditions, directly affecting the energy collection process. **The battery** plays the part of an energy storage component and an energy provider. It impacts the node lifetime, and controls to a certain extent the energy management scheme of the node. **The application** sets the WSN objectives, defines requirements, such as the duty cycle, and determines the node energy consumption. Each of these components is characterized by its own set of parameters which have been largely studied in literature [5,10–12]. In this work, we emphasize the existing relation between these components in order to analyze the trade-offs to be considered when deploying

EH-WSNs. By linking energy collection, energy management and energy consumption, that are respectively dictated by the environment, the battery and the application, we elaborate an energy harvesting analytical model for a single node in a WSN. Using our model, the operational limits of a component can be specified given information on the other two.

In the following, we detail each component, and the corresponding use cases adopted in the numerical analysis.

3   The Environment Characterization

3.1   Energy Collection Model

A PV panel absorbs natural or artificial light, and converts it into electrical energy that powers the node. Hence, it is the light intensity that determines the amount of energy that can be harvested by the node.

Without loss of generality, we consider in the rest of this study that time is slotted into timeslots of duration T hours each. We denote by $\mathcal{T}$ the set of all timeslots τ, for the whole observation period. The harvested energy during a timeslot τ, denoted by $E_h(\tau)$, serves as input to our model. $E_h(\tau)$ depends on several factors such as the environment-specific light conditions, and the PV panel characteristics. We denote by S_{PV} the PV panel illuminated area in m^2, and by η_{PV} the PV cell efficiency. Light intensity is modeled by the global horizontal irradiance, $I_{gh}(\tau)$, measured in W/m^2, representing the total amount of shortwave radiation received by the PV panel. The harvested energy by the node during timeslot τ is measured in Joules, and computed as follows:

$$E_h(\tau) = I_{gh}(\tau) \cdot S_{PV} \cdot \eta_{PV} \cdot T \tag{1}$$

3.2   Use Cases

In our analysis, we consider deployment periods of approximately one year, divided in hourly timeslots, i.e. $T = 1$ h. For the PV panel characteristics, we suppose an illuminated area $S_{PV} = 10$ cm^2, having an efficiency $\eta_{PV} = 0.25$ [13].

We consider in our analysis both outdoor and indoor deployment scenarios, covering a wide range of lighting conditions. We determine the harvested energy at each timeslot $E_h(\tau)$ by using real-world datasets providing measurements of the hourly $I_{gh}(\tau)$. For the outdoor scenario, we use measurements collected in Los Angeles during the year of 2014, provided by the U.S. Department of Energy [14]. The average daily irradiation is month-dependent, and varies between 39.04 J/cm^2/day in December, and 109.56 J/cm^2/day in June. For the indoor scenario, we consider datasets provided by Gorlatova et al. [15] presenting measurements in office buildings in New York City. We select two particular indoor locations with significantly different lighting conditions. Location A is poorly lighted, with window shading used at all times, and an average daily irradiation of 1.3 J/cm^2/day. Conversely, location B is very well lighted, it has large windows with unobstructed view, and an average daily irradiation of 63 J/cm^2/day.

4   The Battery Characterization

4.1   Battery Model

In harvesting-based WSNs, nodes are usually equipped with a rechargeable battery capable of storing the harvested energy, as well as powering the node. Rechargeable batteries are mainly characterized by their capacity E_{max}, and their charge/discharge efficiency $\eta_{bat} < 1$ that causes energy loss when the battery is used. It should be noted that battery characteristics may vary under different operating temperatures and battery age [16]. Without loss of generality, we consider a constant discharge efficiency representing the average of the charge losses a battery may suffer from under different conditions.

In our model, we assume the presence of a power manager responsible of delivering the necessary amount of energy to the node. A node is powered, through the power manager, from the harvested energy alone, from the energy stored in the battery, or from both. The power manager has an output regulator, with an output efficiency $\eta_{out} < 1$. This means that, when energy is delivered to the node, a certain amount of this energy is lost. Consequently, if we denote by $E_c(\tau)$ the energy consumption of a node during a timeslot τ (later detailed in Sect. 5), the power manager must deliver $E_c(\tau)/\eta_{out}$ Joules to the node in order to supply its demand [11].

To model the battery level variation as a function of time, we adopt the energy management model presented by Taneja et al. [11]. We extend their model by adding the battery charge/discharge efficiency η_{bat}. We denote by $E_r(\tau)$ the battery residual energy level at the beginning of timeslot τ. At each timeslot, the node power source is determined by the power manager by comparing the amount of harvested energy $E_h(\tau)$, and the energy to be delivered to the node $E_c(\tau)/\eta_{out}$. Two cases unfold, determining the battery behavior at each timeslot:

– $E_h(\tau) \geq E_c(\tau)/\eta_{out}$: in this case, the harvested energy is enough to solely supply the node. The remaining amount of energy, unused by the node, is stored in the battery, causing the battery to charge according to Eq. 2:

$$E_r(\tau + 1) = \min\left(E_r(\tau) + \eta_{bat} \cdot \left(E_h(\tau) - \frac{E_c(\tau)}{\eta_{out}}\right), E_{max}\right) \quad (2)$$

– $E_h(\tau) < E_c(\tau)/\eta_{out}$: in this case, the harvested energy is not enough to solely supply the node. The missing amount of energy must be provided by the battery. Consequently, two possibilities open up:

1. The amount of harvested energy, combined with the amount of the battery residual energy is enough to supply the node energy demand. The node is powered, and the battery discharges according to Eq. 3:

$$E_r(\tau + 1) = E_r(\tau) + \frac{1}{\eta_{bat}} \cdot \left(E_h(\tau) - \frac{E_c(\tau)}{\eta_{out}}\right) , \text{ if } E_h(\tau) + \eta_{bat} \cdot E_r(\tau) \geq \frac{E_c(\tau)}{\eta_{out}} \quad (3)$$

2. The amount of available energy is not enough to power the node. In this case, the node is in power outage situation, and cannot be properly powered. The node must cease its activity, and we refer to it as *non-operational*. Let $\mathcal{T}_{no}$ be the subset of timeslots τ during which the node is non-operational:

$$\tau \in \mathcal{T}_{no} \subseteq \mathcal{T} \iff E_h(\tau) + \eta_{bat} \cdot E_r(\tau) < \frac{E_c(\tau)}{\eta_{out}} \tag{4}$$

If power outage is tolerated, the nodes remain non-operational for a limited duration only, since the battery can still be recharged by the harvested energy. Once the available energy is sufficient again, a non-operational node resumes its activity. We consider that the tolerance for temporarily non-operational nodes depends on the specific application requirements. In this work, we study two classes of applications: *outage intolerant* and *outage tolerant*, further defined in detail in Sect. 5.

4.2 Use Cases

We consider in our analysis a variety of rechargeable batteries, with different characteristics. The most common rechargeable technologies are Nickel Metal Hydride (NiMH) and Lithium Ion (Li-ion). NiMH batteries suffer from low charge/discharge efficiency; however, they have a simple charging method, lowering their cost. Li-based batteries have higher charge/discharge efficiency, but a more complex charging method [5]. The set of batteries we consider in our analysis covers a wide range of characteristics, summarized in Table 1. Regarding the output regulator, we consider an efficiency $\eta_{out} = 0.8$ [11].

Table 1. Comparison of rechargeable batteries characteristics [6,17].

Model	Type	Volume (cm^3)	E_{max} (J)	Capacity (mAh)	η_{bat}	Notation
AA	NiMH	7.7	10800	2500	0.66	AA-Ni
AAA	NiMH	3.8	5625	1250	0.66	AAA-Ni
AA	Li	7.7	9857	740	0.99	AA-Li
Ultrathin_200	Li	2.7	2664	200	0.99	U-Li-200
Ultrathin_100	Li	1.3	1332	100	0.99	U-Li-100
Ultrathin_43	Li	0.6	573	43	0.99	U-Li-43
Ultrathin_10	Li	0.6	133	10	0.99	U-Li-10

5 The Application Characterization

5.1 Energy Consumption Model

The power consumption differs from one application to another, depending on the sensing activity by the sensor, the computation performed by the CPU, and the communication scheme of the radio transceiver. In this study, we consider that the radio transceiver is the main energy consumption source [12], disregarding other consumption sources. However, including CPU or sensing consumption in the model is straightforward. We suppose that during each timeslot τ of duration T, a duty cycle $DC(\tau)$ is forced on the radio transceiver. The latter will be awake, and consuming power, during each timeslot, for a time equal to $DC(\tau) \cdot T$. For the remaining time, the node is asleep, with a negligible power consumption. When awake, the transceiver switches between three states: receive, transmit, and listen. The power consumed in these states is very similar, as proven by experimental studies [12], and transceiver datasheets [18]. We consider that the power consumed by the radio while active is equal to the average of the power consumption of these three states, denoted by P_{avg}. Thus, during a time period T, the energy $E_c(\tau)$ consumed by the transceiver is dictated by the duty cycle $DC(\tau)$, such as:

$$E_c(\tau) = P_{avg} \cdot DC(\tau) \cdot T \tag{5}$$

5.2 Application Classes

The application determines the tolerance for having a temporarily non-operational node. In this study, we consider the following application classes:

Outage Intolerant: For these applications, a node cannot enter the non-operational state. The node is required to sustain its activity without suffering from a power outage, meaning that continuous operation is a requirement. We denote by $\mathcal{T}_{co}$ the set of consecutive timeslots during which the node is operational up until the first power outage takes place. We formally define continuous operation (*c.o.*) as:

$$c.o. \iff E_h(\tau) + \eta_{bat} \cdot E_r(\tau) > \frac{E_c(\tau)}{\eta_{out}}, \quad \forall \tau \in \mathcal{T}_{co} \subseteq \mathcal{T} \tag{6}$$

The node lifetime in this case is equal to the duration during which the node is continuously operational, up until the first power outage takes place. We refer to it as the continuous operation lifetime. With timeslots of duration T, the continuous operation lifetime is equal to $|\mathcal{T}_{co}| \cdot T$.

Outage Tolerant: For these applications, the node can be temporarily non-operational. A node will remain non-operational as long as the amount of harvested energy combined with the battery residual energy are below the node energy requirement. Once the delivered energy is enough, the node resumes its activity. During a non-operational timeslot, the harvested energy is only used to recharge the battery:

$$E_r(\tau + 1) = \min(E_r(\tau) + \eta_{bat} \cdot E_h(\tau), E_{max}) \tag{7}$$

In this work, we consider the duty cycle to be the percentage of time during which a node is active in the timeslots where the node is operational. In non-operational timeslots, where the node is inactive, the duty cycle is null:

$$DC(\tau) = \begin{cases} DC, & \forall \tau \in \mathcal{T} \setminus \mathcal{T}_{no} \\ 0, & \forall \tau \in \mathcal{T}_{no} \end{cases} \tag{8}$$

We limit our study to static duty cycles, such as the duty cycle is set to a fixed value DC in the operational timeslots. Although dynamic duty cycles are out of the scope of this paper, expanding the framework to include adaptive duty cycles in each timeslot is straightforward.

In outage tolerant applications, it is possible that the node frequently switches between the non-operational and the operational state. We define n_c as the consecutive number of timeslots during which the node is non-operational. Each time the node enters the non-operational state, a new instance $n_c(\tau)$ of the counter n_c is created. We denote by $n_{c_{max}}$ the maximum number of consecutive non-operational timeslots:

$$n_{c_{max}} = \max_{\tau \in \mathcal{T}_{no}} n_c(\tau) \tag{9}$$

Finally, we define r_{no} as the percentage of time during which the node is non-operational out of the whole observation period $\mathcal{T}$:

$$r_{no} = 100 \times \frac{|\mathcal{T}_{no}|}{|\mathcal{T}|} \tag{10}$$

5.3 Use Cases

In our analysis, we do not limit the study to specific applications, but we study different application requirements, which can be matched to a target application afterwards. Our use cases are representative of several WSN applications. We study outage intolerant applications, where continuous operation is required. We also consider several cases of outage tolerant applications with different values of $n_{c_{max}}$ and r_{no}. Regarding energy consumption, based on the CC2420 transceiver datasheet [18], we fix the energy consumption during a one hour timeslot to $P_{avg} \cdot T = 180\,\mathrm{J}$.

6 Outage Intolerant Applications

In this section, we consider outage intolerant applications, where continuous operation is required (Eq. 6). Using our model, we are able to determine the continuous operation lifetime, i.e. the time duration before the first power shortage takes place. With timeslots of duration one hour, the continuous operation lifetime is equal to $|\mathcal{T}_{co}|$ hours. For comparison, we use the metric r_{co}, which represents the percentage of the continuous operation out of the whole observation period, such as $r_{co} = 100 \times \frac{|\mathcal{T}_{co}|}{|\mathcal{T}|}$.

6.1 Outdoor

Figure 1 shows a comparison of r_{co} for an outdoor, one year deployment. We compare the results for the deployment starting at different months of the year. The fixed duty cycle values are such that $DC \in \{20\,\%, 33\,\%, 50\,\%\}$. The choice of such high duty cycles is not uncommon in harvesting-based WSNs, where energy constraints are alleviated [7]. Having higher duty cycles is beneficial, since it allows lower communication delays, and higher throughput.

For $DC = 20\,\%$ (Fig. 1a), all the batteries are capable of guaranteeing continuous operation during the whole year, except for the smallest Li batteries. In fact, with the 10 mAh Ultrathin battery, the node is continuously active for 2 h only. For $DC = 33\,\%$ (Fig. 1b), continuous operation is achieved for a whole year for all the Li batteries, except the smallest ones, no matter the deployment month. However, for the NiMH batteries, continuous operation is only achieved for the AA-sized battery having the largest capacity, and only if the deployment starts at January. For $DC = 50\,\%$ (Fig. 1c), none of the batteries are capable of achieving continuous operation for a whole year, no matter the month of deployment. These results suggest that Li-based batteries generally outperform NiMH batteries. This is due to the high charging efficiency of Li batteries, as opposed to the low efficiency of NiMH batteries causing them to charge less and discharge more aggressively. On the other hand, the smallest Li batteries (43 and 10 mAh) are limited by their relatively small capacity. This suggests that much lower duty cycles are recommended when using these batteries.

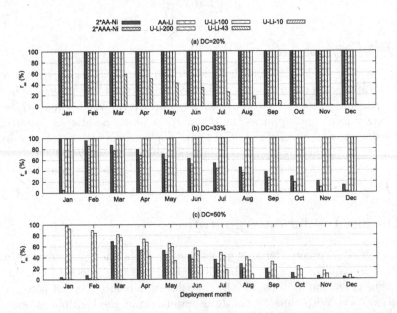

Fig. 1. Percentage of the continuous operation out of the whole observation period - Outdoor. (Color figure online)

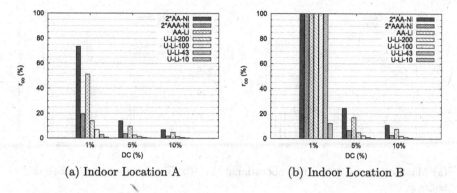

(a) Indoor Location A   (b) Indoor Location B

Fig. 2. Percentage of the continuous operation out of the whole observation period - Indoor.

A pattern concerning the effect of the deployment month on the achieved lifetime emerges. This is due to the significant difference in the amount of harvestable energy from one month to another. It is generally around November and December that the aggressive discharge will take place considering the few recharge opportunities. Thus, it is more likely that a node will enter the non-operational state during these months.

6.2 Indoor

Figure 2 shows a comparison of r_{co} for the indoor locations. Since harvestable energy indoor is far less available than outdoor, smaller duty-cycles are considered: $DC \in \{1\%, 5\%, 10\%\}$. Unlike the outdoor scenario, we do not have the deployment month dependency. At the well lighted location B, Fig. 2b shows that, with $DC = 1\%$, all the batteries, except the 10 mAh Ultrathin, are capable of achieving continuous operation for the whole observation period. This means that any of these batteries can be used with a duty cycle $DC = 1\%$ or lower. Figure 2a shows that, at the poorly lighted location A, the longest time for which the node is capable of sustaining its continuous operation is equal to $r_{co} = 73\%$ of the observation period, for $DC = 1\%$ and two AA-sized NiMH batteries. The results show that duty cycles on the order of $DC = 5\%$ are relatively high for indoor applications, since no battery is capable of guaranteeing continuous operation for longer than $r_{co} = 24\%$ of the observation period. Another observation in Fig. 2a is that the Ultrathin 43 mAh and 10 mAh batteries are only suitable for applications requiring duty cycles even lower than $DC = 1\%$.

7 Non-operational Time in Outage Tolerant Applications

In this section, we consider outage tolerant applications, where continuous operation is not a requirement. The node may become non-operational for a certain

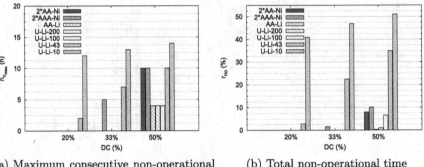

(a) Maximum consecutive non-operational hours

(b) Total non-operational time

Fig. 3. Outage tolerance parameters - Outdoor.

duration before regaining enough energy to resume its activity. Some applications set regulations for the non-operational period, by bounding the outage tolerance parameters $n_{c_{max}}$ (Eq. 9) and/or r_{no} (Eq. 10). Using our model, we compute $n_{c_{max}}$ and r_{no} for a given fixed duty cycle DC.

7.1 Outdoor

Figure 3a and b show values of $n_{c_{max}}$ and r_{no}, respectively, during a one year deployment in the outdoor location, for different battery models, and for different static duty cycle values: $DC \in \{20\%, 33\%, 50\%\}$. Since continuous operation is no longer a target, the month of deployment does not affect the output. Tolerance for non-operational time and the required duty cycle, set by the application, are used to determine the suitable battery. For example, a node powered by a 43 mAh Ultrathin battery, with $DC = 20\%$, is in power outage for two consecutive hours maximum, and remains operational for a total of 97% of the whole deployment period. For applications tolerating these conditions, a node can be equipped with this type of battery. In an outage intolerant application, such batteries were not even capable of surviving one month under the same conditions.

7.2 Indoor

Figure 4 shows values of $n_{c_{max}}$ and r_{no}, for indoor locations A and B, for different batteries, and for different static duty cycle values: $DC \in \{1\%, 5\%, 10\%\}$. In location A, Fig. 4a and b show that even with the relatively low $DC = 1\%$, the tolerance for non-operational time must be high. As the duty cycle increases and the battery capacity decreases, the total non-operational time increases significantly. The values of the total non-operational time reached in location A (Fig. 4b), are highly impractical in real-life deployments. It is useless to have the node non-operational for 80% of the time, for example, as required by most batteries. In location B, we can see in Fig. 4d that there is no power outage for

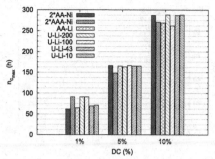
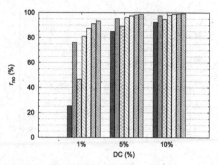

(a) Location A - Maximum consecutive non-operational hours

(b) Location A - Total non-operational time

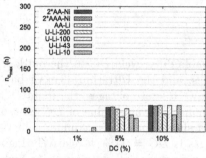

(c) Location B - Maximum consecutive non-operational hours

(d) Location B - Total non-operational time

Fig. 4. Outage tolerance parameters - Indoor locations A (a,b) and B (c,d). (Color figure online)

$DC = 1\%$, except with the Ultrathin 10 mAh battery. This is in accordance with the results shown in Fig. 2b. With the Ultrathin 10 mAh battery and $DC = 1\%$, for an outage intolerant application, the node is active for only 30% of the observation period. However, for outage tolerant applications, results show that the node is active, with $DC = 1\%$, for 96.6% of the time, while staying in the non-operational state for a maximum of 10 consecutive hours. For the higher duty cycles, total non-operational time increases significantly.

8   Duty Cycle Dimensioning in Outage Tolerant Applications

We consider now the complementary problem of the previous section. Given predetermined boundaries on $n_{c_{max}}$ and r_{no}, we compute the maximum achievable DC value.

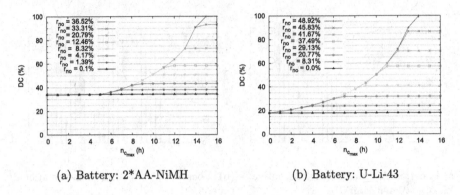

(a) Battery: 2*AA-NiMH   (b) Battery: U-Li-43

Fig. 5. Maximum achievable duty cycle for pre-determined conditions - Outdoor.

8.1   Outdoor

Figure 5a and b show the maximum achievable duty cycle DC in the outdoor location, with the AA-sized NiMH batteries, and the Ultrathin 43 mAh battery, respectively. As anticipated, the more we tolerate outages, the more DC increases. Since r_{no} is bounded, the maximum achievable value of DC is also bounded. With the NiMH batteries, Fig. 5a shows that the maximum achievable duty cycle with no tolerance for outages is $DC = 34\%$, a relatively high value. As the tolerance for non-operational time increases, duty cycles as high as $DC = 93\%$ can be achieved, with the node being non-operational for a maximum of 15 consecutive hours, and for 33% of the observation period. However, conditions are more strict for the Ultrathin 43 mAh battery, as shown in Fig. 5b. The maximum achievable duty cycle with no tolerance for outages is $DC = 18\%$.

8.2   Indoor

Figure 6a and b show the maximum achievable DC in indoor locations, A and B, respectively, with AA-sized NiMH batteries. The difference is clear in terms of the achieved duty cycle between these two locations. In the poorly lighted location A, the duty cycle reaches a maximum of $DC = 1.3\%$ being non-operational for 43% of the time, with 200 maximum consecutive non-operational hours. In the well lighted location B, the achieved duty cycle for the same non-operational conditions is $DC = 5\%$. Higher duty-cycles are achieved at the costly expense of having a large amount of time during which the node is non-operational. Some of the non-operational constraints may be impractical, meaning that applications requiring higher duty cycles as well as short outage time are difficult to accommodate to indoor environments.

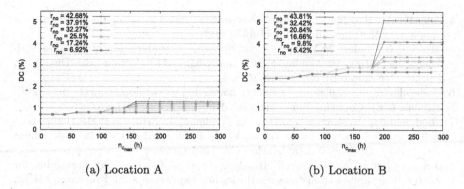

(a) Location A   (b) Location B

Fig. 6. Maximum achievable duty cycle for pre-determined conditions with 2*AA-NiMH batteries - Indoor.

9   Conclusion

In this paper, we present a PV energy harvesting model for a single node in a WSN. We build our model around three major components: the environment, the battery, and the application. Based on our model, given two of the three components, we can characterize the third one. We define several use cases using real-world datasets of light intensity, in both indoor and outdoor scenarios, a wide range of rechargeable batteries, and various application parameters. We show that, given particular application parameters, we are able to determine which batteries are suitable, given the node duty cycle, and vice versa. Furthermore, the feasibility of an application in a specific environment, or for a specific battery, can be studied. Results show that tolerating short power outage periods allows extending the node lifetime, with higher duty cycle values.

For future work, it is of utmost importance to further expand the model from a single wireless node to the whole network. Collaboration between the nodes, and relaying functionalities may influence the node behavior. Moreover, the outage tolerant applications raise new challenges on how to deal with the unexpected "disappearance" of a node when non-operational, followed by its "reappearance" in the network.

References

1. Yick, J., Mukherjee, B., Ghosal, D.: Wireless sensor network survey. Comput. Netw. **52**(12), 2292–2330 (2008)
2. Lee, S.H., Lee, S., Song, H., Lee, H.S.: Wireless sensor network design for tactical military applications: remote large-scale environments. In: Proceedings IEEE Milcom 2009, Boston, MA, USA (2009)
3. Chen, D., Liu, Z., Wang, L., Dou, M., Chen, J., Li, H.: Natural disaster monitoring with wireless sensor networks: a case study of data-intensive applications upon low-cost scalable systems. Mob. Netw. Appl. **18**(5), 651–663 (2013)

4. Oliveira, L., Rodrigues, J.: Wireless sensor networks: a survey on environmental monitoring. J. Commun. **6**(2), 143–151 (2011)
5. Basagni, S., Naderi, M.Y., Petrioli, C., Spenza, D.: Wireless sensor networks with energy harvesting. In: Mobile Ad Hoc Networking: The Cutting Edge Directions, pp. 701–736, 2nd edn. Wiley (2013)
6. Sudevalayam, S., Kulkarni, P.: Energy harvesting sensor nodes: survey and implications. IEEE Commun. Surv. Tutorials **13**(3), 443–461 (2011)
7. Raghunathan, V., Kansal, A., Hsu, J., Friedman, J., Srivastava, M.: Design considerations for solar energy harvesting wireless embedded systems. In: Proceedings ACM/IEEE IPSN 2005, Los Angeles, CA, USA (2005)
8. Vijayaraghavan, K., Rajamani, R.: Active control based energy harvesting for battery-less wireless traffic sensors: theory and experiments. In: Proceedings IEEE ACC 2008, Seattle, WA, USA (2008)
9. Cammarano, A., Spenza, D., Petrioli, C.: Energy-harvesting WSNs for structural health monitoring of underground train tunnels. In: Proceedings IEEE Infocom Student Workshop, Turin, Italy (2013)
10. Gorlatova, M., Wallwater, A., Zussman, G.: Networking low-power energy harvesting devices: measurements and algorithms. IEEE Trans. Mob. Comput. **12**(9), 1853–1865 (2013)
11. Taneja, J., Jeong, J., Culler, D.: Design, Modeling, and capacity planning for microsolar power sensor networks. In: Proceedings of ACM/IEEE IPSN 2008, St Louis, MO, USA (2008)
12. Prayati, A., Antonopoulos, C., Stoyanova, T., Koulamas, C., Papadopoulos, G.: A modeling approach on the telosb wsn platform power consumption. J. Syst. Softw. **83**(8), 1355–1363 (2010)
13. Green, M., Emery, K., Hishikawa, Y., Warta, W., Dunlop, E.: Solar cell efficiency tables (v. 45). Prog. Photovoltaics Res. Appl. **23**(1), 1–9 (2015)
14. Andreas, A., Wilcox, S.: Solar Resource & Meteorological Assessment Project (SOLRMAP). Rotating Shadowband Radiometer (RSR), NREL Report No. DA-5500-56502, Los Angeles, CA, USA (2012)
15. Gorlatova, M., Zapas, M., Xu, E., Bahlke, M., Kymissis, I., Zussman, G.: CRAWDAD Dataset Columbia/Enhants (2011). http://crawdad.org/columbia/enhants/20110407. Accessed Apr 2016
16. Buchmann, I.: Batteries in a Portable World: A Handbook on Rechargeable Batteries for Non-engineers, 2nd edn. Cadex Electronics, Richmond (2001)
17. Ultrathin Rechargeable Lithium Polymer Batteries from Powerstream. http://www.powerstream.com/thin-lithium-ion.htm. Accessed Apr. 2016
18. Texas Instruments, Chipcon CC2420 Datasheet. http://www.ti.com/lit/ds/symlink/cc2420.pdf. Accessed Apr. 2016

Short Paper: Structural Network Properties for Local Planarization of Wireless Sensor Networks

Florentin Neumann$^{(\boxtimes)}$, Daniel Vivas Estevao, Frank Ockenfeld,
Jovan Radak, and Hannes Frey

Institute for Computer Science, University Koblenz-Landau, Koblenz, Germany
{fneumann,dvivas,fockenfeld,radak,frey}@uni-koblenz.de

Abstract. Network graphs satisfying *redundancy* and *coexistence* can be planarized by means of local algorithms without making assumptions on the nodes' communication ranges. We generalize these properties to *k-weak* and *k-strong* coexistence and redundancy. Subsequently, we give empirical evidence that these properties are present in simulated wireless sensor networks as well as in actual testbed deployments, especially if looking at the graph's *c*-hop clustered overlay representations, for $c \leq 2$.

Keywords: WSN · Planarization · Local algorithm · Geometric graph · Redundancy · Coexistence · Log normal shadowing · Testbed measurements

1 Introduction

In *wireless sensor networks* (WSNs) there are various problems, e.g., *guaranteed delivery georouting* [1] and *tracking of mobile objects* [7], that can be solved efficiently by means of local algorithms, provided that the network graph is *planar* (i.e., its straight line drawing in the plane is free of edge intersections). Typically wireless network graphs are not planar. Therefore, local planarization algorithms are required. These algorithms often assume perfect or almost perfect unit transmission ranges of nodes. However, such assumptions lack practical relevance and often reduce mathematical reasoning to problems in circle geometry. The local algorithm LLRAP [5] does not require such assumptions. It correctly outputs a connected, planar subgraph of a given network graph, if this network graph satisfies the structural properties *redundancy* and *coexistence* (see Definitions 1 and 2).

In this preliminary study, we make two novel contributions: Firstly, we generalize the said terms to *k-strong/k-weak redundancy* and *coexistence*. Secondly, we investigate empirically to which extent these properties are present in actual sensor network deployments, as well as in simulated large-scale networks. Moreover, we study if *c*-hop clustered overlay representations ($c \leq 2$) of these graphs are more likely to satisfy these properties. Our results show that even for small constants k and c, respectively, the properties are almost always present.

© Springer International Publishing Switzerland 2016
N. Mitton et al. (Eds.): ADHOC-NOW 2016, LNCS 9724, pp. 229–233, 2016.
DOI: 10.1007/978-3-319-40509-4_16

2   Generalization of Redundancy and Coexistence

Definition 1 (*k*-strong and *k*-weak redundancy property). *A geometric graph* $G = (V, E)$ *satisfies the k-strong [k-weak] redundancy property, if for any pair of intersecting edges* $uv, xy \in E$ *it holds that at least one of the nodes in* $\{u, v, x, y\}$ *is connected in G to both nodes [at least one node] from the intersecting edge by a path of length at most k-hops. The* redundancy property *defined in [5] is equivalent to the 1-strong redundancy property defined here.*

Definition 2 (*k*-strong and *k*-weak coexistence property). *A geometric graph* $G = (V, E)$ *satisfies the k-strong [k-weak] coexistence property, if for any pairwise connected node triple* $u, v, w \in V$ *and all nodes* $x \in V$ *located inside the triangle* $\triangle(uvw)$ *it holds that x is connected in G to all nodes [at least one node] from* $\{u, v, w\}$ *by a path of length at most k-hops. The* coexistence property *defined in [5] is equivalant to the 1-strong coexistence property defined here.*

Any *unit disk graph* (UDG) satisfies 1-strong redundancy [2, Lemma 4.1] and 1-strong coexistence [5, Lemma 1]. Moreover, any *d-quasi unit disk graph* (QUDG) [3], for $d \geq 1/\sqrt{2}$, satisfies 1-weak redundancy [3, Lemma 8.1] and 1-weak coexistence (we omit a formal proof of this last claim due to space restrictions). Also note that for an arbitrary geometric graph G with *graph diameter* δ it trivially holds that G satisfies δ-strong redundancy and δ-strong coexistence.

3   Empirical Analysis

3.1   Simulation Setup and Testbed Data

Simulation. We generated 56 large-scale wireless network graphs as follows. On a 2600×2600 square area we randomly placed nodes using a *spatial homogeneous Poisson process* with rate $\lambda = 2000$. *Path loss* between two nodes at distance d is modeled according to the *Log Normal Shadowing* (LNS) model [6, Eq. (3.69a)], using a *path loss at reference distance* d_0 of 39.15 dB, *path loss exponent* $\gamma = 3$, and *shadowing variance* $\sigma^2 = 5$. Given the path loss, we computed SNR[dB] at the receiver assuming a *transmit power* of 1 dBm and a *noise floor* of -100 dBm. Lastly, we obtained a *probability of bit error* $P_e = 10^{-SNR[dB]/10}$ according to [6, Eq. (5.162)] assuming FSK modulation and maintained the link only if $P_e < 1/2$. On average, these graphs consist of 2018 nodes and have a node degree of 8.

Testbed. The wireless sensor network *CitySee* [4] is deployed in the urban area of Wuxi City, China. The data available to us is a three days measurement in a network composed of 302 wireless sensor nodes (based on TelosB, equipped with CC2420 RF transceivers [4]). During the measurements, ca. every 10 minutes each node gathered its neighborhood tables including *received signal strength indicator* (RSSI) values. The size of a node's neighborhood table was deliberately restricted to at most 10 neighbors (the criteria used for limitation are unknown to us). For obtaining snapshots of this network graph, we split the series into chunks of 10 min intervals and obtained the respective adjacency matrices. In total 360 snapshot graphs were obtained consisting of 297 nodes on average.

Clustering. We employ the following randomized c-hop cluster scheme to obtain the clustered overlay representation $G_c = (V_c, E_c)$ for a given graph $G = (V, E)$. Initially all nodes are *unclustered*. While there is an unclustered node, one such node u is chosen uniformly at random and added to V_c. Node u and its unclustered c-hop neighbors form a cluster and are marked as *clustered*. Once all nodes are clustered, any two nodes $u, v \in V_c$ are connected by an edge in E_c, if there is an edge $e \in E$ which connects cluster members of u and v. We use a randomized technique instead of some more advanced one to avoid a bias of the results.

3.2 Results

Table 1 shows the averaged results as percentages for CitySee and LNS graphs, as well as their 1-hop and 2-hop clustered overlays. The last three columns refer to the *average* and *maximum graph diameter*, and list the *average number of events* (redundancy: two edges intersect; coexistence: node is contained in a triangle).

In the unclustered CitySee graphs, for $k = 3$ all properties but k-strong coexistence are satisfied in more than 99 % of cases; at $k = 4$ all properties reach the 99 %-level. The 100 %-level for the weak and strong properties are reached

Table 1. Avg'ed results of CitySee and LNS graphs including clustered representations

		Prop.	$k=1$	$k=2$	$k=3$	$k=4$	$k=5$	$k=6$	$k=7$	δ_{avg}	δ_{max}	#Events$_{\text{avg}}$
CitySee	uncl.	k-w-red	0.55829	0.96406	0.99920	0.99983	0.99998	1.0	1.0	6.039	10	42310.3
		k-s-red	0.20119	0.86075	0.99750	0.99945	0.99990	0.99999	1.0			
		k-w-coe	0.64319	0.94977	0.99679	0.99931	1.0	1.0	1.0			3249.7
		k-s-coe	0.08831	0.74183	0.97082	0.99794	0.99931	1.0	1.0			
	1-hop	k-w-red	0.94352	0.99863	0.99991	1.0	1.0	1.0	1.0	4.086	8	9878.3
		k-s-red	0.72847	0.99650	0.99893	0.99992	1.0	1.0	1.0			
		k-w-coe	0.71016	0.99417	0.99973	1.0	1.0	1.0	1.0			4443.8
		k-s-coe	0.15149	0.90471	0.99863	0.99974	1.0	1.0	1.0			
	2-hop	k-w-red	0.99459	0.99985	1.0	1.0	1.0			2.983	5	153.7
		k-s-red	0.94208	0.99983	1.0	1.0	1.0					
		k-w-coe	0.80770	0.99991	1.0	1.0	1.0					113.4
		k-s-coe	0.19222	0.97859	1.0	1.0	1.0					
LNS	uncl.	k-w-red	0.99998	1.0	1.0					48.821	55	13406.6
		k-s-red	0.97901	1.0	1.0							
		k-w-coe	1.0	1.0	1.0							3583.9
		k-s-coe	0.91705	1.0	1.0							
	1-hop	k-w-red	1.0	1.0	1.0					24.571	28	112.3
		k-s-red	0.99260	1.0	1.0							
		k-w-coe	1.0	1.0	1.0							37.1
		k-s-coe	0.74459	1.0	1.0							
	2-hop	k-w-red	1.0	1.0	1.0					15.661	19	30.4
		k-s-red	0.99633	1.0	1.0							
		k-w-coe	0.99881	1.0	1.0							14.0
		k-s-coe	0.65195	0.99881	1.0							

at $k = 6$ and $k = 7$, respectively. This is well below the maximum diameter, but matches its average. Similar results can be observed for the 1-hop and 2-hop clustered overlays, although for a smaller k. The 100 %-level is reached for all properties at $k = 5$ and $k = 3$ regarding 1-hop and 2-hop clustered graphs, respectively. The data also reveals a huge improvement between $k = 1$ and $k = 2$: whereas for $k = 1$ some properties are rarely satisfied, at $k = 2$ all properties are satisfied in the vast majority of cases. However, we claim that these results are negatively biased, due to the neighborhood size restriction in the available data.

The analysis of the large-scale LNS graphs shows that all properties are satisfied in 100 % of cases independently of the diameter. In the unclustered graphs, the average diameter is greater than 48, but all properties reach the 100 %-level at $k = 2$. Even for $k = 1$, the strong properties are mainly satisfied. Similar results are obtained for the clustered representations. However, compared with the CitySee data—where for a fixed k the percentages always rise with increasing c-values—a partial reverse of trends regarding clustered representations can be observed. Although this still holds for redundancy, in case of coexistence for $k = 1$ the percentages for 2-hop clustered graphs are smaller compared to the 1-hop and unclustered graphs. One reason for this is that in the clustered LNS graphs the number of events drops vastly and a few deviations have a strong impact.

4   Conclusion

1-strong redundancy and coexistence are sufficient conditions for local planarization of geometric graphs. However, is it reasonable to assume that these properties or their generalizations are actually present in wireless network graphs? Our empirical analysis of simulated and WSN testbed data positively answers this question. Although these properties are not perfectly present when looking directly at the network graph, they are very likely to be present in this graph's c-hop clustered representation, for $c \leq 2$. In conclusion, in WSN algorithm design it is justified and practically relevant to assume and use the presence of the structural network properties redundancy and coexistence. Furthermore, replacing the simplistic unit disk assumptions by these properties helps to detach mathematical reasoning in this field from questions related to circle geometry.

Acknowledgments. The authors would like to thank the CitySee project [4], in particular I. Stojmenović for providing us with the testbed data. This work was supported by the German Research Foundation (DFG), grant "FR 2978/1-2".

References

1. Bose, P., Morin, P., Stojmenović, I., Urrutia, J.: Routing with guaranteed delivery in ad hoc wireless networks. Wirel. Netw. **7**(6), 609–616 (2001)
2. Gao, J., Guibas, L.J., Hershberger, J., Li, Z., Zhu, A.: Geometric spanners for routing in mobile networks. IEEE J. Sel. Areas Commun. **23**(1), 174–185 (2005)

3. Kuhn, F., Wattenhofer, R., Zollinger, A.: Ad hoc networks beyond unit disk graphs. Wirel. Netw. **14**(5), 715–729 (2008)
4. Liu, Y., Mao, X., He, Y., Liu, K., Gong, W., Wang, J.: CitySee: not only a wireless sensor network. IEEE Netw. **27**(5), 42–47 (2013)
5. Mathews, E., Frey, H.: A localized link removal and addition based planarization algorithm. In: Bononi, L., Datta, A.K., Devismes, S., Misra, A. (eds.) ICDCN 2012. LNCS, vol. 7129, pp. 337–350. Springer, Heidelberg (2012)
6. Rappaport, T.: Wireless Communications: Principles and Practice, 2nd edn. Prentice Hall, Upper Saddle River (2001)
7. Tsai, H.W., Chu, C.P., Chen, T.S.: Mobile object tracking in wireless sensor networks. Comput. Commun. **30**(8), 1811–1825 (2007)

Security

A Privacy-Preserving Remote Healthcare System Offering End-to-End Security

Eduard Marin[✉], Mustafa A. Mustafa, Dave Singelée, and Bart Preneel

ESAT-COSIC and iMinds, KU Leuven,
Kasteelpark Arenberg 10, 3001 Leuven-Heverlee, Belgium
{eduard.marin,mustafa.mustafa,dave.singelee,
bart.preneel}@esat.kuleuven.be

Abstract. Remote healthcare systems help doctors diagnose, monitor and treat chronic diseases by collecting data from Implantable Medical Devices (IMDs) through base stations that are often located in the patients' house. In the future, these systems may also support bidirectional communication, allowing remote reprogramming of IMDs. As sensitive medical data and commands to modify the IMD's settings will be sent wirelessly, strong security and privacy mechanisms must be deployed.

In this paper, we propose a user-friendly protocol that is used to establish a secure end-to-end channel between the IMD and the hospital while preserving the patient's privacy. The protocol can be used by patients (at home) to send medical data to the hospital or by doctors to remotely reprogram their patients' IMD. We also propose a key establishment protocol between the IMD and the base station based on a patient's physiological signal in combination with fuzzy extractors. Through security analysis, we show that our protocol resists various attacks and protects patients' privacy.

1 Introduction

Remote healthcare systems usually collect data from medical devices implanted within the patient's body, also known as Implantable Medical Devices (IMDs), several times per day through a base station that is often installed in the patient's house. In the near future, these systems may support bidirectional communication to enable remote reprogramming of the IMD by a doctor in a hospital. While these remote healthcare systems can improve the patients' quality of life and extend the time they can live independently at home, they pose important security and privacy risks. Currently, proprietary protocols are being deployed with limited security measures [7,11]. These insecure wireless protocols may lead to several attacks that can result in fatal consequences for patients.

Remote healthcare systems are typically built such that there is a central node (denoted as data concentrator in the rest of this paper) that collects and redirects all (encrypted) data sent between base stations and hospitals. Context information about these communications, i.e. metadata, can be valuable for

© Springer International Publishing Switzerland 2016
N. Mitton et al. (Eds.): ADHOC-NOW 2016, LNCS 9724, pp. 237–250, 2016.
DOI: 10.1007/978-3-319-40509-4_17

insurance companies or government agencies to monitor public health or compile statistics. However, even if cryptography is used, metadata can leak patients' sensitive information to the data concentrator, which may lead to abuse of data, e.g. denying individuals an insurance contract. To mitigate some of these issues, a trivial solution would be to have several data concentrators (instead of only one) that do not cooperate with each other. This solution would be expensive to deploy and difficult to maintain. In addition, there would not be any guarantee that the data concentrators do not share data with each other. Another solution would be to use a fresh pseudonym in every message. Although this approach would make it more difficult for adversaries to compromise the patient's privacy, this is not sufficient; if the data concentrator knows where the base station is located, then unique patient identifiers may be revealed.

Our first contribution is a user-friendly protocol that allows an IMD to establish an end-to-end secure channel with a hospital while preserving the patient's privacy. The protocol can be used by patients (at home) to send medical data to the hospital or by doctors to remotely reprogram the IMD of their patients. Our protocol makes use of cryptography and an anonymous communication channel to prevent the data concentrator from learning patients' sensitive information. Our second contribution is a key establishment protocol that allows an IMD and a base station to agree on a symmetric session key without using public-key cryptography or any pre-shared secrets between devices. Instead, our protocol uses a patient's physiological signal in combination with fuzzy extractors.

2   Related Work

2.1   Security and Privacy in Remote Monitoring Systems

While much research has focused on making remote monitoring systems more reliable, unobtrusive, energy efficient and scalable, security and privacy have received less attention [10,12]. Ko et al. acknowledged the importance of security and privacy, but they did not provide details about cryptographic mechanisms [9]. Ortiz et al. proposed a protocol to secure the wireless communication between devices in a medical system [13]. However, the protocol does not provide message integrity, and the receiver keeps a list of keys for each transmitter. Transmitters send messages along with their unique device identity (ID) unencrypted to indicate to the receiver which of the keys is used to decrypt the message. This may result in a privacy breach since adversaries can use the unique transmitter ID to track, identify or locate individuals. Perrig et al. presented SNEP, a protocol that provides data confidentiality, integrity, mutual authentication and message freshness [14]. Pre-installed symmetric keys, shared between each device and the base station, are used to derive a new session key every time a device starts communicating with the base station. In contrast to Perrig et al., we also consider privacy. More specifically, our protocol aims to prevent unauthorized entities and adversaries from discovering the patients' real ID, their location or being able to link their messages. Furthermore, the devices are implanted within the patient's body and the base stations do not have any pre-shared keys with them, which makes key management more challenging.

2.2   Key Establishment Protocols

The unique characteristics of IMDs pose novel challenges in the design of key establishment protocols and key management solutions. IMDs are battery-powered and resource-constrained devices in terms of size, memory, processor and energy. The battery typically lasts 7 years. When the battery is drained, surgery is needed to replace the IMD. We note that a trade-off between security and open-access in emergencies is also required. Consider a cardiac patient who is travelling. Although it is clear that no one should be able to access his pacemaker while he is walking on the street, medical staff should have immediate access to his IMD in an emergency situation.

Intuitively, one possible way for the IMD and the base station to establish a key would be to use public-key cryptography. However, IMDs cannot use expensive cryptographic primitives in terms of computations and power consumption. Another possibility would be to pre-install a master (symmetric) key in all IMDs, but this may prevent a patient from receiving care in an emergency situation. A pairing protocol could also be used for establishing a symmetric session key [4,6,18]. Nevertheless, IMDs do not contain a screen, a keyboard or an accelerometer and it is not possible to physically access them once implanted; thus none of the existing solutions can be used. In this paper, we propose a key establishment protocol in which the IMD and the base station measure a patient's physiological signal independently and synchronously to agree on a symmetric key. The protocol uses the time between heart beats, also known as InterPulse Interval (IPI), as the source of randomness, similarly to the touch-to-access protocol proposed by Rostami et al. [15]. Unlike their work, our protocol is more efficient as it only uses symmetric cryptography in combination with fuzzy extractors.

3   Design Preliminaries

3.1   System Model

Our remote healthcare system, similar to existing architectures, consists of the following entities (see Fig. 1). Without loss of generality, in the rest of this paper we will assume that the IMD is a pacemaker.

A *pacemaker* (PM) is a device implanted within a patient's body that is used to monitor and control his heart beat. A *base station* (BS) is an external device installed in a fixed location (e.g. home or a hotel) which collects and forwards medical data to a hospital, and sends commands to PMs as instructed by a doctor in the hospital. BSs have a programming head that incorporates a built-in sensor to read a patient's physiological signal (e.g. the IPI). A *data concentrator* (DC) acts as a bridge between BSs and hospitals, and is in practice

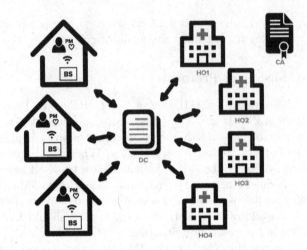

Fig. 1. Our remote healthcare system comprises pacemakers (PMs), base stations (BSs), a data concentrator (DC), hospitals (HOs) and a certification authority (CA).

often managed by the company that manufacturers the PMs and BSs. A *hospital* (HO) is a medical institution where doctors analyse the medical data and send commands to PMs, whereas a *certification authority* (CA) is a trusted entity that issues digital certificates to HOs and the DC.

A BS and a PM can communicate with each other wirelessly using the Medical Implant Communication Service (MICS) band [17], or any other low-energy wireless technology. The communication range between the BS and the PM depends on the wireless communication technology being used, e.g. from two to five meters when using the MICS band. The communication between the BS and the DC takes place over the Internet using a low-latency anonymous communication channel (e.g. a Mix network [16]), whereas the communication between the DC and a HO takes place over the Internet using a standard secure channel, e.g. TLS. To balance the load, multiple DCs are in place; however, for the sake of simplicity, we consider them as one in the rest of this paper.

We consider two possible scenarios depending on whether the doctor is on-line or off-line. For an on-line remote medical check, the patient first makes an appointment with the doctor (e.g. via telephone). At the time of the appointment, the patient sends medical data to the HO through the base station. The doctor then analyses the data and, if required, sends commands to the PM. If the doctor is off-line, the patient can still send medical data to the HO; however, these data will be processed by the doctor at a later stage. We note that in our system the doctor can only reprogram the patient's PM in the on-line scenario.

3.2   Threat Model and Assumptions

Threat Model: PMs and HOs are honest and trusted. PMs follow the protocol specifications as designed by their manufacturers and the U.S.

Federal Communications Commission (FCC). PMs can only establish one communication session with a BS at a time, and do not initiate any communication without receiving a request from a legitimate BS [2]. Adversaries can eavesdrop, modify, inject and jam the messages exchanged between any of the entities. Adversaries can observe only a fraction of the Internet network traffic. This is a common assumption when using low-latency anonymous communication systems. In addition, adversaries might compromise any number of BSs, including the one being used by the patient. The DC is honest but curious; it follows the protocol specifications but it might attempt to discover information about patients by looking at metadata.

Assumptions: We assume that adversaries cannot make physical contact with the patient without this being noticed by the patient. We assume that the communication between the BS and the DC takes place over the Internet using an anonymous communication channel. However, we do not specify which type of anonymous channel is used, since this is out of the scope of this paper. We assume that all entities (except PMs) have the CA's certificate pre-installed. We assume that each HO has a server with a database that contains a list of their patients along with their corresponding PM IDs and cryptographic keys. The server is located in a secure room where it cannot be stolen or tampered with; only authorized medical staff has access to it through appropriate identification, authentication and authorization mechanisms.

3.3   Design Requirements

Our remote healthcare system should satisfy the following functional (F), security (S) and privacy (P) requirements.

   (F1) User-friendly: Reporting medical data and reprogramming the PM should be easy and convenient for patients.

   (F2) BS-independent: Patients should be able to use any legitimate BS, even the ones belonging to other patients.

   (F3) Energy Cost: Computational cost at PMs should be as low as possible to reduce the energy consumption.

   (S1) Mutual Entity Authentication: PMs and HOs should be assured of each other's identity when receiving messages.

   (S2) Message Integrity: PMs and HOs should be assured that the received messages are fresh and have not been altered during transit.

   (S3) Confidentiality of Medical Data and Commands: Only the authorized HO should be able to access patient's medical data and send commands.

   (S4) Availability: Ensures that the system is accessible upon demand by authorised entities.

   (P1) Patient Privacy (minimum data disclosure):

   (P1.1) Patient Identity Privacy: Only HOs should know the identity of the patient who sends medical data.

(P1.2) Hospital Identity Privacy: No unauthorized entity should know to which hospital the patient sends medical data.

(P1.3) Location Privacy: No entity should infer the patient's location.[1]

(P1.4) Session Unlinkability: Only the HO where the patient is registered should be able to link messages sent from the same patient.

4   The Protocol

This section presents our protocol for medical data reporting and remote reprogramming of the patient's PM. It provides end-to-end security (i.e. data confidentiality, integrity, mutual authentication and message freshness) between the patient's PM and the hospital, and protects the patient's privacy. Our protocol is divided into two stages: the medical data reporting stage and the PM reprogramming stage. Prior to the detailed protocol description, we first outline the system initialization process. Table 1 shows the notation used in the paper.

4.1   System Initialisation

Each HO and the DC generate a public/private key pair, PK_{ho_i}/SK_{ho_i} and PK_{dc}/SK_{dc}, respectively. The public keys are signed by the CA. A list of valid HO certificates is stored in the DC, whereas each HO has a valid DC certificate. A group signature scheme is used by BSs to anonymously sign messages on behalf of the BSs group, so that the DC can still verify the authenticity of the message without knowing which specific BS signed the message, thus hiding the ID and location of the patient. All BSs use the same (group) public key, $PK_{bs_{group}}$, and have distinct private keys, SK_{bs_i}. Next, the DC signs the BSs group public key, generates a digital certificate that contains the ID and group's public key, and stores it. The certificate of the DC is pre-installed in all BSs.

During the PM manufacturing process, a symmetric key, $K_{ho\text{-}ps}$, is preinstalled in each PM. $K_{ho\text{-}ps}$ is shared between all PMs and the DC, and used for generating HO pseudonyms. Our protocol uses the same $K_{ho\text{-}ps}$ for all PMs, so that the DC cannot identify the PM that generated the HO pseudonym. In the PM setup phase, two independent symmetric keys, $K_{pm\text{-}ho}$ and $K_{pm\text{-}ps}$, are generated and installed in each PM. The procedure takes place in the HO before the PM is implanted to avoid the PM's manufacturer (often the DC) from learning these keys. $K_{pm\text{-}ho}$ and $K_{pm\text{-}ps}$ are shared between the PM and its corresponding HO. The former is used for encrypting the patient's medical data whereas the latter for generating PM pseudonyms.

Various circumstances may cause the certificates to become invalid before their expiration date. If the private key of any of the entities is compromised, a new public/private key pair is generated (as explained above). The new public key is then signed by the CA and broadcasted to the network. All entities can

---

[1] In an emergency situation doctors have other means (e.g. necklace-based emergency systems) to know the patient's location.

Table 1. Notations.

Symbols	Meanings
d, cmd	medical data of a patient, command sent to the PM
ID_i, PS_i	unique identity of entity i, pseudonym of entity i
K_{pm-ho}	key shared between PM and HO to encrypt/decrypt data/commands
K_{pm-ps}	key shared between PM and HO to create PM pseudonyms (ps)
K_{ho-ps}	key shared between all PMs and DC to generate HO's pseudonyms (ps)
K_s	session key established between PM and BS
N_i, TS_i	nonce generated by entity i, timestamp produced by entity i
msg_{i-j}	message constructed by entity i and intended for entity j
ct_{i-j}	counter used in messages between the entity i and the entity j
C_{i-j}	ciphertext generated by entity i and intended for entity j
PK_i, SK_i	public and private key of entity i
$PRF_K(M)$	pseudorandom function of message M with key K
$AE_K(M)$	authenticated encryption of message M with key K
$E_{PK_i}(M)$	asymmetric encryption of message M with public key of entity i
$Sig_i(M)$	digital signature of entity i on message M

then verify the message authenticity by using the CA's public key. From that point onwards, the old certificate is no longer valid. If the private key of any BS is compromised, the BS is sent to its manufacturer for being reconfigured and replaced. We note that group signature schemes typically allow revocation and addition of new members (i.e. BSs) into the group (for more details, see [3]).

4.2   Medical Data Reporting Stage

After the system initialisation phase, the PM can report medical data to the HO (see Fig. 1). Prior to each reporting stage, a new symmetric session key is established between the PM and the BS for securing the data exchanged over the air. We next describe the proposed key establishment protocol followed by the actual reporting stage more in detail.

IPI-based Key Establishment Protocol: Our protocol requires the IMD and the BS to independently measure the patient's IPI (i.e. time between heart beats) at the same time. These IPI readings, which can be measured anywhere in the patient's body just by touching the patient's skin, are then used for establishing a symmetric key that is valid only for one session. Previous work has shown that the four least significant bits of IPIs are uncorrelated and independently identically distributed (i.i.d) [19]. The IPI cannot be read remotely (e.g. via a webcam), as shown by Rostami et al. [15]. Therefore, based on their results and our assumptions, a remote attacker cannot measure the IPI. Figure 2 shows the IPI-based key establishment protocol.

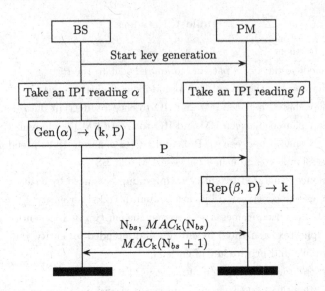

Fig. 2. IPI-based key establishment protocol between the BS and the PM.

To trigger the key establishment procedure, the patient first presses a button on the BS. The PM and BS then take two readings of the patient's IPI at the same time. However, these readings are not equal (but rather similar) due to the noise. Let us denote the reading taken by the BS as α and the reading taken by the PM as β. Fuzzy extractors allow generating a cryptographic key k from α and then successfully reproduce k from β, iff α and β are almost equal [5]. Fuzzy extractors are composed by two functions: generate (Gen) and reproduce (Rep). Gen is executed by the BS with α as an input, and outputs a key $k \in \{0,1\}^l$ and helper data $P \in \{0,1\}^*$. The BS then sends P in order to help the PM to reproduce k. The PM executes Rep with β and P as inputs, and outputs a key k' (if β and α are similar, then k equals k').

To achieve key confirmation, the BS generates a nonce, N_{bs}, and sends it to the PM along with a Message Authentication Code (MAC), $MAC_k(N_{bs})$. Upon receiving the message, the PM verifies $MAC_k(N_{bs})$ using k'. If the MAC is verified correctly, the PM is assured that the BS knows the key; however the BS does not have any assurance that the PM knows the key. For the PM to prove knowledge of the key, it computes $MAC_k(N_{bs} + 1)$ and sends it to the BS. The BS repeats this operation with its own key and checks whether the result of this operation corresponds to the received MAC. If the MAC is verified correctly, both devices can use k, (hereafter denoted as K_s), to securely communicate with each other, otherwise they execute the key establishment protocol again.

Reporting Medical Data to the HO: In this stage the patient's PM sends medical data to the HO. Figure 3 depicts the processes executed by the entities.

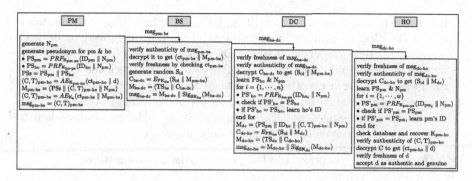

Fig. 3. Medical data reporting protocol.

PM: The PM performs the following steps.

1. It generates a fresh nonce, N_{pm}, that is used to compute two fresh pseudonyms. First, it computes a pseudonym for itself, i.e. $PS_{pm} = PRF_{K_{pm-ps}}(ID_{pm} \parallel N_{pm})$, where PRF is a pseudorandom function (e.g. a secure block cipher) and ID_{pm} is the PM's real identity (e.g. serial number). This pseudonym is only used once (i.e. in the message sent from the PM to the HO). Next, it computes a pseudonym for the HO, i.e. $PS_{ho} = PRF_{K_{ho-ps}}(ID_{ho} \parallel N_{pm})$, where ID_{ho} is the HO's real identity. This pseudonym is used in a pair of messages (i.e. the one sent from the PM to the HO and vice versa). Both pseudonyms protect the patient's privacy while communicating with the HO, i.e. the PM uses PS_{pm} and PS_{ho} instead of ID_{pm} and ID_{ho}.

2. It encrypts the patient's medical data, d, and a counter, ct_{pm-ho}, using the secret key it shares with the HO, K_{pm-ho}, i.e. $(C, T)_{pm-ho} = AE_{K_{pm-ho}}(ct_{pm-ho} \parallel d)$. Since PMs do not contain a precise clock, a counter is used to prevent replay attacks. The counter is initialised to zero every time a new session key between the BS and the PM is established. For better performance, the encryption method used is an authenticated encryption scheme (e.g. AES-CCM [1]), which outputs a ciphertext, C, and an authentication tag, T.

3. It constructs a message, $M_{pm-bs} = PSs \parallel (C, T)_{pm-ho} \parallel N_{pm}$, where $PSs = PS_{pm} \parallel PS_{ho}$, and encrypts it using the session key previously established with the BS, K_s, i.e. $(C, T)_{pm-bs} = AE_{K_s}(ct_{pm-bs} \parallel M_{pm-bs})$.

4. It sends $msg_{pm-bs} = (C, T)_{pm-bs}$ to the BS.

BS: Upon receiving msg_{pm-bs}, the BS performs the following steps.

1. It verifies the authenticity of msg_{pm-bs} and decrypts the ciphertext to obtain $(ct_{pm-bs} \parallel M_{pm-bs})$. It then checks whether the counter, ct_{pm-bs}, is valid, i.e. if it is higher than the counter of the last received message. If this condition is satisfied, the message is accepted, otherwise it is rejected.

2. It generates a random session ID, S_{id}, that is valid for a pair of messages and allows the HO to anonymously send a message back without knowing which specific BS sent the message. Then it encrypts S_{id} along with M_{pm-bs} using the public key of the DC, PK_{dc}, i.e. $C_{bs-dc} = E_{PK_{dc}}(S_{id} \parallel M_{pm-bs})$.

3. It constructs a message, $M_{bs-dc} = (TS_{bs} \| C_{bs-dc})$, where TS_{bs} is a timestamp of the BS used to counter replay attacks. TS_{bs} does not need to be kept secret and is sent unencrypted to the DC; however, its integrity is protected by means of a digital signature. This allows the DC to check the freshness of the message before decrypting the message and verifying its authenticity.

4. It generates a signature on M_{bs-dc} using its private key, $Sig_{SK_{bs}}(M_{bs-dc})$. Then it constructs a message, $msg_{bs-dc} = M_{bs-dc} \| Sig_{SK_{bs}}(M_{bs-dc})$, and sends it to the DC via an anonymous communication channel.

DC: Upon msg_{bs-dc} reception, the DC performs the following steps.

1. It verifies the freshness of msg_{bs-dc} by checking TS_{bs} and the authenticity of the message using the BSs group public key, $PK_{bs_{group}}$. It then decrypts C_{bs-dc} to obtain $(S_{id} \| M_{pm-bs})$, where $M_{pm-bs} = (PS_{pm} \| PS_{ho} \| (C,T)_{pm-ho} \| N_{pm})$.

2. It retrieves ID_{ho} by computing $PS'_{ho} = PRF_{K_{ho-ps}}(ID_{ho_i} \| N_{pm})$ for all HOs, $(ID_{ho_1}, \ldots, ID_{ho_n})$ until the DC finds a match, i.e. PS'_{ho} equals PS_{ho}.

3. It constructs a message, $M_{dc} = (PS_{pm} \| ID_{ho} \| (C,T)_{pm-ho} \| N_{pm})$, in which PS_{ho} is replaced with ID_{ho}. Then it encrypts the message and the session ID, S_{id}, using the public key of the HO, PK_{ho}, i.e. $C_{dc-ho} = E_{PK_{ho}}(S_{id} \| M_{dc})$.

4. It constructs a message, $M_{dc-ho} = (TS_{dc} \| C_{dc-ho})$, where TS_{dc} is a timestamp of the DC, and generates a signature on M_{dc-ho} using its private key, $Sig_{SK_{dc}}(M_{dc-ho})$. Finally, it appends the signature to M_{dc-ho} in order to form a message, $msg_{dc-ho} = M_{dc-ho} \| Sig_{SK_{dc}}(M_{dc-ho})$, and sends it to the HO.

HO: Upon msg_{dc-ho} reception, the HO performs the following steps.

1. It verifies the freshness of msg_{dc-ho} by checking TS_{dc} and the authenticity of the message by checking the validity of $Sig_{SK_{dc}}(M_{dc-ho})$. It then decrypts C_{dc-ho} to obtain $(S_{id} \| M_{dc})$, where $M_{dc} = (PS_{pm} \| ID_{ho} \| (C,T)_{pm-ho} \| N_{pm})$.

2. It retrieves ID_{pm} by computing $PS'_{pm} = PRF_{K_{pm-ps}}(ID_{pm_i} \| N_{pm})$ for all PMs, $(ID_{pm_1}, \ldots, ID_{pm_w})$, where w is the number of patients with a PM registered in the HO, until a match is found, i.e. PS'_{pm} equals PS_{pm}.

3. It searches for ID_{pm} in its database to retrieve K_{pm-ho}. Using this key, the HO verifies and decrypts $(C,T)_{pm-ho}$ to obtain $(ct_{pm-ho} \| d)$. Next, the HO verifies the freshness of the message by checking if the counter, ct_{pm-ho}, is higher than the counter of the previously received message. Only if this condition is fulfilled, the HO accepts the patient's medical data, d, as authentic and genuine.

4.3   PM Reprogramming Stage

After examining the patient's medical data, the doctor can send the necessary command(s) to adjust the PM's settings. This stage, which can only take place if the doctor is on-line, is described next and shown in Fig. 4.

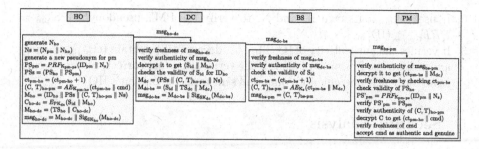

Fig. 4. PM's reprogramming protocol.

HO: The HO performs the following steps.

1. It generates a fresh nonce, N_{ho}, to create a fresh pseudonym for the PM, PS_{pm}. Then it increases the counter $ct_{pm\text{-}ho}$, and encrypts it along with the command, cmd, using $K_{pm\text{-}ho}$, i.e. $(C, T)_{ho\text{-}pm} = AE_{K_{pm\text{-}ho}}(ct_{pm\text{-}ho} \| cmd)$.
2. It constructs a message, $M_{ho} = (ID_{ho} \| PSs \| (C, T)_{ho\text{-}pm} \| Ns)$, where $PSs = (PS_{pm} \| PS_{ho})$ and $Ns = (N_{pm} \| N_{ho})$. Next it encrypts M_{ho} and the session ID using the public key of the DC, i.e. $C_{ho\text{-}dc} = E_{PK_{dc}}(S_{id} \| M_{ho})$.
3. It constructs a message $M_{ho\text{-}dc} = (TS_{ho} \| C_{ho\text{-}dc})$ and generates a signature on it, $Sig_{SK_{ho}}(M_{ho\text{-}dc})$. Then it constructs a message, $msg_{ho\text{-}dc} = M_{ho\text{-}dc} \| Sig_{SK_{ho}}(M_{ho\text{-}dc})$, and sends it to the DC.

DC: Upon $msg_{ho\text{-}dc}$ reception, the DC performs the following.

1. It verifies the freshness and authenticity of $msg_{ho\text{-}dc}$, and decrypts it to obtain $(S_{id} \| M_{ho})$. Once it learns ID_{ho} and S_{id}, it checks if S_{id} is a valid session ID for this HO, i.e. if a message containing this session ID was previously sent to this specific HO. It then constructs $M_{dc\text{-}bs} = (S_{id} \| TS_{dc} \| M_{dc})$, where $M_{dc} = (PSs \| (C, T)_{ho\text{-}pm} \| Ns)$, generates a signature on it, $Sig_{SK_{dc}}(M_{dc\text{-}bs})$, and constructs a message, $msg_{dc\text{-}bs} = M_{dc\text{-}bs} \| Sig_{SK_{dc}}(M_{dc\text{-}bs})$. Finally, $msg_{dc\text{-}bs}$ is sent to the BS over the anonymous communication channel previously used.

BS: Upon $msg_{dc\text{-}bs}$ reception, the BS performs the following.

1. It verifies the freshness and authenticity of $msg_{dc\text{-}bs}$ before checking the validity of S_{id}, i.e. checking if this session ID is the same as one previously generated by the BS. It then increases the counter by one, i.e. $ct_{pm\text{-}bs} = (ct_{pm\text{-}bs} + 1)$. It encrypts $ct_{pm\text{-}bs}$ and M_{dc} using the session key, i.e. $(C, T)_{bs\text{-}pm} = AE_{K_s}(ct_{pm\text{-}bs} \| M_{dc})$, and constructs and sends a message, $msg_{bs\text{-}pm} = (C, T)_{bs\text{-}pm}$, to the PM.

PM: Upon $msg_{bs\text{-}pm}$ reception, the PM performs the following steps.

1. It verifies the authenticity of $msg_{bs\text{-}pm}$, decrypts it to obtain $(ct_{pm\text{-}bs} \| M_{dc})$ and verifies the freshness of M_{dc} by checking $ct_{pm\text{-}bs}$. It then verifies the validity of PS_{ho}, i.e. checks if PS_{ho} has been previously generated at the PM.

2. It uses K_{pm-ps}, its own ID and N_s to verify the PM's pseudonym, $PS'_{pm} = PRF_{K_{pm-ps}}(ID_{pm} \| N_s)$.
3. It verifies the authenticity of $(C, T)_{ho-pm}$, decrypts it to obtain $(ct_{pm-ho} \| cmd)$ and verifies the freshness of cmd. If the verifications are correct, it accepts cmd as an authentic and genuine command sent from the patient's HO.

5 Security Analysis

Message Authenticity: Messages exchanged between any of the entities contain either a digital signature of the message originator or a MAC. Assuming that a standard digital signature scheme (e.g. RSA or Schnorr variant of ECDSA [8]) or a MAC algorithm (e.g. AES CBC-MAC) are used, our protocol guaranties source authentication, message integrity and non-repudiation (only with digital signatures). Thus, attacks that attempt to modify the messages in transit can be detected. All messages include either a counter or a timestamp to ensure freshness, and hence prevent replay attacks (satisfy (S1) and (S2)).

Confidentiality of Medical Data and Commands: Medical data and commands to modify the PM's settings are always encrypted twice (i.e. in two encryption layers). The inner-layer encryption is carried out between the PM and the HO. This provides end-to-end security. The outer-layer encryption is performed between any two communicating entities, and used to hide the pseudonyms from adversaries. In addition, it helps to prevent some types of denial-of-service attacks (satisfy (S4)). Assuming that a standard encryption scheme (e.g. AES) is used, it will be hard for eavesdroppers to learn the content of the messages. Only authorised medical staff will be able to access the medical data or send commands (satisfy (S3)).

Patient's Identity Privacy: Each message exchanged between the PM and the HO contains a distinct pseudonym to hide the PM's real ID. This pseudonym is generated using a PRF, e.g. AES-128, and the symmetric key that is known only to the PM and the authorized HO. AES-128 can be used as a PRF as long as the number of messages for one key is less than 2^{40}, which corresponds to more than 4,000 encrypted messages per second exchanged between the HO and the PM (assuming that the battery lasts 7 years). All unauthorized internal entities (i.e. BSs and DC) as well as external adversaries will not be able to obtain the PM's real ID. Only the authorized hospital can recover the real ID of the PM, and link the medical data to a specific patient (satisfy (P1.1)).

Hospital's Identity Privacy: For each pair of messages exchanged between the PM and the HO, the PM generates a distinct pseudonym to hide the real ID of the hospital where it sends medical data. This pseudonym is generated using a PRF, e.g. AES-128, and the symmetric key shared between all PMs and the DC. As explained above, AES-128 is a secure PRF if the number of encrypted messages is less than 2^{40}. However, external adversaries cannot have access to the pseudonyms produced by the PRF, since pseudonyms are always sent in an encrypted format between the communicating entities (satisfy (P1.2)).

Location Privacy: A low-latency anonymous channel (e.g. a Mix network) between the BSs and the DC in combination with a group signature scheme prevents the DC from learning the location of the BS while being used by the patient (satisfy (P1.3)).

Session Unlinkability: Fresh and random pseudonyms are used in each message exchanged between the PM and the HO. Therefore, by looking at the exchanged messages, no unauthorized entity can link different sessions or learn if two messages have been sent by the same PM (satisfy (P1.4)).

Protection against Stolen BSs or Pre-owned PMs: Since BSs are simply relay devices that do not have any pre-installed shared secrets with PMs, adversaries who get a BS cannot access the content (i.e. medical data and commands) of the messages exchanged between the PM and the HO. In addition, adversaries who obtain a new or a pre-owned PM cannot send data to the HO. Upon a PM replacement, the old PM's ID and the corresponding keys are removed from the database so that the patient is no longer linked to the old PM.

6   Conclusions

In this paper, we proposed a protocol that provides end-to-end security between a PM and a HO while preserving the patient's privacy. Each PM uses two fresh pseudonyms for hiding the unique PM's ID and the HO where the medical data is sent. These pseudonyms allow the DC to forward the medical data to the authorized HO without learning to whom the data belongs to, and prevent adversaries from discovering the PM's real ID and to which hospital the medical data is sent. In addition, all BSs sign their messages using a group signature scheme and send them to the DC over an anonymous channel. This allows (i) the DC to verify the authenticity of the messages and (ii) the HO to link the medical data to the patient without learning which specific BS sent the messages (i.e. the location of the patient). Moreover, we presented an IPI-based key establishment protocol that allows a PM and a BS to agree on a symmetric key without using public-key cryptography or any pre-installed shared secrets.

Acknowledgments. The authors would like to thank George Petrides and the anonymous reviewers for their helpful comments. This work was partially supported by KIC InnoEnergy SE via KIC innovation project SAGA, and the Research Council KU Leuven: C16/15/058.

References

1. RFC3610: Counter with CBC-MAC (CCM). https://tools.ietf.org/html/rfc3610

2. Federal Communications Commission. MICS Medical Implant Communication Services, FCC 47CFR95.601-95.673 Subpart E/I Rules for MedRadio Services
3. Boneh, D., Boyen, X., Shacham, H.: Short group signatures. In: Franklin, M. (ed.) CRYPTO 2004. LNCS, vol. 3152, pp. 41–55. Springer, Heidelberg (2004)
4. Castelluccia, C., Mutaf, P.: Shake them up!: a movement-based pairing protocol for cpu-constrained devices. In: Proceedings of the 3rd International Conference on Mobile Systems, Applications, and Services, NY, USA, pp. 51–64 (2005)
5. Dodis, Y., Reyzin, L., Smith, A.: Fuzzy extractors: how to generate strong keys from biometrics and other noisy data. In: Cachin, C., Camenisch, J.L. (eds.) EUROCRYPT 2004. LNCS, vol. 3027, pp. 523–540. Springer, Heidelberg (2004)
6. Gehrmann, C., Mitchell, C.J., Nyberg, K.: Manual authentication for wireless devices. RSA Cryptobytes 7(1), 29–37 (2004)
7. Halperin, D., Heydt-Benjamin, T.S., Ransford, B., Clark, S.S., Defend, B., Morgan, W., Fu, K., Kohno, T., Maisel, W.H.: Pacemakers and implantable cardiac defibrillators: software radio attacks and zero-power defenses. In: Proceedings of the 29th Annual IEEE Symposium on Security and Privacy, pp. 129–142, May 2008
8. Johnson, D., Menezes, A., Vanstone, S.: The elliptic curve digital signature algorithm. Int. J. Inf. Secur. 1(1), 36–63 (2014)
9. Ko, J., Lim, J.H., Chen, Y., Musvaloiu-E, R., Terzis, A., Masson, G.M., Gao, T., Destler, W., Selavo, L., Dutton, R.P.: Medisn: medical emergency detection in sensor networks. ACM Trans. Embed. Comp. Syst. 10(1), 11:1–11:29 (2010)
10. Malan, D., Thaddeus, F.J., Welsh, M., Moulton, S.: CodeBlue: an ad hoc sensor network infrastructure for emergency medical care. In MobiSys Workshop on Applications of Mobile Embedded Systems, pp. 12–14. ACM (2004)
11. Marin, E., Singelée, D., Yang, B., Verbauwhede, I., Preneel, B.: On the feasibility of cryptography for a wireless insulin pump system. In: Proceedings of the Sixth ACM Conference on Data and Application Security and Privacy, CODASPY 2016, pp. 113–120. ACM, New York (2016)
12. Ng, J.W.P., Lo, B.P.L., Wells, O., Sloman, M., Peters, N., Darzi, A., Toumazou, C., Yang, G.Z.: Ubiquitous monitoring environment for wearable and implantable sensors. In: UbiComp - 6th International Conference on Ubiquitous Computing (2004)
13. Ortiz, A., Munilla, J., Peinado, A.: Secure wireless data link for low-cost telemetry and telecommand applications. In: Electrotechnical Conference. MELECON. IEEE Mediterranean, pp. 828–831. IEEE (2006)
14. Perrig, A., Szewczyk, R., Tygar, J.D., Wen, V., Culler, D.E.: SPINS: security protocols for sensor networks. Wirel. Netw. 8(5), 521–534 (2002)
15. Rostami, M., Juels, A., Koushanfar, F.: Heart-to-heart (H2H): authentication for implanted medical devices. In: Proceedings of the ACM SIGSAC Conference on Computer and Communications Security, CCS, NY, USA, pp. 1099–1112 (2013)
16. Sampigethaya, K., Poovendran, R.: A survey on mix networks and their secure applications. Proc. IEEE 94(12), 2142–2181 (2006)
17. Savci, H., Sula, A., Wang, Z., Dogan, N., Arvas, E.: MICS transceivers: regulatory standards and applications. In: Proceedings of IEEE SoutheastCon, April 2005
18. Stajano, F., Anderson, R.J.: The resurrecting duckling: security issues for ad-hoc wireless networks. In: Proceedings of the 7th International Workshop on Security Protocols, London, UK, pp. 172–194 (2000)
19. Xu, F., Qin, Z., Tan, C.C., Wang, B., Li, Q.: IMDGuard: securing implantable medical devices with the external wearable guardian. In: Proceedings of INFOCOM, pp. 1862–1870, April 2011

DTLS Improvements for Fast Handshake and Bigger Payload in Constrained Environments

Philippe Pittoli[✉], Pierre David[✉], and Thomas Noël[✉]

ICube, Université de Strasbourg, Strasbourg, France
{p.pittoli,pda,noel}@unistra.fr

Abstract. Transport Layer Security (TLS) is a protocol defined by the IETF to secure communications on the Internet, and Datagram Transport Layer Security (DTLS) is its version based on UDP. DTLS is the proposed solution to secure the Internet of Things (IoT). As IoT devices are constrained in memory, in code size and in computation speed, DTLS overhead is a crucial parameter for communication efficiency. The contribution presented in this paper is an improved version of DTLS, with fewer handshake messages and a reduced payload overhead, without compromising security. Fewer handshake messages means a reduced connection delay, with 6 signalling packets instead of 10. Reducing payload overhead improves communication latency and provides more room for application data. As such, our work provides a more efficient connection-based security protocol for the IoT domain.

1 Introduction

Constrained environments are used in a wide range of situations, varying from autonomous wireless sensors for environmental studies to industrial systems with energy supplied devices. Network technologies are equally very diversified, from the multi-hop or single hop wireless network based on IEEE 802.15.4 to a wired network such as Ethernet in industrial context. However, even with such a diversity, there is always a need for end-to-end security in order to protect data from illegitimate eavesdroppers and to protect devices from malicious intrusions.

TLS (Transport Layer Security) is a fundamental cryptographic protocol whose main purpose is to establish a secured connection between two applications. This is accomplished via a negotiation of cryptographic material, enabling the peers to encrypt the whole communication with a shared encryption key. TLS is utilized in various ways on the Internet, to protect web traffic (HTTPS), to secure the e-mail transport protocol (SMTP/TLS) and so on. However, the TCP connection required by TLS is costly, specially in constrained environments such as the Internet of Things (IoT).

Datagram Transport Layer Security (DTLS) [RM12] is a modified version of the TLS protocol to use UDP as the underlying layer. It has been designed

© Springer International Publishing Switzerland 2016
N. Mitton et al. (Eds.): ADHOC-NOW 2016, LNCS 9724, pp. 251–262, 2016.
DOI: 10.1007/978-3-319-40509-4_18

for purposes such as real-time applications. The IETF promotes its use on constrained devices that are utilized in the emerging wireless technologies such as Internet of Things (IoT).

Despite the use of UDP, DTLS is still costly for two main reasons. First, the number of messages transmitted during the handshake delays the first data packet (which is emphasized in duty cycled network [VTW+15]) and energy consumption. Second, the DTLS header limits the room available for application data in a packet.

This article presents an improved version of DTLS for the sake of speed, allowing quicker connection establishments. We reduce the amount of signalling packets and the header overhead, using IETF recommendations as a basis. In order to concentrate on the DTLS layer only, we do not make any assumption on other layers. As a consequence, we reduce the whole protocol stack to the bare minimum as shown in Fig. 1: neither IP nor transport layer nor application layer, in order to measure only the impact of DTLS over a communication network based on IEEE 802.15.4.

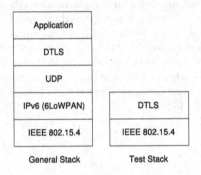

Fig. 1. Classic protocol stack as in deployed networks, and our protocol stack to only measure the impact of the DTLS protocol in communications.

As encryption speed is an important parameter in security protocol efficiency, we quantify in this paper the gain of using hardware encryption rather than software encryption.

We compare our proposal to the original DTLS protocol, and we also compare the software and hardware encryptions in order to measure the improvement of communication speed with a specialized hardware. We focus on the connection speed (handshake and application data exchange) and the memory (RAM and Flash) consumption.

The rest of this paper is structured as follows. In Sect. 2, we introduce DTLS principles. Section 3 describes the objective of this article, and the choice of algorithms. We then explain our improvements over the DTLS protocol in Sects. 4 and 5. Experimental study of performances has been conducted. We describe our hardware platform in Sect. 6, and we summarize performance results in Sect. 8. We describe related works in Sect. 9, before concluding in Sect. 10.

2   Overview of DTLS

The goal of DTLS is to secure a connection between two applications. It is composed of an algorithm and cryptographic material negotiation, then an application data exchange with encryption (confidentiality), integrity checking and signature (authenticity). DTLS is layered above UDP as the transport protocol, where TLS is layered above TCP. This difference implies that DTLS doesn't have a transport layer protocol that handles packet losses and retransmission, reordering and fragmentation itself. Figure 2a shows the different control packets exchanged during connection establishment, which are explained later in this section.

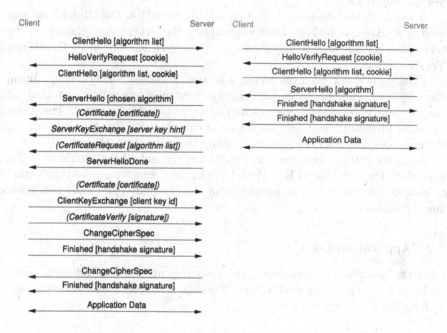

Fig. 2. (a) DTLS handshake. Messages in italics are optional. Payloads of messages are enclosed within square brackets. Context specific messages are enclosed within parentheses. (b) Optimized DTLS handshake.

2.1   Handshake

In order to establish a handshake based on DTLS protocol, multiple packets need to be transmitted, as shown by Fig. 2a. First, the client sends a message **Client Hello** containing the list of algorithms supported by the client to manage the session, and a random number to avoid using the same cryptographic key to encrypt messages on every session. The server replies with a message **Hello Verify Request** containing a cookie which must be sent back by the client with a new message **Client Hello**. This mechanism is needed to mitigate a deny of service attack targetted to overload the server memory by creating multiple sessions. With this mechanism, the server does not keep a state between the first and the second client messages.

Then the server sends a **Server Hello** message with the list of algorithms to use to enforce the integrity, the confidentiality and the authentication, based on the list previously sent by the client. Three optional messages come after: a certificate to authenticate the server to the client, its identity is sent with a **Server Key Exchange** message to help the client choosing the right encryption key, and a message **Certificate Request** if the server requires a certificate to authenticate the client. To conclude this phase, the server sends a message **Server Hello Done**.

If required, the client may send its certificate with a **Certificate** message. Next, it sends a **Client Key Exchange** to help the server choose the appropriate encryption key and then it may send a certificate verification with a **Certificate Verify** message.

Finally the client, then the server, send two messages. The first one, **Change Cipher Spec**, indicates to the peer that the next messages will be protected with the previously negotiated algorithms and cryptographic material. The second message, **Finished**, includes a signature of the whole handshake, to ensure that the messages had not been modified during the connection phase.

Messages within parentheses in Fig. 2a are sent only if the certificate system is needed. The Pre-Shared Key (PSK) mechanism, which assumes that authentication is performed by the knowledge of the pre-shared key, does not involve any certificate.

2.2 Application Data

Once the handshake is completed, the session is up and payload data may be exchanged. The protocol used to carry the data is simple: it only encrypts, signs and verifies the integrity.

3 IoT Application: Key Sharing, Encryption, Certificates

Our main objective will be to be able to establish a secure connection as quickly as possible with a very constrained hardware (from 8 MHz processor) with a constrained link layer (IEEE 802.15.4). To that end we will select the appropriate algorithms and mechanisms, for the key sharing, the encryption, and so on. Choices made for very constrained environments will still be valid for less constrained ones.

There are several ways to share encryption keys. The simplest is to share the key with the devices before the deployment or with an out-of-band communication. This can be acceptable in a sensor network. Sharing encryption keys with a certificate system is a costly mechanism, even with elliptic curves [LN08]: on a device with a 13 MHz processor, there is more than one second to initialize, sign and verify a certificate with resources consuming optimisations (about 1500 bytes of RAM and 10 to 15 KB of Flash memory), and several seconds for a 8 MHz processor.

At the IETF, the DICE working group has defined a DTLS profile suitable for IoT applications [TF15]. It is reasonably implementable on many constrained devices in terms of algorithm complexity and speed. The profile recommends the use of Counter with CBC-MAC (CCM) encryption method, with 8 bytes of Message Authentication Code (MAC) and the AES-128 symmetric encryption algorithm. For key sharing, the profile recommends the use of pre-shared keys or, if possible, Diffie-Hellman exchange for the key agreeing with a certificate authentication system optimized with elliptic curves (Elliptic Curves Digital Signature Algorithm).

4   Improvement: Handshake Message Removal

As shown with Fig. 2a, DTLS is a verbose protocol which was not intended to be used in very constrained networks. In this section, we propose to improve the DTLS handshake by reducing the total number of packets exchanged in pre-shared key (PSK) communication context. We follow the recommendations of the DICE working group for the encryption mechanism: we use AES 128 and CCM 8, and we pre-share the cryptographic keys. We do not use Diffie-Hellman nor certificates because they are costly even with elliptic curves. Thus, the messages required by the certificate system as shown in Fig. 2a are removed. Next, we propose to remove handshake messages which are for signalisation only and that do not contain any useful data (shown by the messages in square brackets): it is possible to do so because we fix the messages sequence, which implies that to indicate the end of a message sequence with a signalling packet becomes non essential. We also remove messages which indicate the identity of a peer, because it is possible to deduce this piece of information in several ways: with a sub-layer protocol (by the MAC or IP address for instance), or from the upper application protocol.

It is important to note that even if the MAC (or IP) addresses are used to identify a peer, it is not possible for an attacker to get any useful data: the attacker may spoof an address, but if it does not know the cryptographic keys, as they are never sent in clear text, the connection cannot be established.

Figure 2b summarizes our proposed DTLS optimizations.

ServerHelloDone and ChangeCipherSpec: these messages are removed because they do not contain any useful data. The **ServerHelloDone** message indicates the end of a message sequence from the server to the client but, in our DTLS version, the server has to send only one message (**Server Hello**) in this sequence. We can further notice that the **ChangeCipherSpec** is about to be removed in TLS 1.3 [Res15].

ServerKeyExchange: this message is non essential if we have only one server and a single key is to use for the server at a time, or if the client already knows the server identity [ET05] (or it can be deduced) and there is only one key at a time.

ClientKeyExchange: this message may be removed in two cases: either the client is located one single hop away to the server, thus letting the server use the MAC (or IP) address to deduce the client identity, or the application protocol is used to communicate the client identity to the server.

With these optimizations, we remove up to five messages during the handshake and settle the number of exchanged messages (no more optional message left). This simplifies the final code, and it is expected to lower the binary size, implementation errors, the maintenance and improve the readability.

The counterpart of these optimizations is that they remove compatibility with current DTLS implementations.

5   Improvements: Redundancy During the Application Data Exchange

Once the connection is established, DTLS transmits the application data using two layers (Record and Application layers) whose headers are shown in Fig. 3. The Record part contains two fields (Epoch and Sequence Number) used to detect packet losses, duplication and unordered packet arrival. When Authenticated Encryption with Associated Data (AEAD) algorithm is used to secure the communication, the Application Data part contains both the payload and an Initialization Vector (IV). The IV has to be unique on every message and is conventionally composed of the Epoch and the Sequence Number from the Record part.

We can collapse this redundant information (8 bytes) in the two layers, reducing the DTLS communication overhead from 29 to 21 bytes. This 8-byte redundancy represents 6 % of the total payload over a IEEE 802.15.4 connection.

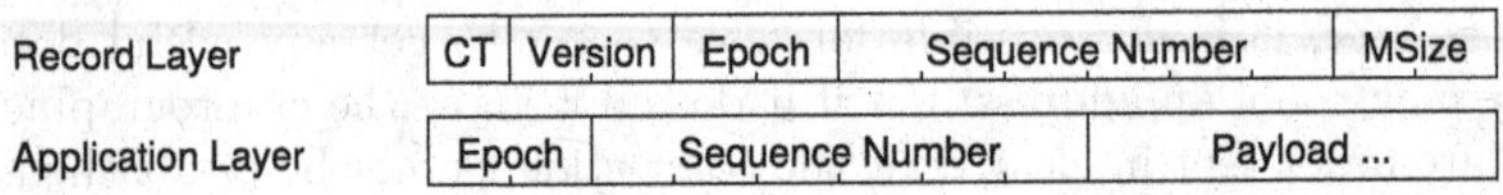

Record Layer	CT	Version	Epoch	Sequence Number	MSize
Application Layer	Epoch	Sequence Number		Payload ...	

Fig. 3. Redundancy in the headers, due to a conventional usage of the initialization vector in AEAD encryption.

The gain is especially the higher packet payload available (97 bytes instead of 89 when used on a IEEE 802.15.4 network, hence 8.2 %). Applications have more room to carry data on a single fly. In addition, the removal of this redundancy is expected to lower the transmission duration.

The counterpart of this optimization is that it removes compatibility with current DTLS implementations.

6   Experiment Environment

We run an experiment in order to measure the improvements brought by our solution in term of memory consumption (RAM and Flash), and connection speed (both handshake and application data exchange).

In this section, we describe the hardware and software environment of our experiment (Sect. 8). Our code is based on TinyDTLS which is a library that implements the current latest DTLS protocol version (1.2).

6.1   Hardware Platform

Our hardware platform is based on ATMega128RFA1 microcontroller (Zigduino) with a native IEEE 802.15.4 connectivity, 16 kB RAM, 128 kB Flash and a 16 MHz CPU frequency. The 802.15.4 connectivity is set in peer-to-peer mode, radio always on in order to avoid the wake-up delay for the link layer. There is only one hop from the client to the server.

The first device acts as a DTLS server which listens to the network, the second acts as a DTLS client which periodically sends a message with an arbitrary length.

6.2   IEEE 802.15.4 and CSMA-CA

To explain the time taken by a packet to reach its destination, we need to know the waiting duration before the transmission. These pieces of information are available from the IEEE 802.15.4 standard [IEE06]. The time spent waiting for CSMA-CA (expressed in symbols) is described with Eq. 1. BE means Backoff Exponent, it is used to fix an upper bound to the waiting time of the CSMA-CA algorithm.

$$T_{\text{wait}_{\text{symbols}}} = \text{random}(2^{\text{BE}} - 1) * \text{unit backoff period} \tag{1}$$

This equation leads to the maximum time spent for CSMA-CA in Eq. 2.

$$
\begin{aligned}
T_{\text{wait}} &= \frac{T_{\text{wait}_{\text{symbols}}}}{\text{nb symbols per second}} \\
&= \frac{\text{random}(2^3 - 1) * 20}{62\,500} \\
&\leq 2.24 \text{ ms}
\end{aligned}
\tag{2}
$$

The maximum waiting time on the first try when we want to send a message is 2.24 ms (Eq. 2), with $BE = 3$ which is the best case (first attempt to access the medium) and the backoff period is 20 (physical unit). If the random function follows a linear distribution, then the average waiting time will be 1.12 ms before sending a message (without interferences).

The minimum link layer has 10 bytes, but we instead use 12 bytes because we add the Personal Address Network identification (PAN id) in addition to the four bytes of MAC addressing (source, destination), in order to accomplish experiments in our laboratory with different sub-networks.

6.3 Network and Transport Layers

As explained in Sect. 1, we do not use network or transport layer since we want to focus on pure DTLS performances and to avoid any unneeded overhead. Network and transport layers are not required for a single hop network since every node in the local network can reach each other, so the addressing is done with the MAC header alone.

7 Hardware Encryption

Since the IEEE 802.15.4 standard requires an AES encryption to secure the communications on the link layer, the AES hardware encryption may be available on many of the IEEE 802.15.4 enabled devices. Dedicated hardware (with specialized instructions) improves an algorithm speed. Table 1 shows the results of a preliminary test of hardware encryption on our microcontroller. The encryption of 16 bytes with AES-128 on specialized hardware is five times quicker than software encryption. The CCM encryption mechanism advised by the DICE profile uses AES intensively, it is interesting to observe the practical gain in connection establishment and transmission durations.

Table 1. Comparison between software and hardware AES encryptions on a 16 bytes block.

	Software encryption	Hardware encryption
16 bytes	484 μs	96 μs

8 Measured Performances

To measure the message exchange duration between two devices (or applications) without reliable time synchronisation, we run an experiment by sending a message from a device to another, then sending it back. The message sizes used are 1 byte (the shortest possible message, such as a temperature without further application protocol like CoAP) and 87 bytes (the largest message on IEEE 802.15.4 with a 12-byte header).

Our experiment is performed with several configurations. The first one is a simple port of DTLS over ATmega128RFA1, the second one is our optimized DTLS exposed in Sects. 4 and 5, the next two are vanilla DTLS and our optimized DTLS with hardware encryption, and the final one is a non-secure exchange (without DTLS nor encryption). The last configuration aims at putting the results into perspective and knowing the cost of security.

Table 2. Comparison between DTLS, optimized DTLS, their hardware encryption variant and without DTLS (nor encryption)

	DTLS	Optimized DTLS	DTLS H.E	Optimized DTLS H.E	Without security
RAM (bytes)	11 216	10 388	7 120	6 292	897
Flash (bytes)	48 966	46 610	42 208	39 852	8 000
median handshake dur. (μs)	233 214	186 660	220 840	173 628	
median dur. (1 byte) (μs)	18 720	18 248	11 220	10 748	4 780
median dur. (87 bytes) (μs)	43 664	43 664	21 488	21 168	10 600

8.1 Flash Memory and RAM

Table 2 summarizes the observed consumption results. First, the cost in RAM for the security is more than 10 kB, and almost 41 kB in Flash. The optimized version consumes less of these ressources (7.3 % less RAM and 5 % less Flash), which is explained by the code removal due to the unneeded messages. The biggest difference is between software and hardware encryption: the encryption library (from OpenBSD) is speed-optimized by unrolling loops which implies a larger amount of instructions. On the other hand, the AES algorithm is removed with hardware encryption, which reduces the binary size of 4 kB, it is useful when the application code is large and this should be considered for complex applications over very constrained devices.

8.2 Handshake Performances

The handshake duration is measured on the client, from the first message of the connection to the handshake final (Finished) message. The median of observed handshake durations done over 100 connections is about 233 214 μs with DTLS, and 186 660 μs with our optimized version of DTLS. The gain is 46.5 ms, 19.9 % of the duration of the handshake: the message removal lowers the number of exchanges, thus lower the total duration of CSMA-CA waiting.

8.3 Data Exchange Performances

The removal of the Initialization Vector (IV) explained in Sect. 5 brings several improvements: transmission is quicker because the message is smaller, leaving more room for application data and decreasing the use of fragmentation.

Figure 4 shows our results over 3000 measurements for each configuration. The difference between DTLS and our optimized version is pronounced with the single byte message because difference of the message size to transmit is proportionally more important, the gain is about 2.6 %. The difference between DTLS (optimized or not) and a non protected exchange is measured with a non-secure exchange, to put the cost of security into perspective. The message

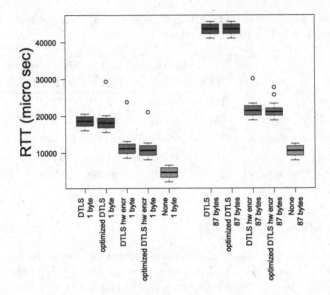

Fig. 4. Sending a message and returning it back, with only one byte then 87 bytes message: DTLS, optimized DTLS, then their hardware encrypted versions and without DTLS (Color figure online).

protection implies a RTT 3.9 times higher than without protection. Encryption, plus the overhead in the packet payload for DTLS, significantly slows down the connection.

Secondly, the message payload can be 8 bytes higher, thus the number of messages required for sending a certain amount of data is lower. For instance to send 1700 bytes, which represents 18 messages with the optimized version of DTLS, it is required to send 20 messages of the original DTLS.

The different improvements can be meaningful depending on the application. For instance, the handshake optimization lowers the connection duration: this helps connection with highly mobile nodes (drones, cars, etc.). There is only a slight improvement on the latency (on small packets) due to the IV removal, so it is useful mainly to reduce fragmentation.

9   Related Work

Lithe [RSH+13] takes another approach to reduce DTLS overhead: it relies upon 6LoWPAN to compress DTLS informations in order to gain up to 8 bytes during data exchange in the optimal case, where Epoch and Sequence Number fields (in the Record header) are truncated respectively to 8 and 16 bits (instead of 16 and 48 bits). This reduced width limits the number of exchanged messages in a single connection. Lithe does not remove redundant informations between the Record and Application layers. Furthermore, Lithe is tied to 6LoWPAN and is not independant from the lower layers.

Robust Header Compression (ROHC) is a method described in several RFCs to compress headers, such as IP and UDP ones (amongst others). The key principle is not to send an header if it is the same than the previous one, or just to send the difference. This may be used to limit the protocol payload, and can be used with our optimized DTLS if the IP context can't be avoided.

Elliptic curves cryptographic performances on sensors are tested on [LN08] through a number of evaluations with a library called TinyECC. This library lets the developers turn on and off several optimizations of elliptic curves algorithms, and can be used to improve the speed performances and the power consumption. Thanks to this work, we concluded that elliptic curves are one order of magnitude slower than symmetric cryptography on constrained devices.

Finally, the number of messages exchanged during the handshake induces degraded performance in duty-cycled networks, this is covered by [VTW+15]. In those networks the handshake duration can be more than 30 seconds, but our optimization can greatly helps to boost the performances.

10   Conclusion and Future Work

Our work leads to an improvement of the time spent on the handshake, of the RAM and Flash memories used by removing several messages. The most significant improvement is the larger payload of application data, due to the removal of Initialization Vector for AEAD algorithms, such as suggested by the DICE IETF working group. This improvement is at the cost of an incompatibility with current DTLS implementations.

As future work, we could experiment new standardized algorithms for encryption (ChaCha20) and for integrity checking (Poly1305) to see the differences in memory, code size and processing time. Furthermore, the IETF working group on TLS 1.3 tends to change completely the protocol, leading to a simpler and faster handshake, new mandatory-to-implement algorithms and to even change the fields.

References

[CCCP15] Capossele, A., Cervo, V., De Cicco, G., Petrioli, C.: Security as a coap resource: an optimized DTLS implementation for the IoT (2015)

[ET05] Eronen, P., Tschofenig, H.: Pre-Shared Key Ciphersuites for Transport Layer Security (TLS). RFC 4279 (Proposed Standard), December 2005

[HGmH+] Heer, T., Garcia-morchon, O., Hummen, R., Keoh, S.L., Kumar, E.S., Wehrle, K.: Security challenges in the IP-based internet of things. Wirel. Pers. Commun. **61**(3), 527–542 (2011)

[IEE06] IEEE. 802.15.4 (2006). http://standards.ieee.org/getieee802/download/802.15.4-2006.pdf

[LN08] Liu, A., Ning, P.: TinyECC: A configurable library for elliptic curve cryptography in wireless sensor networks. In: Proceedings of the 7th International Conference on Information Processing in Sensor Networks, IPSN 2008, pp. 245–256. IEEE Computer Society, Washington, DC, USA (2008)

[Res15] Rescorla, E.: The Transport Layer Security (TLS) Protocol Version 1.3. Internet-Draft draft-ietf-tls-tls13-07.txt, IETF Secretariat, July 2015

[RM12] Rescorla, E., Modadugu, N.: Datagram Transport Layer Security Version 1.2. RFC 6347 (Proposed Standard), January 2012

[RSH+13] Raza, S., Shafagh, H., Hewage, K., Rene, H., Voigt, T.: Lithe: lightweight secure CoAP for the internet of things. IEEE Sens. J. **13**(10), 3711–3720 (2013)

[TF15] Tschofenig, H., Fossati, T.: TLS/DTLS Profiles for the Internet of Things. Internet-Draft draft-ietf-dice-profile-17.txt, IETF Secretariat, October 2015

[VTW+15] Vučinić, M., Tourancheau, B., Watteyne, T., Rousseau, F., Duda, A., Guizzetti, R., Damon, L.: DTLS Performance in Duty-Cycled Networks. In: International Symposium on Personal, Indoor and Mobile Radio Communications (PIMRC - 2015). IEEE, Hong-Kong, China, August 2015

A Novel Algorithm for Securing Data Aggregation in Wireless Sensor Networks

Haythem Hayouni$^{(\boxtimes)}$ and Mohamed Hamdi

Higher School of Communication of Tunis (Sup'Com),
University of Carthage, Tunis, Tunisia
haythem.hayouni@supcom.tn

Abstract. Data aggregation is an important method to reduce the energy consumption in wireless sensor networks (WSNs). However, in hostile environments, the aggregated data can be subject to several types of attacks, and provide security is necessary. Recently, several secure data aggregation schemes based on homomorphic encryption have been proposed for wireless sensor networks. These data aggregation schemes provide better security compared with traditional aggregation since the aggregator can directly aggregate the ciphertexts without decryption. Based on our survey of existing research for ensuring secure data aggregation in WSNs, an Efficient Secure data aggregation scheme based on fully Homomorphic Encryption (E-SHE), is proposed. This scheme can protect end-to-end data confidentiality and support arbitrary aggregation operations over encrypted data. In addition, by utilizing message authentication codes (MACs), we can also verify data integrity during data aggregation and forwarding processes so that false data can be detected. Our experiments show that E-SHE requires less computation overheads and communication than previously methods and can effectively preserve data integrity, and precise data aggregation rate while consuming less energy to extend network lifetime.

Keywords: Wireless sensor networks · Data aggregation · Homomorphic encryption · MAC · Integrity

1 Introduction

Wireless sensor networks consist of a set of devices having limited computing resources. This type of network has attracted much attention in recent years, not only in academia but also in industry, for the study and development of a number of potential applications. However, the resource constraint [1] is the most important feature of this network. Indeed, the wireless sensors are limited in terms of calculation, storage, battery, etc. Therefore, a solution that aims to conserve these resources is widely desired.

However, compared to conventional computer networks, implementing security is not easy in wireless sensor networks due to limited processing power,

© Springer International Publishing Switzerland 2016
N. Mitton et al. (Eds.): ADHOC-NOW 2016, LNCS 9724, pp. 263–276, 2016.
DOI: 10.1007/978-3-319-40509-4_19

storage, bandwidth, and energy of sensor nodes. In addition to security, limited battery power and bandwidth of sensor nodes make it a challenging task to provide efficient solutions to data gathering problem. Therefore, in order to reduce the power and bandwidth consumption of wireless sensor networks, several mechanisms are proposed such as data aggregation [2].

Data aggregation is a technique widely used because it reduces the number of messages transmitted in the network, and therefore reduce energy consumption and improve the life of the network. However, Sensor networks are typically deployed in hostile environments, therefore, the aggregated data are vulnerable to attacks by adversaries, where a compromised sensor node can either illegally report arbitrary values in the result data. Therefore, an adversary can attack both the confidentiality and the integrity of the data by capturing several number of aggregator nodes that are near the base station. Encryption is the only answer to this problem, but aggregation of encrypted data is needed.

Secure data aggregation schemes can be categorized as hop-by-hop encryption and end-to-end encryption [3]. In hop-by-hop encryption, aggregator nodes must decrypt all sensor data they receive, aggregate the data according to the corresponding aggregation function, and encrypt the aggregation result before sending it to next hop node. In end-to-end encryption schemes, one intermediate node receives the ciphertexts from leaf nodes and then aggregates them with its own encrypted sensor data; the result will finally be sent to a next node. In recent years, some schemes have been proposed focusing on guaranteeing the data privacy during data aggregation phase, but these do not protect the integrity of aggregation data sent to the Base station. For this reason, Message Authentication Code (MAC) protocols are often used to detect false data and protect data integrity.

To solve the problems mentioned above, this paper introduces a novel way to provide confidential and integrity preserving aggregation in wireless sensor networks. The proposed scheme use Elliptic Curve Elgamal to check confidentiality. Inside, the encrypted data is combined with homomorphic MAC to achieve the aggregation data integrity in WSNs.

The main contributions of this paper may be summarized as follows:

- To address the drawbacks of privacy homomorphic cryptography, we focus on the investigation of providing fully homomorphic encryption for end-to-end data confidentiality in WSNs.
- We apply MACs to protect data integrity confidentially and conveniently. In our scheme, sensor nodes compute MACs for the aggregated data and sent it to the aggregator, which can verify the integrity of data through computing and comparing the MACs directly. If the aggregator is captured, the attacker will not know the data.
- The encrypted data is combined with homomorphic MAC to achieve the aggregation data integrity in WSNs.

The rest of the paper is organized as follows: Sect. 2, reports on the related work in the field on secure data aggregation. Section 3 introduces the problem statement. We describe our proposed scheme in Sect. 4, followed by its security analysis and performance evaluation in Sects. 5 and 6. Finally, we summarize our work and conclude the paper and propose some future work in Sect. 7.

2 Related Work

In this section, we present some secure the aggregate schemes.

Katz and Lindell [4] after giving a formal definition of used primitives, presents the following MAC aggregation scheme: each node i sends a message and a tag $tag_i = MAC_K(m_i)$ to aggregator node, which itself concatenates the messages and calculates the XOR of received tags in addition to his. The authors also show the security of their scheme and measure the impact of the tag length on the verification time of a simple message.

He et al. [5] present a reliable data aggregation forwarding scheme for WSNs, based on cluster formation. The overall effect of all measures taken at each phase is clearly visible on the energy spent in the setup step of the scheme. Confidentiality is ensured since all messages are encrypted and there is a group updating key. However, the integrity of personal data is not provided since the attacker can easily modify data between nodes and the aggregator.

Ozdemir and Cam [6] propose a data aggregation protocol based on false data in some sensor node, to ensure the validation of the data sent from the aggregator node to the base station. To ensure the result of the aggregate, the aggregator must provide evidence that get from several witnesses (with false data) nodes. A witness node is a dedicated node for monitoring and also makes him aggregation, but does not transmit the result to the base station.

Ozdemir and Xiao [7] present an hierarchical data aggregation scheme. The authors use an algorithm based on elliptic curve cryptography to ensure the confidentiality. Each cluster has a different public key of the other clusters, to encrypt data. After receiving all messages from members of his cluster, the aggregator aggregates the different encrypted data according to homomorphic property and calculates a MAC on the result.

Papadopoulos and Yun [8] propose a scheme which provides both confidentiality and integrity by combination of homomorphic encryption and secret sharing. It can shield great aggregates and return exact results. This scheme defines an aggregate authentication process to provide integrity and securing against some attacks.

Engouang et al. [9] propose an efficient aggregation of encrypted data. This solution is designed to provide efficiency and confidentiality in WSNs. The authors propose an homomorphic hash while providing security of aggregated data. The basic idea is to replace the XOR operation by a simple modular addition. This solution is robust against passive attacks.

Parmar and Jinwala [10] propose an efficient data aggregation solution based Message Authentication Code (MAC) to provide authentication. The solution use Aggregate MAC (AMAC) to reduce the transmission cost incurred by MAC.

The authors present a cluster based scenario where they can apply AMAC to reduce the number of bits transmitted for authentication.

Table 1 presents a comparison between these schemes and our proposed scheme.

Table 1. The comparison of different schemes with respect security requirements.

Scheme	Confidentiality	Integrity	Integrity method	Encryption type
[4]	No	Yes	MAC	–
[7]	Yes	Yes	MAC	End-to-End
[5]	No	No	–	End-to-End
[9]	Yes	Yes	Hash	End-to-End
[6]	Yes	Yes	MAC	Hop-by-Hop
[8]	Yes	Yes	MAC	End-to-End
[10]	Yes	Yes	MAC	End-to-End
E-SHE	Yes	Yes	Homomorphic MAC	End-to-End

3 System Model and Assumptions

3.1 Network Model

As shown in Fig. 1, a clustering hierarchical heterogeneous WSN architecture consists of three kinds of nodes, i.e. sensors nodes, Aggregator (Cluster Head) and base station (Sink).

Sensors monitor interesting events and periodically send raw readings to the storage nodes with a secure communication channel. The design objective of our scheme is to provide precise data aggregation with modest extra communication overhead to preserve data integrity.

3.2 Adversary Model

We assume that the attacker tries to compromise all aggregators to capture the data and read it. In other words, the attacker tries to break the confidentiality and integrity.

To describe the security of different schemes presented above and our scheme, we define four attacks [11]:

- Unauthorized aggregation: an unauthorized node can perform communications with the network nodes and aggregate more ciphertexts with false data and send them into network.
- Malleability: this attack allows the adversary to alter the contents of an valid encrypted data to deceive the base station.

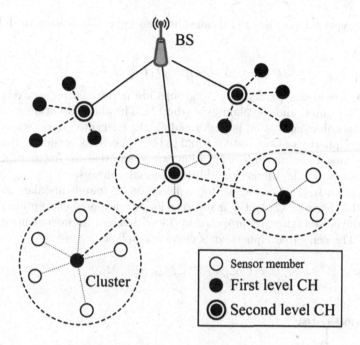

Fig. 1. Network model of our scheme.

- Node compromise: we can have physical access to the sensors due to their dispersion in nature often. An adversary can therefore possibly extract information from these sensors as its physical layout.
- Replay attack: in this case, an adversary save a part of the network traffic, without even understanding the content and replay it later to introduce aggregation error, and therefore, the aggregate results will be affected.

3.3   Assumptions

To preserving privacy, we use homomorphic encryption techniques [12].

♦ Homomorphic encryption

We have seen that the solutions end-to-end would be preferable, because compromise node would not provide any information about the aggregate or data sent by other nodes in the system. The ideal would be to have a solution where each node encrypt data with a key that he shared with the aggregator and with which the intermediate nodes would handle encrypted data without access to the plaintext. To achieve this, we need a homomorphic encryption functions. In homomorphic encryption, aggregation functions (sum, average, etc.) can be applied on the ciphertext. When receiving an encrypted data, each node apply the aggregation function on this ciphertext.

An encryption algorithm $E()$ is called homomorphic if it satisfies the following property:

$$D\left(E(a)\,\Delta_x\,E(b)\right) = D\left(E(a\,\Delta_y\,b)\right) \tag{1}$$

Where operations on Δ_x and Δ_y groups are performed respectively in the ciphertexts group C and the plaintexts group M. The above equation means that the homomorphic encryption is performed on the encrypted data, where in the sum of the ciphertext of two data is equal to the ciphertexts version of the sum of these two data. Homomorphic encryption can be additive and/or multiplicative. In our scheme, article, we use an additive homomorphisms, $\oplus$.

Encryption step for our algorithm is based on 32 rounds unbalanced Feistel network [13]. Figure 2 shows the main steps for our proposed encryption process. (We use the Feistel structure proposed in [13] with major improvements in round process). The generated ciphertext is defined as follow:

$$C = Concat(M_{33}, M_{34}, M_{35}, M_{36}) \tag{2}$$

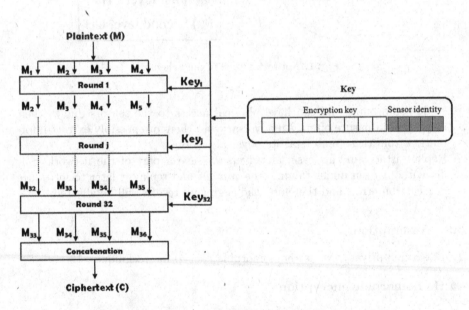

Fig. 2. Encryption process with improvement of that in [13].

♦ Homomorphic MAC

Homomorphic MAC scheme enables a sensor node to use his secret key for generating a tag which authenticates a message M. Given a set of tags tg to authenticate the different parts of message $M : m_1, m_2, , m_n$, the sensor node can generate an homomorphic MAC for this different tag, to generate one output MAC for the message M.

Specifically, given n MACs, $MAC_1...MAC_n$, the aggregated MAC can be computed as:

$$MAC_{agg} = MAC_1 \oplus MAC_2 \oplus ... \oplus MAC_n \tag{3}$$

4 Proposed Scheme

Based on our survey of existing research for ensuring secure data aggregation in WSNs, an Efficient Secure data aggregation scheme based on fully Homomorphic Encryption (E-SHE), is proposed based on scheme [14] with some improvements. E-SHE contains four process: *Key Generation, Sign-Encrypt, Aggregate* and *Verify*.

4.1 Key Generation

Given $\alpha \in Z$, the tuple (q_1, q_2, q_3, E) is generated, where q_1, q_2, q_3 are large primes, E is the set of elliptic curve points. Then, take three points (g_1, g_2, g_3) randomly from E. Compute points $P = q_2q_3*g_1$, $Q = q_1q_3*g_2$, and $H = q_1q_2*g_3$. Such that the order of P, Q and H is q_1, q_2, and q_3 respectively. The generated encryption key Y is:

$$Y = (E, P, Q, H) \tag{4}$$

4.2 Encrypt-Sign Process

Encryption step for our algorithm is based on 32 rounds unbalanced Feistel network [13]. Figure 2 shows the main steps for our proposed encryption process. The generated ciphertext is defined as follow:

$$msg = Concat(M_{33}, M_{34}, M_{35}, M_{36}) \tag{5}$$

After, the sensor computes the signature t_i of (x_i, id_i, msg_i). Our contribution presented as follow:

To formally our schemes, message msg is formed as s segments of l bits. Let $r = 2^l$, then the message space is F_r^s. All contributors and verifiers share one global MAC key that consists of (key_1, key_2). Let K_1 and K_2 denote the key spaces of key_1 and key_2 respectively, and $\mathcal{I}$ denote the space of node identities. Two pseudo random functions are required: $Rd_1 : K_1 \rightarrow F_r^s$ and $Rd_2 : (K_2 \times \mathcal{I}) \rightarrow F_r$.

t_i is computed as follow:

$$a = Rd_1(key_1) \tag{6}$$

$$b_i = Rd_2(key_2, id_i) \tag{7}$$

$$t_i = a.msg_i + b_i \tag{8}$$

At the end, the sensor node i sends the couple $(\mathbf{E(msg_i)}, \mathbf{t_i})$ to aggregator.

4.3 Aggregate Process

The aggregator aggregates $((msg_1, t_1, w_1), ..., (m_j, t_j, w_j))$ as follow:

(i): Aggregated Ciphertext

$$msg' = \sum_{i=1}^{j} w_i.(R_i, T_i) \tag{9}$$

$$= \sum_{i=1}^{j} w_i.msg_i \in F_r^s \tag{10}$$

(ii): Aggregated MAC

$$t' = \sum_{i=1}^{j} w_i.t_i \in F_r \tag{11}$$

At the end, the aggregator sends the aggregated result (msg', t') to the base station BS.

4.4 Verify Process

The Base station (BS) decrypts the aggregate result using its private key, where it needs to inverse the mapping from the point on the elliptic curve to the aggregate result.

While the BS receives (msg_i', t_i') from $Aggregator_i$, it can recover and verify each sensing data as follows:

- BS obtains the m_i by decrypting msg_i'.
- BS verify (x_i, id_i, m_i, t_i'):

$$a = Rd_1(key_1) \in F_r^s \tag{12}$$

$$b = \sum_{i=1}^{j} (w_i.Rd_2(key_2, id_i)) \in F_r \tag{13}$$

If $a.m_i + b = t_i'$ then the result is accepted, else the result is refused.

5   Security Analysis

In this section, we analyze our scheme and some schemes against the attacks mentioned above.

5.1   Security Requirements

In WSNs using data aggregation, end-to-end security services, namely, confidentiality and integrity are widely desired.

✣ Data confidentiality

Denote B the Network security factor. In our scheme, each MAC generated by the aggregator has the size of $32/(B+1)$ bits (the data packet reserves 4 bytes for MAC), and $B+1$ aggregators MACs calculated by $B+1$ sensor nodes form a MAC. If an attacker find all the $B+1$ aggregators MACs, he can successfully forge a valid MAC with the low probability of 1 in $2^{32/(B+1)}$ for each aggregator MAC. Thus, in E-SHE the confidentiality is achieved without the need to use any other function.

✣ Data integrity

End to end integrity is provided by using the Homomophic MAC aggregation. While authors in [6,9], show that the active adversary can make multiple attacks especially in the case of a homomorphic encryption. Indeed, it is known that this type of encryption is very vulnerable to active attacks because of its homomorphic properties.

Theorem 1. The tag MAC transmitted toward the second step is secure.

Proof. The security is provided assuming that the MAC are affected. Suppose that E-SHE uses an unforgeable MAC solution, adversary cannot successfully forge messages in transmission step of homomorphic tags. Thus, E-SHE preserve integrity by forwarding selective packets. Beside, the uses of homomorphic MAC provide the freshness of different messages after defined time interval. When the aggregator receives a false data, it fails the MACs verification process to aggregator for lower levels. Furthermore, BS can verify the integrity of all messages and authenticates transmitters using its private key, which that never expires, compared to other solutions.

5.2   Security Against Attacks

✣ Unauthorized aggregation

An adversary can also inject bad data by replacing existing data by forged data. E-SHE uses different keys for each message with a homomorphic MACs. Therefore, the adversary can not generate an encrypted data and a valid MAC without knowing the current key.

✤ Malleability

Malleability allows the adversary to alter a valid encrypted data to deceive the base station. Indeed, the adversary can modify encrypted data without necessarily know the contents and this can be done by simply adding a number to an encrypted data in our case. E-SHE provides end-to-end integrity allowing the base station to verify individual data and authenticate the transmitters. Therefore, if an encrypted data is altered, the verification will fail and the corresponding aggregated data will be rejected. The base station starts the procedure for identifying the malicious sensor node.

✤ Node compromise

In E-SHE, the base station is able to detect the malicious node through the identification process for compromise nodes. If the aggregator is captured, the later can not access to plaintexts data of its members, and can not deceive the base station by a malicious message because the corresponding keys are unknown to the adversary. In other words, a compromised aggregator can only edit his own message to try to deceive the BS.

✤ Replay attack

The replay attack is to send a valid packets already transmitted to produce an unauthorized action. E-SHE uses different encryption and authentication keys from one message to another. For example, after a number of aggregation levels L, the adversary replays the packet of round i of the aggregation phase. It is obvious that the verification process fails because the base station uses the key of round $(L+i)$ to verify the data, because the key to the i-th round i is expired. Beside, E-SHE is vulnerable against this attack, thanks to the homomorphic encryption based MAC and the factorization of large integers.

6   Numerical Results and Discussion

In our experiments, we use TinyOS 2.0 simulator (TOSSIM) [15]. Energy is an important constraint in WSNs, so we use TinyOS simulator PowerTOSSIM [16] a power modeling extension to TOSSIM. The implementation was done on the MicaZ mote, which is a typical device for WSNs that equipped with 8-bit processor. We use the TinyECC library [17], which is implemented in nesC programming language.

The simulation parameters are presented in Table 2. We evaluate our algorithm against iPDA [5], EIRDA [9], and AMAC [10] schemes in terms of communication overhead and energy consumption.

Table 2. Simulation parameters.

Parameter	Value
Number of sensor nodes	20–300
Transmission range	30 m
Area size	400 m × 400 m
Transmit power	0.720 mw
Receiving power	0.405 mw
Initial energy	6.3 J
Packet size	45
Noise floor	−105 dB
Simulation time	500 s

6.1   Communication Overhead

Communication overhead shows the execution time for the sensors to generate ciphertexts and MAC before transmit it to the next level. Aggregation delay is also evaluated by measuring delay indented on aggregating ciphertexts and MAC. We compare the communication overhead among our E-SHE, EIRDA, iPDA, and AMAC. In order to make the comparison clearly, we evaluate the communication overhead by the sum of sending bytes of all sensor nodes during aggregation phase for these four schemes. We run every experiment over 20 times and check the average value, in order to reduce the relative error. The results are shown in Fig. 3, depending on the number of sensor nodes, which shows that when the number of transmissions at sensor nodes increased, the efficient of data aggregation of E-SHE is increased compared to the three others schemes. This is

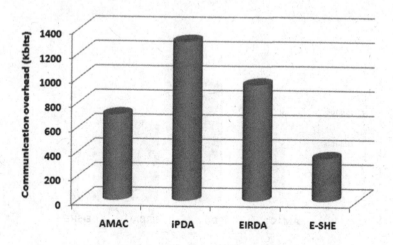

Fig. 3. Communication overhead.

due to the fact that if the density of the network increases, each data aggregator is able to aggregate data for more sensors. For example, for 100 sensors, the base station must wait about 0.9 s to receive the aggregated encrypted data using EIRDA, 0.65 s using AMAC and then it must only wait 0.2 s with E-SHE. In addition, the homomorphic MAC records 1000 bytes less memory compared to that used in EIRDA.

6.2   Energy Consumption

Energy consumption is the central issue in WSNs. The calculation and communication are two aspects that have a direct impact on energy consumption and therefore the life of the sensor nodes. The energy consumption of our cryptographic functions can be calculated and are shown in Table 3.

Table 3. Estimated energy for E-SHE operations.

Operation	Energy consumption (mj)
Key generation	$10.82\,mj$
Encryption	$0.547\,mj$
Homomorphic primitives	$0.02\,mj$

These measures are added to our energy model and we conduct our simulations for the four studied solutions. It is clear from Fig. 4 that E-SHE offers a significant reduction in energy consumption compared to other solutions. This gain can be explained by the fact that far fewer computational load is engaged in our algorithm, because of the use of Feistel encryption primitives with some

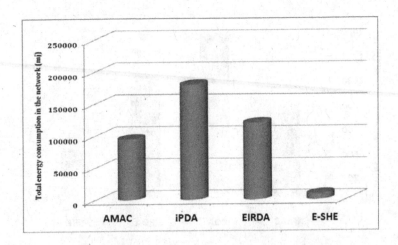

Fig. 4. Total energy consumption in the network.

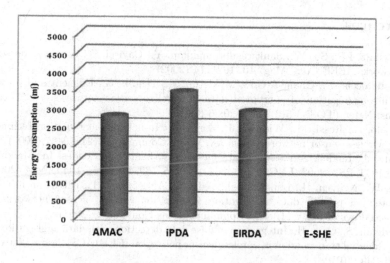

Fig. 5. Energy consumption by sensor nodes.

improvements. Therefore, for the same level of provided security, the network lifetime is significantly improved. Figure 5 shows that the average amount of energy consumed by sensor nodes is considerably reduced in our algorithm. This is due to that costly operations, in E-SHE, are only performed in the transmission phases. Furthermore, in addition to participating in the Feistel process, the sensor nodes perform the different functions. In EIRDA and AMAC, this operation requires operations on the elliptic curve as it only requires a small amount of calculations in E-SHE (modular addition).

7   Conclusions and Future Work

Data aggregation is an important method to reduce the energy consumption in wireless sensor networks (WSNs). However, in hostile environments, the aggregated data can be subject to several types of attacks, and provide security is necessary. The focus of our work is on a scheme for a confidential and integrity data exchange in WSNs that support data aggregation.

In this paper, we present secure data aggregation scheme that provides secure data integrity and security assumptions. The proposed scheme is based on an additive homomorphic encryption algorithm that allows aggregation on encrypted data, combined with homomorphic MAC. However, confidentiality and data integrity are provided by this scheme. The security analysis and performance evaluation shows that our scheme is able to resist against various attacks such as compromise node attacks and Unauthorized aggregation. In our future work, several improvements can be provided and could lead to a further reduction in power consumption.

References

1. Akyildiz, I.F., Su, W., Sankarasubramaniam, Y., Cayirci, E.: A survey on sensor networks. IEEE Com. Mag. **40**, 102–114 (2002)
2. Castelluccia, C., Chan, A.C.F., Mykletun, E., Tsudik, G.: Efficient and provably secure aggregation of encrypted data in wireless sensor networks. ACM Trans. Sens. Netw. (TOSN) **5**(3), 20:1–20:36 (2009)
3. Fasolo, E., Rossi, M., Widmer, J., Zorzi, M.: In-network aggregation techniques for wireless sensor networks: a survey. Wirel. Commun. **14**(2), 70–87 (2007)
4. Katz, J., Lindell, A.Y.: Aggregate message authentication codes. In: Malkin, T. (ed.) CT-RSA 2008. LNCS, vol. 4964, pp. 155–169. Springer, Heidelberg (2008)
5. He, W., Nguyen, H., Liu, X., Nahrstedt, N., Abdelzaher, T.: iPDA: an integrity-protecting private data aggregation scheme for wireless sensor networks. In: Proceedings of IEEE Military Communications Conference (2008)
6. Ozdemir, S., Cam, H.: Integration of false data detection with data aggregation and confidential transmission in wireless sensor networks. IEEE/ACM Trans. Netw. **18**, 736–749 (2010)
7. Ozdemir, S., Xiao, X.: Integrity protecting hierarchical concealed data aggregation for wireless sensor networks. Comput. Netw. **55**, 1735–1746 (2011)
8. Papadopoulos, S., Kiayias, A., Papadias, D.: Exact in-network aggregation with integrity and confidentiality. IEEE Trans. Knowl. Data Eng. **24**(10), 1760–1773 (2012)
9. Engouang, T.D., Yun, L.: Aggregate over multi-hop homomorphic encrypted data in wireless sensor networks. In: Proceedings of the 2nd International Symposium on Instrumentation and Measurement, Sensor Network and Automation (2013)
10. Parmar, K., Jinwala, D.C.: Aggregate MAC based authentication for secure data aggregation in wireless sensor networks. In: Huang, D.-S., Jo, K.-H., Wang, L. (eds.) ICIC 2014. LNCS, vol. 8589, pp. 475–483. Springer, Heidelberg (2014)
11. Perrig, A., Stankovic, J., Wagner, D.: Security in wireless sensor networks. Commun. ACM **47**, 53–57 (2004)
12. Parmar, K., Jinwala, D.C.: Symmetric-key based homomorphic primitives for end-to-end secure data aggregation in wireless sensor networks. J. Inf. Secur. **6**(1), 38–50 (2015)
13. Shi, Y., He, Z.: A lightweight white-box symmetric encryption algorithm against node capture for WSNs. In: 2014 IEEE Wireless Communications and Networking Conference (WCNC) (2014)
14. Zhou, Q., Yang, G., He, L.: An efficient secure data aggregation based on homomorphic primitives in wireless sensor networks. Int. J. Distrib. Sens. Netw. (2014)
15. Levis, P., Lee, N., Welsh, M., Culler, D.: TOSSIM: accurate and scalable simulation of entire TinyOS applications. In: Proceedings of the 1st ACM International Conference on Embedded Networked Sensor Systems (SenSys) (2003)
16. Shnayder, V., Hempstead, M., Chen, B., Welsh, M.: PowerTOSSIM: efficient power simulation for TinyOS applications. In: Proceedings of ACM International Conference on Embedded Networked Sensor Systems (SenSys) (2004)
17. Liu, A., Ning, P.: TinyECC: a configurable library for elliptic curve cryptography in wireless sensor networks. In: Proceedings of the 7th International Conference on Information Processing in Sensor Networks (IPSN) (2008)

VANET and ITS

SADP: A Lightweight Beaconing-Based Commercial Services Advertisement Protocol for Vehicular Ad Hoc Network

Kifayat Ullah[1(✉)], Luz M. Santos[1,2],
João B. Ribeiro[1], and Edson D.S. Moreira[1]

[1] ICMC, University of São Paulo (USP), São Carlos Campus, São Paulo, Brazil
{kifayat,lsantosj,edson}@icmc.usp.br, joao.b@usp.br
[2] University of Pamplona, Pamplona, Colombia

Abstract. Vehicular Ad hoc Network (VANET) is an auspicious technology for the future Intelligent Transportation System (ITS). A large number of safety applications are envisioned for VANET. Nonetheless, non-safety applications, more specifically advertisement of roadside commercial services, still need a great deal of attention. The classical methods of advertisements (e.g., billboards) are not adequate for the business owners (for instance: hotel, restaurant, supermarket and gas station) to promote their commercial services to the nearby travellers (drivers or passengers) on highways. We address this problem by proposing SADP—a lightweight beaconing-based commercial services advertisement protocol for VANET. Our solution is based on the Wireless Access in Vehicular Environments (WAVE) architecture. SADP performance was evaluated in a congested multi-lane highway scenario, by an extensive set of experiments. Simulation results conclude that: (a) the Road Side Unit (RSU) broadcast frequency and vehicles speed affect the reception of advertisement beacon packets; (b) a frequency of 10 advertisement beacon packets per second assures 100 % reception of all the advertised services by all vehicles and could be the basis for a premium-like type of service advertisement plan for the business owners.

Keywords: WAVE · IEEE 802.11p · IEEE 1609.4 · VANET · I2V communication · Beaconing · Services advertisement

1 Introduction

Vehicular Ad hoc Network (VANET) is a special kind of Mobile Ad hoc Network (MANET) [14], with distinct features like dynamic network topology, restricted and high mobility of vehicles, short inter-contact time between vehicles and roadside infrastructure, different operational scenarios and real-time data exchange requirements. The communication among the vehicles is named Vehicle to Vehicle (V2V) communication. Moreover, the communication between the roadside infrastructure and the vehicles is called Infrastructure to Vehicle (I2V) communication.

© Springer International Publishing Switzerland 2016
N. Mitton et al. (Eds.): ADHOC-NOW 2016, LNCS 9724, pp. 279–293, 2016.
DOI: 10.1007/978-3-319-40509-4_20

In the last two decades, VANET gains a great deal of attention from various standardization organizations, network communities, government departments and auto-mobile industries. In the U.S. the Institute of Electrical and Electronics Engineers (IEEE) has developed Wireless Access in Vehicular Environments (WAVE) architecture for VANETs [16]. Furthermore, in Europe, the European Telecommunications Standards Institute (ETSI) standard for VANET is known as ETSI ITS-G5 [4]. Additionally, in Japan the VANET related standardization efforts appear in the Association of Radio Industries and Businesses (ARIB) standard called ARIB STD-T75 [1].

VANET is playing a major role in deployment of a broad range of safety applications. The primary objective of these applications is to assure the safety of travellers,[1] by eliminating the hazard of road accidents. Moreover, to provide comfort and entertainment to the passengers and to make their journeys more pleasurable, some non safety applications has been envisioned. Nevertheless, the dissemination of roadside commercial services to the drivers is an unexplored and challenging task. The Business Owners (BOs) of hotels, supermarkets, petrol stations, coffees shops and restaurants would be interested in advertising their services to the drivers for making more revenues. On the other side, the drivers would like to discover these commercial services during their trips to other cities and states.

In order to address this problem, we proposed a novel lightweight beaconing-based application layer Service Advertisement Protocol (SADP). We believe that BOs would benefit from our solution by promoting their commercial services to on-road customers in a cost efficient way.

The reminder of this paper is organized as follows. Section 2 gives an overview of the related work. Our proposed commercial services advertisement protocol is detailed in Sect. 3. Section 4 explains the system model with its main entities. The simulation tools, parameters and scenario are explained in Sect. 5. Section 6 illustrates the results and discussions. Finally, Sect. 7 concludes this paper with work in progress and some future directions.

2   Related Work

In this section, work related to services advertisement, discovery and incentive mechanisms in VANET is reviewed.

In [19], the authors proposed a multi-hop advertisement dissemination and forwarding scheme. They used incentive mechanism to motivate selfish nodes and dynamic window technology to prevent malicious nodes from launching Denial of Service (DoS) attacks. The business owners need to offer virtual cash to all the advertisement disseminating vehicles, which would be impractical and expensive. They used NS-2 simulator to evaluate the performance. However, they did not mention the parameters and matrices used. Moreover, they did not consider WAVE protocols stack for their implementation.

---

[1] We use the words travellers, drivers and passengers interchangeably.

The authors in [10], presented Address Based Service Resolution Protocol (ABSRP). ABSRP is an Road Side Unit (RSU) dependent service discovery protocol for VANET. Vehicle asking for a service sends a request towards its leader RSU, which would send it towards the target, only if it knows the IP address. Otherwise, the RSU broadcast the request to other RSUs using Internet. The responsible RSU then reply with response message, either by using VANET or backbone Internet. ABSRP would be infeasible for high-speed vehicles having short inter-contact time with RSUs. Moreover, the solution is based on UDP protocol (and not WAVE Short Message Protocol) at the transport layer. Additionally, they used IEEE 802.11a interface card for wireless communications.

In [9], FleaNet—A virtual flea market to buy and sell goods over VANET— was presented. RSU or vehicles propagate the advertisements, using IEEE 802.11b interface cards. On the other hand, vehicle looking for a service will periodically send query message toward its neighbor vehicles. Upon receiving this query message, vehicle search its local database. In case a match is found, the originator vehicle of the query message is informed. Like previous solutions, FleaNet does not consider the VANET standard architecture.

In [5], the authors proposed a greedy algorithm for selecting Roadside Access Points (RAPs) by the business owners. The business owners rent RAPs for advertising their services to the drivers. These RAPs are installed and controlled by RAP Service Provider (RSP). The main goal of this work was the reduction of advertisement cost by selecting least number of RAPs. The correctness of the algorithm was verified by theoretical analysis and the performance was evaluated using simulations. However, it was not mentioned which wireless technology was used for communications.

A secure incentive-based advertisement dissemination framework called Signature Seeking Drive (SSD) was proposed in [8]. They used Public Key Infrastructure (PKI) for V2V communications among vehicles. This PKI is provided to each vehicle by an authorized certificate authority. Upon successful forwarding of the advertisements to neighbour vehicles, a digitally signed receipt from the vehicle is received. The drivers could use these receipts as a virtual cash at locations such as petrol stations. The main disadvantage of the proposed incentive scheme is the cost. It would be expensive to offer incentives to each advertisement disseminating and receipt providing vehicle. Furthermore, they do not considered the VANET standard protocols.

In a recent study in [15], the problem of roadside services advertisement and discovery in VANET was partly addressed. The authors proposed a beaconing-based scheme to discover the advertised services in a V2V communication scenario. The solution is based on three phases, i.e., service advertisement phase, service query phase and service response phase. They evaluated the performance of these phases using simulations. For service advertisement phase, they increased the number of RSUs offering the services (from 1 RSU to 7 RSUs). Additionally, for service query and response phases, they increased the number of vehicles asking and replying for querying (from 1 vehicle to 21 vehicles). Authors used the WAVE architecture for VANET, but they did not defined the proper packet format for advertisement, query and response packets. Moreover, they did not consider the multi-channel operation of IEEE 1609.4 protocol [3].

Aside from [15], none of the cited works considered the WAVE protocols standard. Hence, our proposed solution is different in various ways: (a) we add our solution on top of WAVE protocols stack, while considering the IEEE 802.11p and IEEE 1609.3 standard protocols; (b) we take into consideration the multi-channel operation of IEEE 1609.4 standard protocol [7]; (c) our protocol is based on text-based beaconing technology and is, therefore, a lightweight solution; (d) we specify the proper format of the advertisement beacon packet and define all the required parameters for repeating the experiments in the future. In this context, the main contributions of our work are:

1. The design of a lightweight beaconing-based application layer commercial services advertisement protocol for VANET.
2. Implementation of SADP in a congested highway scenario, while considering the WAVE architecture and standard VANET protocols.
3. Performance evaluation of SADP by an extensive set of simulations.

3   Service Advertisement Protocol (SADP)

Advertisement is the most popular form of marketing. The statistics portal reported that in 2015 the global investment on advertisements was almost 600 billion U.S. dollars, and the prediction for 2018 is 667.65 billion U.S. dollars [13]. Driven by today's modern advertisement methods (e.g., email marketing, web banners, pop up ads, mobile advertisement and in-store advertisement), we introduce the concept of a novel beaconing-based commercial services advertisement protocol for VANET. Some key features of our protocol are discussed in this section.

3.1   Based on IEEE WAVE Architecture

SADP is based on the IEEE WAVE standard protocols stack. At the transport and network layers, it makes use of WAVE Short Message Protocol (WSMP) [16]. WSMP is an alternative for the TCP/UDP and IP traffic. It is defined in the IEEE 1609.3 standard. Additionally, at the MAC and PHY layers, SADP benefits the IEEE 802.11p and IEEE 1609.4 standard protocols.

3.2   Beaconing-Based Technology

In the context of VANET, a beacon refers to a short text-based message, transmitted by each node (RSU or Vehicle) using single-hop communication at a regular interval of time. Generally, a beacon contain status information such as speed, heading, acceleration and position of a node. Other terms used for beacon are heartbeat message, Wave Short Message (WSM) and Cooperative Awareness Message (CAM). Beacons are initiated by the application layer. While, at the transport and network layers beacons are handled by the WSMP. For convenience, we referred to these beacons as Safety Beacons (SBs). Likewise, we introduce –for the first time– the concept of Advertisement Beacon (AB). RSU would broadcast the registered commercial services using AB packets at a regular interval.

3.3 Multi-channel Environment

In the United States, the Federal Communications Commission (FCC) has allocated 75 MHz of bandwidth in the 5.9 GHz band for the exclusive use of VANET. The allocated spectrum consists of 5 MHz guard band and seven channels of 10 MHz each. The most important channel is the Control Channel (CCH). The remaining six channels are termed as Service Channels (SCHs). Each channel is labelled with a number. The channel number 178 (i.e., CCH) is devoted to the use of safety applications, including WAVE Service Advertisement (WSA) [7]. In accordance with IEEE1609.4 standard, we adopted the channel number 178 for the exchange of SB packets (10 times per second). On the other side, we use the channel number 174 for the broadcast of AB packets (Fig. 1). Hereof, our protocol supports the multi-channel operation. More details on this are provided in the IEEE 1609.4 standard [6].

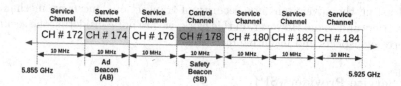

Fig. 1. Multi-channel operation

3.4 Advertisement Beacon (AB) Packet Format

The information about the advertised commercial services is encapsulated in the AB packet. This information encompasses the fields: message type, identity (ID), category, Time To Live (TTL), name, description, longitude, latitude and address. The general format and fields of the AB packet is shown in the Fig. 2. The size of the packet was limited to 256 Bytes.

1	1	1	1	256 Bytes				
				56	100	4	4	88
Message type	ID	Category	TTL	Name	Description	Longitude	Latitude	Address

Fig. 2. Advertisement Beacon (AB) packet format

4   System Modelling

In this section, we explain the main entities of our proposed system.

4.1   Business Owner (BO)

Business Owner (BO) is an entity owning a business (e.g., a restaurant, hotel, gas station, coffee shop, supermarket and gift centre) along highways. The BO demands to advertise its commercial services (such as food menu, price list, hotel facilities, special offers, discounts and new arrivals) to the nearby passengers with a third-party broker. However, only text-based commercial services would be permitted to advertise using SADP.

4.2   Service Advertising Agency (SAA)

The Service Advertising Agency (SAA) is a third party broker. The SAA would be responsible for the instalment of RSUs in its vicinity. The BO interested to advertise its services would require to register it with the SAA using Service Level Agreements (SLAs). Besides, the SAA would prepare different advertisement policies for the BOs. These policies include the number and location of RSUs to advertise the commercial services, lifetime of the advertised services, frequency of broadcasting the advertisements, security of transactions, incentive mechanism for the users and so on. However, in this work, we are not considering the security and privacy aspects and incentive mechanism.

4.3   Service Provider (SP)

In our model, the RSU[2] performs like a Service Provider (SP). SP is the central component, responsible for the broadcast of commercial services advertisement, using AB packets in an I2V fashion. We adopt push-based communication strategy for this purpose. The SPs would be installed alongside the roads and highways, under the administration of the SAA. We assumed that these SPs have storage and processing capabilities. Furthermore, SP would be equipped with two interfaces: (a) a fixed interface connected with the wired Internet; (b) IEEE 802.11p based wireless interface for I2V communication. Using the wireless interface, the SP would broadcast the AB packets to its neighbour vehicles, several times per second.

4.4   Service User (SU)

The travellers (drivers and passengers) on highways are termed as Service Users (SUs). We supposed that all the SUs has smarts vehicles, equipped with telematics and navigation system (for example: GPS). Furthermore, there are built-in On Board Units (OBUs) with storage and processing capabilities. Moreover, for

---

[2] We use the words RSU and SP interchangeably.

communicating with the SP, all the vehicles are equipped with IEEE 802.11p based wireless interface cards. Thereafter, entering into the coverage area of a SP, the vehicle starts receiving the advertised commercial services (AB packets). The SU might avail a service or it would store the AB packets, for a pre-defined Time to Live (TTL).

5   Simulation Setup

This section explains the simulation scenario, tools and parameters.

5.1   Simulation Scenario

For the purpose of performance evaluation of SADP, we assume a multi-lane double-sided congested highway scenario. The length of the highway is 1 Km. A total of 384 vehicles (approximately 96/km) was allowed to take part in each experiment. We assumed that all the vehicles travel at a constant speed in two different directions (East-West and vice-versa). Moreover, the vehicles depart from two different origins in a uniform fashion. Likewise, they leave the scenario at two different destinations. We install a single SP (RSU) in the middle of the scenario having a communication range of approximately 400 m. The simulated scenario is depicted in Fig. 3. The SP is connected to the SAA using wired Internet. Additionally, it regularly communicates with vehicles using I2V communications. We registered 80 different commercial services of five different categories (i.e., hotel, restaurant, coffee shop, gas station and supermarket) with the SP. The SP uses a single AB packet to advertise each of the registered service. The SP broadcast these services on channel number 174 (i.e., SCH) at a regular time interval. When a SU (with a smart vehicle) enters into the coverage area of a SP, it starts to receive the advertised AB packets. The vehicle would stores these AB packets in the local cache for a predefine TTL value. However, in case of a duplicate AB packet the vehicle will simply discard the packet. Additionally, vehicles send and receive SBs (10 times per second) to other vehicles in a V2V fashion on channel number 178 (i.e., CCH).

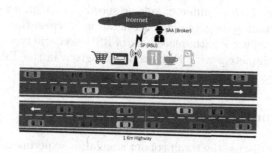

Fig. 3. Multi-lane highway scenario

5.2 Simulation Tools

In the domain of computer networks, the use of simulators is highly applicable for implementation, comparison, testing and performance evaluation of protocols. Consequently, to implement and evaluate the performance of SADP, we rely on Simulation of Urban Mobility (SUMO) [2]. SUMO is an open source microscopic vehicular mobility simulator. SUMO generates realistic traffic flows on a microscopic level. Realistic traffic flows are very important in VANET, as the performance evaluation is highly influenced by the mobility of vehicles.

Apart from SUMO, we also use an open source discrete event network simulation called Objective Modular Network Testbed in C++ (OMNET++) for communication purpose [17]. Inside OMNET++, each vehicle and SP is represented by a node. Likewise, a good choice for performing VANET simulations is Vehicles in Network Simulation (Veins) framework [11]. Veins uses Traffic Control Interface (TraCI) [20] to connect OMNET++ with SUMO using a TCP connection. The IEEE WAVE protocols stack (e.g., WSMP, IEEE 802.11p, IEEE 1609.3 and IEEE 1609.4) is available inside Veins. Figure 4 shows these simulations components.

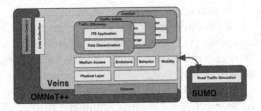

Fig. 4. Simulation tools [18]

5.3 Simulation Parameters

Parameters related to vehicles properties and highway scenario were defined in SUMO. One of the utmost important factor is the vehicle speed. We carried out our experiments with three different vehicle speeds, i.e., high, medium and low. Additional important parameters are acceleration, deceleration, minimum gap between vehicles, vehicle length, placement of RSU, scenario type and length. All SUMO related parameters with their values are given in the Table 1. Likewise, parameters relevant to OMNET++ are shown in Table 2.

6   Results and Discussions

In this section, we discuss the results of our simulation experiments. Experiments were divided into three parts and henceforth, we explain them separately in the following subsections. Additionally, each of the experiment was further divided into four cases. In the first case, we used a broadcast frequency of 0.5/s. In the

Table 1. SUMO parameters and their values.

Parameter	Value
Scenario type	Multi-lane double highway
Highway length	1 km on each side
Number of vehicles	384 vehicles
Vehicles density	Congested (96 vehicles/km)
Vehicles speed	High = 120 km/h, Medium = 90 km/h, Low = 60 km/h
Vehicles length	5 m
Vehicles acceleration	2.6 m/s
Vehicles deceleration	4.5 m/s
Minimum gap	2.5 m
SP(RSU) position	500 m (middle of scenario)

Table 2. OMNET++ parameters and their values.

Parameter	Value
Simulation time	300 s for each experiment
Number of runs	5 runs for each experiment
AB packet size	256 Bytes
AB packet broadcast frequency	0.5/s, 1/s, 5/s and 10/s
SB packet broadcast frequency	10/s
Number of services	80 services (5 categories)
Transport &network layer protocol	WSMP
MAP &PHY layer protocol	IEEE 802.11p
Channel bandwidth	10 MHz
Frequency band	5.9 GHz
Communication range (RSU &Vehicles)	400 diameter
Radio propagation model	Two Ray Interference Model [12]
Transmission rate	6 Mbps
Transmission power	3.4 mW
Channel priorities	CCH(178) = 3, SCH(174) = 2
Channel used	178 for SB packets & 174 for AB packets

second case, we used 1/s. In the third case, we used 5/s. While, in the final case, we used a broadcast frequency of 10/s. We determined the percentage of vehicles that received the AB packets from RSU. Besides, we also examined the effect of vehicles speed and broadcast frequencies of AB packets on the performance.

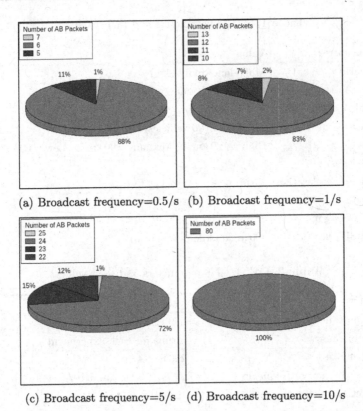

(a) Broadcast frequency=0.5/s   (b) Broadcast frequency=1/s

(c) Broadcast frequency=5/s   (d) Broadcast frequency=10/s

Fig. 5. Percentage of vehicles with received AB packets in a high-speed scenario (Color figure online)

6.1 Evaluation of AB Packets in a High-Speed Vehicles Scenario

In the first series of experiments, we evaluated the performance of our protocol in terms of AB packets reception in a high-speed (120 km/h) vehicles scenario under four broadcast frequencies (i.e., 0.5/s, 1/s, 5/s and 10/s). The results collected are shown in Fig. 5(a)–(d).

For a broadcast frequency of 0.5/s (i.e., sending an AB packet after every two seconds), about 88 % of the vehicles received only 6 AB packets (out of 80 broadcast AB packets). It means that a maximum of 6 different roadside commercial services were received by most of the vehicles. The results of the first case is illustrated in Fig. 5(a). In Fig. 5(b), we observed 100 % performance improvement for the second case (i.e., for a broadcast frequency of 1/s). A total of 83 % of vehicles received 12 AB packets. The performance further improves after increasing the broadcast frequency of AB packets to 5/s (third case). As depicted in Fig. 5(c), most of the vehicles (72 %) received 24 AB packets. However, the AB packets reception, in all these three cases, is still low. The reason for this low performance is the time spend by RSU between sending AB packets. On the other side, the high-speed of vehicles (120 km/h) also affect the performance, as the vehicles has small inter-contact time with the RSU. In the final case, we further increase the broadcast frequency of AB packets (i.e., 10/s). As a result, 100 % of vehicles received all the advertised AB packets (80 advertisements). This is shown in Fig. 5(d). The reason for this outstanding performance is obviously the frequent broadcasts of AB packets by the RSU. In this case, vehicles has more chances to receive the AB packets, even if they missed it in the first chance, for instance because of congestion at MAC layer.

6.2   Evaluation of AB Packets in a Medium-Speed Vehicles Scenario

Likewise the first set of experiments, we evaluated the performance by considering the same broadcast frequencies for AB packets (0.5/s, 1/s, 5/s and 10/s). However, we changed the vehicles speed to 90 km/h (medium-speed vehicles). The results collected are shown in Fig. 6(a)–(d).

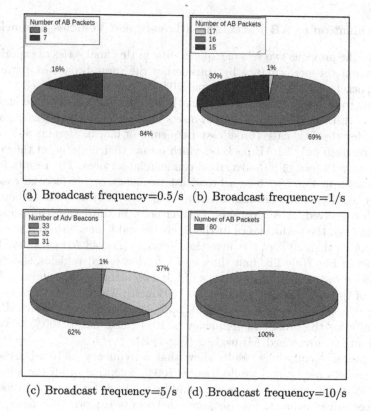

(a) Broadcast frequency=0.5/s   (b) Broadcast frequency=1/s

(c) Broadcast frequency=5/s   (d) Broadcast frequency=10/s

Fig. 6. Percentage of vehicles with received AB packets in a medium-speed scenario (Color figure online)

In the first case (broadcast frequency of 0.5/s), 84 % of the vehicles successfully received only 8 AB packets, which means most of the vehicles received only few of the advertised services. This case is shown in Fig. 6(a). However, for the second case (i.e., broadcast frequency of 1/s), we observed a 100 % performance improvement and 69 % of vehicles received 16 AB packets. The results are depicted in Fig. 6(b) accordingly. Furthermore, after increasing the broadcast frequency of AB packets to 5/s (third case), about 62 % of the vehicles received 31 AB packets. This case is illustrated in Fig. 6(c). As compared to the previous set of experiments, the performance (in terms of AB packets reception) is higher (even for same broadcast frequencies). The main reason for this performance improvement is the speed of the vehicles. In this case, all the vehicles move with a uniform speed of 90 km/h (medium-speed) and has more inter-contact time with RSU. We further increase the broadcast frequency of AB packets to 10/s (forth case). Likewise, the forth case of previous experiment, herein 100 % of vehicles received all the advertised AB packets (80 advertisements). The results of this case is shown in the Fig. 6(d).

6.3   Evaluation of AB Packets in a Low-Speed Vehicles Scenario

Similar to the previous two sets of experiments, in this final series of experiments we evaluated the performance by considering the same broadcast frequencies for AB packets (i.e., 0.5/s, 1/s, 5/s and 10/s). However, we changed the vehicles speed to 60 km/h (low-speed vehicles). The results collected are shown in Fig. 7(a)–(d). Likewise the previous two sets of experiments, the performance is very low for the first case (broadcast frequency of 0.5/s). Herein, 86 % of the vehicles received only 12 AB packets, which means that majority of the vehicles received only 12 (out of 80) advertised commercial services. The results for this case is shown in Fig. 7(a). For a broadcast frequency of 1/s (second case), we observed 100 % performance improvement (Fig. 7(b)) over first case. About 73 % of vehicles received 24 AB packets. Furthermore, in third case (broadcast frequency of 5/s), the performance further increase and hence, 45 % of the vehicles received more than 50 % of the advertised services (i.e., 48 AB packets). This is illustrated in Fig. 7(c). The first three cases, for low-speed vehicles, show better performance over the first three cases of last two series of experiments. This is because of the low speed of vehicles (i.e., 60 km/h), which allow them to spend more time in the coverage range of RSU and as a result receive more AB packets. Finally for the broadcast frequency of 10/s (forth case), 100 % of vehicles received all the advertised AB packets (Fig. 7(d)).

The overall simulations results show that a frequency of 10 advertisement beacon packets per second would assures 100 % reception of all the advertised services by all the vehicles and hence could be the basis for a premium-like type of services advertisement. Furthermore, the results explore how the speed of the vehicles and the broadcast frequencies affect the success rate of AB packet receptions.

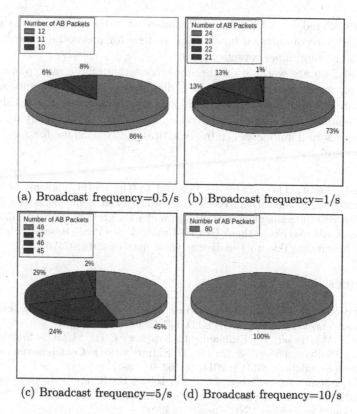

(a) Broadcast frequency=0.5/s   (b) Broadcast frequency=1/s

(c) Broadcast frequency=5/s   (d) Broadcast frequency=10/s

Fig. 7. Percentage of vehicles with received AB packets in a low-speed scenario (Color figure online)

7   Conclusion and Future Work

In this paper, we proposed SADP—a novel lightweight beaconing-based road-side commercial services advertisement protocol for VANET. SADP operates in a multi-channel VANET environment and take advantage of beaconing technology and store-carry-and-forward mechanism. The main entities of the system are: business owners, service advertising agency, service provide and service user. We also reveal a complete list of simulation parameters for repeating the experiments in the future. We implement SADP on top of WAVE architecture and perform extensive simulations using SUMO, OMNET++ and Veins simulators. We evaluate the performance of the protocol in a congested highway scenario under different parameters. From the results we noticed that all the vehicles received 100 % (80 different services about five categories) of the advertised services, for a broadcast frequency of 10/s. This is important for the BO, as they would be interested that none of the driver has missed the opportunity to receive their advertised services. Additionally, it gives a basis for the entities, involved in roadside commercial services advertisement, to assess the several new possibilities offered by VANET and formulate new business plans to explore them. SADP is recommended for business owners for promoting their roadside services and earning more revenues.

Currently, we are working on the implementation of the next phase of the protocol i.e., commercial roadside services discovery using V2V communications. In the future, we would study incentives schemes to motivate the selfish nodes in the dissemination of commercial services. Moreover, we are studying the use of pseudonyms and public key infrastructure (PKI) schemes for making our protocol secure.

Acknowledgement. The authors are thankful to The World Academy of Sciences (TWAS), Italy and to the national council for scientific and technological development (CNPq), Brazil for providing financial support under the grant number 190275/2011-1. The authors would also like to thank CEPID-CeMEAI, Sao Paulo Research Foundation (FAPESP) for funding this work under the grant number 2013/07375-0.

References

1. Association of Radio Industries and Businesses: Dedicated short-range communication system, version 1.0, ARIB STD-T75 (2001)
2. Behrisch, M., Bieker, L., Erdmann, J., Krajzewicz, D.: SUMO - Simulation of Urban MObility: an overview. In: The Third International Conference on Advances in System Simulation, SIMUL 2011, pp. 63–68 (2011)
3. Chen, Q., Jiang, D., Delgrossi, L.: IEEE 1609.4 DSRC multi-channel operations and its implications on vehicle safety communications. In: 2009 IEEE Vehicular Networking Conference (VNC), pp. 1–8 (2009)
4. European Telecommunications Standards Institute: Intelligent Transport Systems; European profile standard on the physical and medium access layer of 5 GHz ITS, Version 1.1.0, ES 202 663 (2009–2011)
5. Hu, Y., Xiao, M., Huang, L., Cheng, R., Mao, H.: Nearly optimal probabilistic coverage for roadside advertisement dissemination in Urban VANETs. arXiv preprint arXiv:1504.03824 (2015)
6. IEEE Standard for Wireless Access in Vehicular Environments (WAVE)-Multi-channel Operation: IEEE Std 1609.4-2010 (Revision of IEEE Std 1609.4-2006) pp. 1–89 (2011)
7. Jiang, D., Delgrossi, L.: IEEE 802.11p: towards an international standard for wireless access in vehicular environments. In: Vehicular Technology Conference, VTC Spring 2008, pp. 2036–2040. IEEE (2008)
8. Lee, S.B., Park, J.S., Gerla, M., Lu, S.: Secure incentives for commercial ad dissemination in vehicular networks. IEEE Trans. Veh. Technol. **61**(6), 2715–2728 (2012)
9. Lee, U., Lee, J., Park, J.S., Gerla, M.: FleaNet: a virtual market place on vehicular networks. IEEE Trans. Veh. Technol. **59**(1), 344–355 (2010)
10. Mohandas, B., Nayak, A., Naik, K., Goel, N.: ABSRP- a service discovery approach for vehicular ad hoc networks. In: Asia-Pacific Services Computing Conference, APSCC 2008, pp. 1590–1594. IEEE (2008)
11. Sommer, C., German, R., Dressler, F.: Bidirectionally coupled network and road traffic simulation for improved IVC analysis. IEEE Trans. Mob. Comput. **10**(1), 3–15 (2011)

12. Sommer, C., Joerer, S., Dressler, F.: On the applicability of two-ray path loss models for vehicular network simulation. In: 4th IEEE Vehicular Networking Conference (VNC 2012), pp. 64–69. IEEE, Seoul, Korea, November 2012
13. The Statistics Portal: http://www.statista.com/statistics/273288/advertising-spending-worldwide
14. Ullah, K., Santos, L., Michelin, J., dos Santos Moreira, E.: Extended abstract: file transfer in vehicular ad-hoc networks. In: 2013 III Brazilian Symposium on Computing Systems Engineering (SBESC), pp. 175–176 (2013)
15. Ullah, K., Santos, L., Yokoyama, R., dos Santos Moreira, E.: Advertising roadside services using vehicular ad hoc network (VANET) opportunistic capabilities. In: 2015 4th International Conference on Advances in Vehicular Systems, Technologies and Applications, VEHICULAR, pp. 7–13 (2015)
16. Uzcategui, R., Acosta-Marum, G.: WAVE: a tutorial. IEEE Commun. Mag. **47**(5), 126–133 (2009)
17. Varga, A., Hornig, R.: An overview of the OMNeT++ simulation environment. In: Proceedings of the 1st International Conference on Simulation Tools and Techniques for Communications, Networks and Systems & Workshops, Simutools 2008, pp. 60:1–60:10 (2008)
18. Veins Framework: http://veins.car2x.org/documentation/veins-arch.png
19. Wang, Z., Li, Y., Zhu, J., Duo, X.: A new model for advertisement dissemination in ad hoc networks. In: 2010 6th International Conference on Wireless Communications Networking and Mobile Computing (WiCOM), pp. 1–4 (2010)
20. Wegener, A., Piórkowski, M., Raya, M., Hellbrück, H., Fischer, S., Hubaux, J.P.: TraCI: an interface for coupling road traffic and network simulators. In: Proceedings of the 11th Communications and Networking Simulation Symposium, CNS 2008, pp. 155–163. ACM, New York (2008)

Experimental Evaluation of a Low-Cost Digital Sign-Posts Architecture for ITS Applications

Carlos Fernandez-Laguia, Juan-Carlos Cano,
Carlos T. Calafate, and Pietro Manzoni[⊠]

Universitat Politècnica de València, Camino de Vera, s/n, Valencia, Spain
pmanzoni@disca.upv.es
http://www.grc.upv.es/

Abstract. Integrating road signs information is becoming a critical goal for Intelligent Transportation Systems (ITS). Unlike other driving automation features, this capacity requires not only the vehicle, but also posts and infrastructure to be adapted thus involving an investment that can only be justified by a substantial number of users.

In this paper we describe an architecture that aims to facilitate the introduction and deployment of this technology based on low cost devices as the digital sign-posts and the integration of smartphones as an alternative in-vehicle user-interface. Wireless communications based on IEEE 802.11 is used for the basic connectivity requirements.

From the results obtained through an experimental evaluation, we show that, despite the smartphone constraints, we can achieve successful detection and recognition experiences at up to 90 km/h. Ultimately the experiment described confirms that the use of smartphones represents an opportunity to expand wireless technology in the traffic sign digitalisation area.

Keywords: Digital traffic signs · Smartphones · Intelligent transportation systems · Vehicle-to-infrastructure communications

1 Introduction

Automotive industry has made considerable advances over the last years in the fields of safety and comfort. Some of the main contributors have been the features categorized as Advanced Driver Assistance Systems (ADAS), a large number of which are already commercially available. Despite ADAS has been proved as a good opportunity to decrease the number of road accidents [1], usage rate is still significantly low. One of the reasons is that although automotive development cycles have been reduced in the last years, and in-cycle actions represent now a major update on the product concept, the vehicle replacement time is still lengthy. In the European Union the average for a passenger vehicle replacement is approximately 9.5 years while in the United States the figure exceeds 11 years. On the other hand recent surveys pointed out that a lack of perceived usefulness is one of the main reasons why awareness of ADAS is still significantly higher

© Springer International Publishing Switzerland 2016
N. Mitton et al. (Eds.): ADHOC-NOW 2016, LNCS 9724, pp. 294–307, 2016.
DOI: 10.1007/978-3-319-40509-4_21

than its usage rate [2]. Therefore new vehicle features do require a considerable amount of time to become a technology of common use.

This is especially critical when the technology update does not only involve the vehicles, but the roadway infrastructure. This is the case for applications like CoMoSeF [3] or traffic management systems [4] that are design to exchange information with roadside stations. In this category of communication systems, which is known as Vehicle-to-Infrastructure (V2I), only a significant amount of users can justify the investment to adapt the road according to the requirements. In particular using wireless technologies to exchange data with road infrastructure requires each traffic sign (or group of them under certain conditions, like its relative position) to be equipped with a network adapter.

In this paper we describe an architecture that aims to facilitate the introduction and deployment of this type of service which is based on low cost devices as the digital sign-posts and the integration of smartphones as an alternative in-vehicle user-interface. Wireless communications based on IEEE 802.11 is used for the basic connectivity requirements. To the best of our knowledge, there is no previous empirical experiences introducing the smartphone as a key element in the V2I strategy.

The paper is organized as follows. Section 2 describes the relevant related work on the topic. Section 3 details the proposal and Sect. 4 presents the experimental set-up details. Finally, Sect. 5 describes the evaluation results and Sect. 6 the final conclusions.

2 Related Work

New automobile safety features like Intelligent Speed Adaptation (ISA) or Traffic Sign Recognition (TSR) are increasingly highlighting the importance of bringing traffic signing into the digital domain in an efficient manner. This trend will only be reinforced in a future where transportation is outlined by more connected and more autonomous vehicles.

The problem has been broadly addressed in the literature using different approaches. Among them, the most extended proposal exploit enhanced vision techniques that have been adapted to road environments. Another system that has made its way to real applications is based on digital maps that contain relevant data and location references, so the client can query speed limits and other traffic information based on a known position. Although these solutions have reached a certain degree of popularity, there are some intrinsic limitations that make them not suitable for a driverless scenario. For instance light conditions and partial occlusions are two important handicaps for camera-based systems. Further, in the case of digital maps, an accurate location awareness and database update frequency are two major concerns to be addressed.

The solution that has been recently identified as a better candidate applies wireless technologies to exchange road information with vehicles. This approach appears for first time in the literature on [5]. Yoshimi Sato and Koji Makanae took the opportunity that camera-based systems were not yet extended in Japan and proposed a different solution: equipping the road with general-purpose RFID tags that contains relevant traffic sign information. The main issue observed by other researches about Satos' work is the position of the RFID tags and the antenna. The communication range of the type of RFID devices used by Sato is approximately 40 cm.

Later proposals achieved better results and versatility, for example on [6], the authors increased the communication range up to approximately 30 m, allowing an easier installation of the communication devices on both sides, infrastructure and vehicles. However the approach of wireless technologies had not been extensively studied until the vehicle-to-vehicle communications (V2V) concept appeared in the literature. Along with the vehicles, traffic signs have also been considered to be equipped with a wireless interface [7]. A new 802.11 amendment was published in 2010 motivated by a quicker data exchange between mobile nodes within a maximum range of one kilometer [9].

Despite the progress in this area, when we compared the cited proposals for wireless data exchange with other techniques they all share two common disadvantages: (1) the requirement of dedicated equipment and (2) the infrastructure investment. Vision systems have already established a strong position in the automotive market, therefore the adoption of a more expensive technology is not an easy challenge. Moreover this cost is not only applied to the vehicle price, but it also requires a major change in the infrastructure, which can only be justified by an important number of users. In this work we proposed an alternate platform to the in-build capture device that aims closing the gap between technical prerequisites and the requirement of potential users to easy an infrastructure change.

3   Proposed Architecture

Following our main objective, which is providing an efficient solution at a low cost, we have based our proposal on extended wireless communication standards rather than specialised protocols. The idea behind is creating a basic model that could be implemented on different platforms. We set our target on fulfilling major requirements to transmit road signs information via wireless media and support future updates. Consequently we have chosen to draft a proposal that allows high compatibility at different stack layers.

The communication model is the classical client-server scheme where road signs act as service providers, offering traffic data to the various clients concurrently upon request. The communications topology has been conceived as a traditional infrastructure network under a transport layer where the messages are transmitted via TCP. Finally, a REST-like protocol has been drafted to exchange information at the application layer.

The Medium Access Control (MAC) services are exploited at the client application level, managing the connection among different traffic signs. The mobile device inside the vehicle listens to the wireless medium for beacons. Each traffic sign, which has been previously equipped with the proposed server solution, periodically sends beacons with the relevant information for the connection. When the vehicle approaches a traffic sign, the device receives this beacon and immediately sends an authentication/association request (see Fig. 1). After the connection is established, the client can access the relevant data from the traffic sign and its location. Because the mobile device continuously monitors the signal strength received from the traffic sign, these values can be used as a reference to determine whether the vehicle is approaching or leaving a particular area. The reliability of this strategy still needs be studied under certain conditions like shielding or multi-path; however based on our experience, we believe that this data can be exploited for that purpose on its own or to complement other real-time location systems.

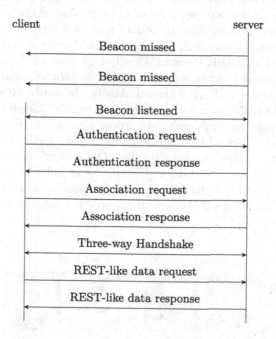

Fig. 1. Proposed communication flow diagram

It should be underlined that the proposed architecture is intended to be implemented with only two requirements: (1) a platform able to exchange data and manage the connection with different nodes in a specific order and (2) a network adapter that handles the connection at a lower level. This is specially important at the client side, which aims at gaining availability by integrating the complete solution on a smartphone.

On the server side, this architecture propitiates the convenient conditions to reduce the requirements for computing power, and therefore to lower the battery consumption. Consequently there could be a significant cost reduction if we use a low-cost solution to equip traffic signs with a wireless interface.

4   Experimental Set-Up

The idea that we propose could potentially pull ahead the benefits of wireless communications applied to traffic sign recognition by offering the driver a interim solution to adopt this technology regardless of the vehicle age. In order to demonstrate that the proposal accomplishes the basic requirements we have built an experimental solution, which has been used for the field application test that is explained on Sect. 5.

In order to keep costs low we chose to base our experimental solution on well-proven open source software. Figure 2 shows a simplified representation of the different software packages used on the field application test. Both sides of the systems have been implemented on Linux-kernel-based operating systems. In particular, the server side uses a *Debian* distribution where we have installed a *hostapd* daemon that handles the 802.11 standard services to connect with the different clients. The network layer is managed by the *dnsmasq* application, which is used as a light DHCP server. Finally high layer messages are managed by our own script, which implements the REST-like protocol, and incorporates a Python library for the TCP/IP standard. Among the hardware solutions in the market that fulfill our requirements, we have chosen the *Raspberry Pi*. Its wide range of connectors, specially general-purpose inputs/outputs (GPIO), along with the fact that it uses a Linux-kernel-based operating system makes it a versatile and robust solution for our objective. The network adapter is marketed under the name RTL8188CUS-GR, an USB device included in most of the *Raspberry Pi* development kits, which provides an inexpensive and complete solution for the 2.4 GHz band. For the trials we configured the network manager to work on 802.11n mode.

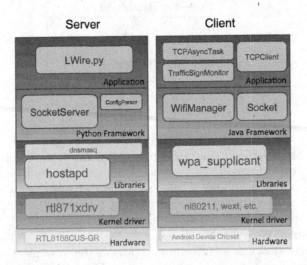

Fig. 2. Simplified network stack representation based on open source software

The client side implementation relies on the *Android* API (Application Programming Interface) to communicate with the *wpa_supplicant* at the user space. The supplicant handles the connection with the different traffic signs including applying communication filters based on the beacon information. This is significantly important because the supplicant should dismiss any traffic signs that does not contain relevant information for the driver. We have considered this topic as a subject of future work, hence it is not expound it further in this text.

The model we presented is not restricted to be implemented on a smartphone and even less on a specific operating system. We have considered that the smartphone has become a commodity of widespread adoption among drivers, hence a solution based on smartphones could significantly increase the number of users, if does not cause any negative effect on the device performance. Additionally as an electronic platform, it offers flexibility and portability at a relatively low cost. Among the mobile device operating systems we chose *Android* because it led the market share of 2015. Its availability and familiarity among the potential users makes it a convenient platform to speed up the adoption of this technology. It should be noted that despite its benefits we are sometimes limited by the *Android* interface as it is described in the next section.

(a)   (b)

Fig. 3. (a) Smartphone attached to the vehicle windshield during the experiment. (b) *Raspberry pi* solution installed on a zebra crossing sign.

For further analysis we have developed an application that does not only retrieve information from the traffic signs in the surroundings, but also implements a minimal user interface, monitors the smartphone performance and measures the process cycle time. *Android* has been designed as an operating system for touch screen devices, hence the importance of the user interfaces. This might be a problem executing time-consuming network commands. To overcome this situation we have implemented the main logic of the application on a separate thread designed as a finite-state-machine as we can see on Fig. 4.

We installed the application on a *GT-I9195*, also known as *Samsung Galaxy S4 mini*. This smartphone with a dual core at 1.73 GHz and 1331 MB of RAM belongs to the *Samsung* lower cost lineup that reached a certain degree of popularity in 2013 because of its price. The device is shipped with a Li-Ion battery,

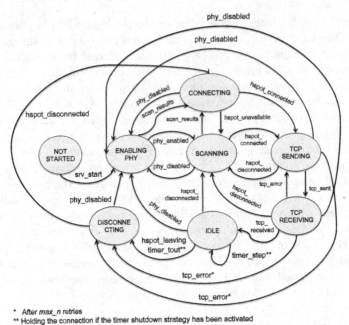

Fig. 4. Our proposal application state machine diagram

which offers autonomy of 1900 milliamp-hour. The operating system running on the device is *Android 4.4.2*, labeled as *KitKat*. This version of the system is the last one that *Google* released with the virtual machine *Dalvik*, however it is still the most extended one among the *Android* users (87.6 % according to official sources).

For our experiment we installed the smartphone on a car mount attached to the vehicle windshield where driver can see it without affecting his/her driving ability as shown in Fig. 3. Also we attached a *Raspberry Pi*, which had been previously set up with a wireless network adapter and an external battery, to a traffic sign at approximately 1.5 m from the ground. Along with our application we additionally downloaded and installed other three traffic *Android*-based solutions that helped us to create a more complete overview on the efficiency of our application by doing separate trials on each one.

The applications installed were: *myDriveAssist* from *Robert Bosch GmbH*, which uses the built-in camera from the smartphone to detect traffic signs; *aCo-Driver* from *Evotegra*, an independent company that combines camera and GPS (Global Positioning System) data to retrieve the speed limits from the road; and finally *Michelin Navigation* a navigation system, which implements a mode where only speed limits and not directions are prompted to the driver relying on a remote database (Fig. 5).

Fig. 5. Satellite image from area where experiments have been performed. (source: *Google Earth*)

5   Results

The evaluation for our proposal has been carried out by two different experiments. The first one aims at proving that our solution is able to detect, register and prompt relevant data to the driver before the vehicle overtakes the traffic sign. In the second one we wanted to establish whether the proposal was efficient from the perspective of a resource-constrained device.

For the first experiment we assumed the maximum communication range between traffic sign and client invariable. The distance travelled by the smartphone when the information is prompted to the driver would then be determined by the vehicle speed and the required time for a complete data exchange between client and server. The distance is given as:

$$d = R - \frac{s}{t_{detection}} \tag{1}$$

where R is the maximum communication range and s the vehicle speed. On the one hand the speed entirely depends on the driver and the traffic conditions; therefore we have considered different speed values for our experiment. On the other hand we wanted to establish the magnitude of the effect that this speed has on the data exchange time; with this purpose the client application logs time stamp and location at the different stages of the process (vehicle moving, traffic sign detected, etc.). The results from the first experiment are shown in Fig. 6.

From them we can conclude that higher speed values do not necessarily increase the time that the smartphone needs to complete traffic sign detection, so for simplicity we can assume that its influence is limited to the amount of time available for detection. However the detection time shows a wide variability regardless the vehicle speed. In order to understand the requirement on cycle time for the data exchange, we modified the source code on the *Android* application to create a log file with the timestamps at critical process steps like request for traffic sign discovery, authentication, connection, etc. We installed the modified application on the *GT-I9195* and run a trial on the simulated environment

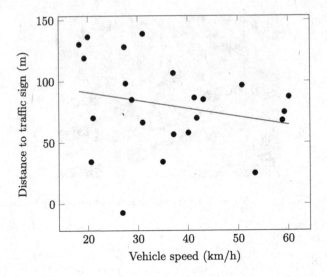

Fig. 6. Relation between vehicle speed and traffic-sign distance.

in the lab. Both devices, smartphone and digital post were placed approximately one meter from each other.

The results in Table 1 shows that after approximately one hour execution 74 detections have been successfully achieved and more than two thirds of the average time per cycle was spent on the discovery phase. This phase gets initiated when the application calls the `startScan()` method and ends either when the network driver performs a full scan or a known profile is available for connection.

Table 1. Detection time break down

Execution phase	Average time (ms)
Scan	2859
Authentication	134
Association	830
Rest-like	483
Disassociation	18

From the *wpa_supplicant* documentation[1] we have learnt that every time the application calls the `startScan()` method the network driver is requested to perform a full scan, which in the 2.4 GHz band consists of 13 channels spaced 5 MHz apart. The driver is usually designed to stay on each channel a certain amount of time before switching to the next one. The reason is that access points

---

[1] https://w1.fi/cgit/hostap/plain/wpa_supplicant/.

send periodically a beacon with relevant information for the connection, without that beacon the communication cannot be established. The full-scan latency is therefore determined by:

$$scan\_latency = N \times beacon\_interval \tag{2}$$

This beacon interval is typically set up to 100 ms on the access point (AP) [10], however most of stations (STAs) use a more conservative value to ensure that the beacon does not get missed if the antenna is physically located within the communication range. For example on *WEXT*, one of most extended kernel drivers for Linux, this value is defined as 250 in the source code by the constant WEXT_CSCAN_PASV_DWELL_TIME, therefore a full scan could take up to 3250 ms.

The *wpa_supplicant* and most popular drivers for *Android* devices allow minimising the scan delay by reducing the number of channels, however the *Android* API accepts no parameter for the startScan() method. This is, indeed, an important inefficiency compared with other solutions. Dedicated applications, like the ones based on RFID technologies, do not really require a full scan since transmitter and receiver can be designed to exchange data only in a preset frequency band. Despite it defines seven channels of 10 MHz bandwidth, the 802.11p standard solves the scan-latency problem restricting the number of channels to discover devices to only one, which is dedicated to periodical dissemination of this control information [9].

Another possible method that could potentially overcome this issue is a similar approach to the one used by the passive RFID solutions. The transceiver does not broadcast periodic frames, but it sends the relevant information for connection after receiving the request from the receiver. A similar method is included in the 802.11 specification with the name of active scanning. Like the single-channel scanning, this service is supported by the *wpa_supplicant* and the analysed drivers, but not by the *Android* API. The problem has been previously studied for a different application in [11]. Unfortunately the proposed solution inevitably means a more complicated installation process, including in some cases root access to the system. These enhancements can discourage potential users from adopting this technology, which goes against one of our main objectives. We should, therefore, accept that the use of an *Android* device as an alternate platform implies a maximum scan delay that is determined by Eq. (2).

After analysing this limitation we can state that the integrity of the system could be affected by the vehicle speed. The maximum value that guarantees the information prompted before the traffic sign is overtaken can be defined as:

$$s_{max} = R \times t_{max} \tag{3}$$

Although we have initially considered a constant value for R, the experience shows that the maximum distance to establish a safe and durable connection between the smartphone and the traffic sign is significantly reduced when the experiment is carried out in an urban environment. We have actually moved from 100-meters communication ranges in open spaces, like a highway, to a maximum distance of 40 m in narrow streets with presence of different obstacles (e.g. tall buildings, parked vehicles, trees, etc.). Despite these limitations, when we consider that the maximum speed allowed in urban areas is 50 km/h, our solution is still able to exchange information before the vehicle overtakes the transceiver. Applying Eq. 3 to the highway conditions, the maximum speed obtained is approximately 96 km/h, which means that the reliability of the system can be compromised above that vehicle speed.

On a different theme, we consider that the smartphone is not a dedicated piece of equipment and our solution must avoid any adverse effect on the usual device performance. In order to have a better picture of how our proposal performs on a smartphone, we have run a second experiment with each of the other applications installed on the device and collected the following indicators: CPU usage, memory allocated and battery status. The test consisted of a one-hour trip on an itinerary that was 30 % urban and 70 % freeway, covering approximately a distance of 50 km. We aimed to establish whether the application could have negative effects on the smartphone performance.

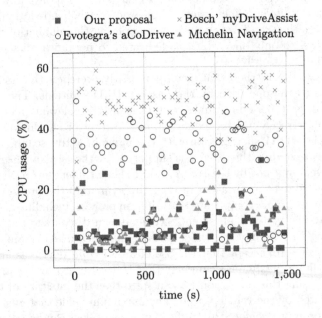

Fig. 7. *GT-I9195* 1.7 GHz dual-core CPU utilisation by application.

From the captured data, the two applications that use the camera for sign detection are the ones making the most extensive use of the CPU as we can see on Fig. 7. The other two, *Michelin Navigation* and our proposal, rely on the network interface as the main input. Unlike the camera, the network adapter is used based on the application demand, and not continuously, hence the CPU load should be in general less uniform, but lower than the camera-based systems.

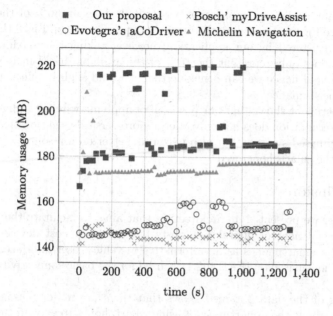

Fig. 8. *GT-I9195* RAM usage by application.

As a result the battery consumed during the test trial was also more representative on applications that use the in-build camera than on network-based ones.
In particular *myDriveAssist* had already consumed 44 % of the battery charge

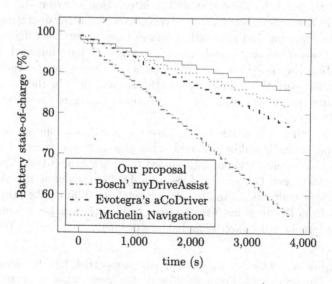

Fig. 9. Real discharge for the Li-Ion 1900 mAh removable battery by application.

after one hour, while our application was still showing an 86 % of the battery available (see Fig. 9). Regarding the memory requirements, on Fig. 8 the highest figures are registered by our application; however taking the maximum value recorded, it does only represent the 16 % of the available RAM memory of the smartphone, and hence we can conclude that it would slightly affect negatively the device performance.

The experiment shows that our proposal of applying wireless technologies to traffic sign recognition does not only offer a more versatile solution than camera or data-map based systems, but also performs better on a resource-constrained device like the smartphone used.

6   Conclusion

In this paper we presented an architecture that aims to facilitate the introduction and deployment of digital sign-posts based on low cost devices and the integration of smartphones as an alternative in-vehicle user-interface. Wireless communications based on IEEE 802.11 is used for the basic connectivity requirements.

Analysis of the data obtained shows that there are major reasons for further studies about the opportunity of using smartphones to expand the number of potential users of this technology. With a low budget we have implemented an application that is able to run for a long time on any *Android* device without potential CPU or memory capacity issues and reliably detect traffic signs equipped with our wireless solution at any vehicle speed up to 96 km/h.

The main limitations we have discovered during the implementation are introduced by the APIs involved in wireless communication for *Android*. The *Wifi-Manager* class, which handles most of the interaction between the application and the Wi-Fi controllers, is very limited compared with other operating systems. Moreover the scanning and association phases are not very well documented in the official *Android* reference, and most of premises we have followed during the implementation are based on our own experience. Nevertheless *Android* is an on-going project and these limitation might be overcome in the future, which would enhance the system response and, therefore, increase the effective maximum vehicle speed.

Despite its current limitations, we have compared our solution with others *Android* applications based on different strategies, and this proposal proves the benefits of using wireless technologies among the other exploit techniques underlined in this document. Our system does not only offer the possibility of detecting a wider range of traffic signs and have additional capabilities like providing live information from the area, but also we have verified that makes a more efficient use of the smartphone resources like battery or CPU.

Acknowledgments. This work was partially supported by the *Ministerio de Economía y Competitividad, Programa Estatal de Investigación, Desarrollo e Innovación Orientada a los Retos de la Sociedad, Proyectos I+D+I 2014*, Spain, under Grant TEC2014-52690-R.

References

1. Alkim, T.P., Bootsma, G., Hoogendoorn, S.P.: Field operational test the assisted driver. In: IEEE 2007 Intelligent Vehicles Symposium, pp. 1198–1203. IEEE, Istanbul (2007)
2. Trübswetter, N., Bengler, K.: Why should i use ADAS? Advanced driver assistance systems and the elderly: knowledge, experience and usage barriers. In: Proceedings of the 7th International Driving Symposium on Human Factors in Driver Assessment, Training and Vehicle Design, pp. 495–501. Bolton Landing, New York (2013)
3. Maenpaa, K., Sukuvaara, T., Ylitalo, R., Nurmi, P., Atlaskin, E.: Road weather station acting as a wireless service hotspot for vehicles. In: 2013 IEEE International Conference on Intelligent Computer Communication and Processing (ICCP), pp. 159–162. IEEE, Cluj-Napoca (2013)
4. Djahel, S., Smith, N., Wang, S., Murphy, J.: Reducing emergency services response time in smart cities: an advanced adaptive and fuzzy approach. In: 2015 IEEE First International Smart Cities Conference (ISC2), pp. 1–8. IEEE, Guadalajara (2015)
5. Yoshimichi, S., Koji, M.: Development and evaluation of in-vehicle signing system utilizing RFID tags as digital traffic signs. Int. J. ITS Res. 4(1), 53–58 (2006)
6. Pérez, J., Seco, F., Milanés, V., Jiménez, A., Díaz, J.C., De Pedro, T.: An RFID-based intelligent vehicle speed controller using active traffic signals. Sensors 10(6), 5872–5887 (2010)
7. Naja, R.: Wireless Vehicular Networks for Car Collision Avoidance. Springer, New York (2013)
8. Huang, W., Zhongdong, Y., Zhu, F., Yang, L., Wang, F. Y.: Applicability of short range wireless networks in V2I applications. In: 2013 16th International IEEE Conference on Intelligent Transportation Systems-(ITSC), pp. 231–236. IEEE, The Hage (2013)
9. ETSI, TCITS: Intelligent Transport Systems (ITS); European profile standard on the physical and medium access layer of 5 GHz ITS. Draft ETSI ES 202.663 (2009): V0
10. Murray, D., Dixon, M., Koziniec, T.: Scanning delays in 802.11 networks. In: The 2007 International Conference on Next Generation Mobile Applications, Services and Technologies (NGMAST 2007), pp. 255–260. IEEE, Cardiff (2007)
11. Brouwers, N., Zuniga, M., Langendoen, K.: Incremental wi-fi scanning for energy-efficient localization. In: 2014 IEEE International Conference on Pervasive Computing and Communications (PerCom), pp. 156–162. IEEE, Budapest (2014)
12. Choi, P., Gao, J., Ramanathan, N., Mao, M., Xu, S., Boon, C.C., Fahmy, S.A., Peh, L.S.: A case for leveraging 802.11 p for direct phone-to-phone communications. In: Proceedings of the 2014 International Symposium on Low Power Electronics and Design, pp. 207–212. ACM, La Jolla (2014)

Robots and MANETs

Asynchronous Gathering in Rings with 4 Robots

François Bonnet[1]([⊠]), Maria Potop-Butucaru[2,3], and Sebastien Tixeuil[2,3,4]

[1] Graduate School of Advanced Science and Technology, JAIST, Nomi, Japan
f-bonnet@jajst.ac.jp
[2] UPMC Sorbonne Universités, LIP6-CNRS, 7606 Paris, France
[3] CNRS, LIP6-CNRS, 7606 Paris, France
[4] Institut Universitaire de France, Paris, France

Abstract. In this paper we consider the gathering of oblivious mobile robots in a n-node ring. In this context, the single class of configurations left open in the most recent study [2] is $\mathcal{SP}4$ (a special class of configurations with only four robots).

We present an algorithm to solve some of the most intricate configurations in $\mathcal{SP}4$, those that can lead to a change of the axis of symmetry. Our approach lays the methodological bases for closing the remaining open cases for $\mathcal{SP}4$ solvability.

1 Introduction

The Distributed Computing community, motivated by the variety of tasks that can be performed by autonomous robots and their complexity, started recently to propose formal models for these systems and to design and prove protocols in these models. The seminal paper by Suzuki and Yamashita [16] proposes a robot model, two execution models, and several algorithms (with associated correctness proofs) for gathering and scattering a set of robots. In their model, robots are identical and anonymous (they execute the same deterministic algorithm and they cannot be distinguished using their appearance), robots are oblivious (they have no memory of their past actions) and they have neither a common sense of direction, nor a common handedness (chirality). Furthermore robots do not communicate in an explicit way. However they have the ability to sense the environment and see the position of the other robots, which lets them find their way in their environment. Also, robots execute three-phase cycles: *Look*, *Compute* and *Move*. During the *Look* phase robots take a snapshot of the other robots' positions. The collected information is used in the *Compute* phase in which robots decide to move or to stay idle. In the *Move* phase, robots may move to a new position computed in the previous phase. The two execution models are denoted (using recent taxonomy [8]) FSYNC, for fully synchronous and SSYNC, for semi-synchronous. In the SSYNC model, an arbitrary non-empty subset of robots execute the three phases synchronously and atomically. In the FSYNC model all robots execute the three phases synchronously. Since the initial introduction of this distributed computing model, an asynchronous model (denoted by ASYNC) has also been suggested [8]. In the ASYNC model there is

© Springer International Publishing Switzerland 2016
N. Mitton et al. (Eds.): ADHOC-NOW 2016, LNCS 9724, pp. 311–324, 2016.
DOI: 10.1007/978-3-319-40509-4_22

no synchronization between the various phases that are executed by the robots (that is, a robot may be looking while the other is moving).

A recent trend, motivated by practical applications such that exploration or surveillance, is the study of robots evolving in a discrete space with a *finite* number of locations. This discrete space is modeled by a graph, where nodes represent locations or sites, and edges represent the possibility for a robot to move from one site to the other.

One of the benchmarking [8] problems for mobile robots evolving in a discrete space is that of *gathering*. Regardless of their initial positions, robots have to move in such a way that they are eventually located on the same location, not known beforehand, and remain there thereafter. The case of ring networks is especially intricate, since its regular structure introduces a number of possible symmetric situations, from which the limited abilities of robots make it difficult to escape. A particular disposal (or configuration) of robots in the ring is *symmetrical* if there exists an axis of symmetry, that maps single robots into single robots, multiplicities (a.k.a. towers, more than one robot on the same node) into multiplicities, and empty nodes into empty nodes. A symmetric configuration can be edge-edge, node-edge or node-node symmetrical if the axis goes through two edges, through one node and one edge, or through two nodes, respectively. A *periodic* configuration is a configuration that is invariant by non-trivial rotation. Configurations that are neither symmetrical nor periodic are called *rigid*.

On the negative side, it was shown in [13] that gathering is impossible when the algorithm run by every robot is deterministic and there are only two robots, or if the initial configurations are periodic, or edge-edge symmetric, or if the ability for a robot to detect multiple robots on a single location (denoted as *multiplicity detection*) is not available. Running a probabilistic algorithm [15] permits to start from an arbitrary initial configuration (including periodic and edge-edge symmetric) but still requires multiplicity detection and the SSYNC execution model.

On the positive side, a number of deterministic ring gathering algorithms have been proposed in the literature [1–5,7,9–13] for the cases left open by impossibility results, focusing on the problem solvability for different initial configurations and different values for the size of the ring and the number of robots. The first two solutions [12,13] are complementary: [13] is based on breaking the symmetry whereas [12] takes advantage of symmetries. Nevertheless, the case of an even number of robots proved difficult [3,11] as more symmetric situations must be taken into account. When multiplicity detection is only available on the current position of each robot, more involved and specific approaches [5,9–11] are needed. When the robots are able to detect whether there is one or multiple robots in each location (capability known as global weak multiplicity), a unified deterministic strategy is available [1,2] for the most general execution model, ASYNC.

1.1 Motivation

Our work is motivated by the recent unified results proposed in [2]. Following their taxonomy, the set $\mathcal{I}$ of possible initial configurations (*i.e.* configurations without multiplicities) can be partitioned into the following three sets:

- $\mathcal{NG}$, Non-Gatherable configurations; configurations that are non gatherable (two robots, periodic configurations, edge-edge symmetrical configurations),
- $\mathcal{SP}4$, SPecial configurations with four robots (defined below),
- $\mathcal{A} = \mathcal{I}\backslash(\mathcal{NG} \cup \mathcal{SP}4)$, Admissible configurations that are known to be gatherable; there exists a unified deterministic algorithm [2] that allows to solve the gathering problem provided that the initial configuration is in $\mathcal{A}$.

In [2] the configurations in $\mathcal{SP}4$ are defined as: *"symmetric configurations of type node-edge with 4 robots and the odd interval cut by the axis bigger than the even one, with an interval being a maximal set of empty consecutive nodes."*

Figure 1 represents the configurations in $\mathcal{SP}4$ where the odd block is of size b, the even block is of size c, and $b > c$.

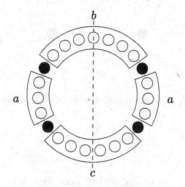

Fig. 1. $\mathcal{SP}4$ configuration (a, b, a, c); b is odd, c is even, and $b > c$.

Notation. In the remaining of the paper, $\mathcal{SP}4$ configurations will be uniquely represented by quadruplets (a, b, a, c) where b is odd, c is even, and $b > c$.

Remark. For any configuration of $\mathcal{SP}4$, according to the definition, the size n of the ring satisfies the equation $n = 4 + 2a + b + c$ where b and c are respectively odd and even. Therefore all configurations of $\mathcal{SP}4$ contain four robots on a ring of odd size.

Specificity of $\mathcal{SP}4$ Configurations. The only case that it is still open, with respect to the gathering problem, is the set of configurations $\mathcal{SP}4$. The difficulty of addressing the class $\mathcal{SP}4$ is stated in [2] as follows: *"The main difficulty faced when dealing with configurations in $\mathcal{SP}4$ comes from the fact that among the* *two intervals cut by the axis, the odd one is bigger than the even one. Intuitively the middle node of the odd interval is the only possible candidate to finalize the gathering [...]. Hence, when robots move towards such a node to make a multiplicity, it may happen that only one of the two symmetric robots allowed to move effectively moves. The subsequent configuration contains now two intervals of even size corresponding to those intervals originally cut by the axis of symmetry. Possibly they can be of the same size and hence they may induce different symmetries with respect to the original one."*

In the following we exemplify the above statement starting from the configuration depicted in Fig. 1 where we try to move the two upper robots towards the top. If only one of them moves (per scheduler choice), the system reaches the configuration depicted in Fig. 2 that contains a new symmetrical axis.

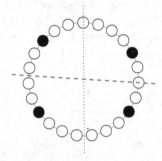

Fig. 2. Why it is complex to solve $\mathcal{SP}4$ configurations.

1.2   Contributions

In this paper we investigate the gathering problem in rings of size n with n odd and $n > 5$, $k = 4$ robots in the standard settings (ASYNC execution model, global weak[1] multiplicity detection). More precisely, we investigate the open class $\mathcal{SP}4$. For some of the most problematic configurations $C \in \mathcal{SP}4$ (those that may introduce a change of the symmetry axis), we propose an algorithm that gathers robots when starting in C.

Remark. A recent study proves that $\mathcal{SP}4$ configurations are ungatherable for rings of size 7 or 9, and also conjectures that these configurations are ungatherable for any size of ring [6]. This is not a contradiction with our proposed results. In that paper, configurations are considered globally, while here we try to analyze each of them independently. Note that, even if it was possible to solve independently each of them (which is probably not true), it would not imply a possible solution to solve all of them globally.

---

[1] In our context of four robots, strong and weak multiplicity are equivalent. Given the knowledge of four robots, when observing a tower (multiple robots on the same node), one can always deduce the exact number of robots.

1.3   Roadmap

Section 2 defines the model. Section 3 describes and proves an algorithm solving the gathering problem when the initial configuration is known. Section 4 explains how to combine algorithms of Sect. 3, when it is possible. Section 5 concludes the paper. In Appendix A we instantiate the results for some rings.

2   Model

2.1   Robots on Graphs

We consider an undirected n-sized ring-shaped graph. The graph is *anonymous*, that is, there exists no labeling to distinguish nodes or edges. The robots are *identical*, *i.e.*, they cannot be distinguished and all execute the same protocol. Moreover, the robots are *oblivious* and *disoriented*, meaning that they have no memory of past actions, and they share no common handedness, *i.e.*, they do not agree on a common right (or, left) side. Robots are placed on the graph nodes, and a link between two nodes denotes the ability for a robot to move from one node to the other. Several robots can be located at the same node concurrently and they form a *tower*.

The robots cannot explicitly communicate, but have the ability to sense their environment and see the position of the other robots, in their local coordinate system. The sensing capabilities of the robots permits them to know whether there is one or multiple robots in each location (that is, we use the *global weak multiplicity* detection model).

Robots operate in cycles of three *deterministic* phases: Look, Compute, and Move. During its Look phase a robot takes a snapshot of the graph together with all robots' positions. The collected information (position of the other robots in the egocentric view) is used in the Compute phase during which the robot may decide to move or stay. In the Move phase, the robot may move to one of the adjacent nodes, as computed in the previous phase.

The computational model we consider is the ASYNC model [8] in a discrete setting. It means that, the start and duration of each Look-Compute phases and the start of each Move phase of each robot are arbitrary and determined by an adversary. However, when a robot takes a snapshot of the graph and other robots' positions, it sees the other robots on nodes only (that is, we assume that the move phases are instantaneous). In particular, this execution model permits a robot to make a move based on a previously observed configuration (*e.g.* if its Look phase occured before the Move phase of another robot). A configuration at a given time is defined by the positions of all robots at that time.

2.2   The Gathering Problem

The general problem considered in our work is the gathering problem, where starting from any arbitrary gatherable configuration (that is, a configuration from which gathering is not proved impossible by a deterministic algorithm),

k robots have to gather in one location not known in advance before stopping there forever.

In this paper, we focus on the remaining open case for gathering with 4 robots, starting from an initial configuration in $\mathcal{SP}4$.

3   Gathering from a Known Initial Configuration of $\mathcal{SP}4$

In this section, we propose an algorithm that solves the gathering problem from a known initial configuration. The given algorithm is parametrized by this initial configuration. In Sect. 4, we combine these specific algorithms to obtain a unified algorithm. Using the same notations as in Fig. 1, let us define the subset $\mathcal{SP}4_g$ of $\mathcal{SP}4$ as the gatherable $\mathcal{SP}4$ configurations:

$$\mathcal{SP}4_g = \left\{ (a,b,a,c) \in \mathcal{SP}4 \text{ such that } \begin{array}{|l} a \text{ is odd} \\ c = b - 1 \\ a \neq b - 2 \end{array} \right\}$$

We do not claim that $\mathcal{SP}4_g$ contains all gatherable configurations of $\mathcal{SP}4$, but it contains all the ones that we solve in this paper. Note that there are already an infinite number of configurations in $\mathcal{SP}4_g$.

From any configuration of $\mathcal{SP}4_g$, it is possible to gather robots and our algorithm works in two phases. In Phase 1 starting from a configuration in $\mathcal{SP}4_g$ the algorithm (described in Fig. 3) brings the system in a not $\mathcal{SP}4$ configuration. In Phase 2, the algorithm (described in Fig. 5) starts in a configuration output of Phase 1 and brings the system in the gathered configuration (i.e. all robots share the same position).

3.1   Phase 1: Leaving $\mathcal{SP}4$

Figure 3 describes the first phase of our algorithm. It explains how to reach a "good" symmetrical configuration from a configuration of $\mathcal{SP}4_g$. The dotted lines in Fig. 3 depict the symmetrical axes that exist. Black arrows correspond to the moves that should be *computed* by robots if they *look* at the current configuration (note that symmetrical robots have always symmetrical moves). Let us emphasize that our algorithm is entirely represented by these black arrows. Other arrows (explained in next paragraph) describe only possible executions of this algorithm.

Dashed arrows correspond to potential pending moves (*computed* from a previous configuration). Double arrows indicate all possible transitions between the configurations. Due to the ASYNC model assumption, only a subset of robots can be scheduled to move; all cases are considered and distinguished with labels "both" (two robots move), "one" (a single robot moves), "dashed" (a single robot moves according to a pending move), "both bold" (two robots move according to currently computed moves), ...

From the initial configuration of $\mathcal{SP}4_g$ (top-left), the system always reaches a configuration outside of $\mathcal{SP}4$ (top-right or bot-mid) which does not contain any potential pending move.

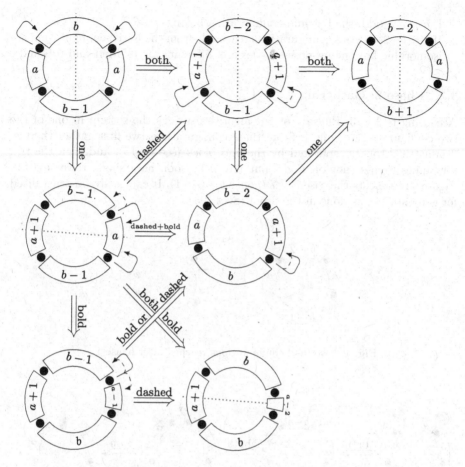

Fig. 3. Algorithm – Phase 1: leaving $\mathcal{SP}4$.

Notation. Let us denote $\mathcal{L}_1$ the list[2] of configurations used during this phase 1. $\mathcal{L}_1$ includes all configurations depicted on Fig. 3.

Specific values of a or b. Figure 3 is inaccurate for the following cases, but can be easily adapted:

- $b = 1$; Instead of having $b - 2$ empty nodes, there should be a tower.
- $a = 1$; Instead of having $a - 2$ empty nodes, there should be a tower.
- $b = a$; The two final configurations are identical.

Remark. The constraints defining $\mathcal{SP}4_g$ are vital for the well behavior of our algorithm. It will appear clearly in the proof, but is already noticeable from the algorithm description:

---

[2] We use here the word *list* instead of *set* because we do not assume uniqueness of each configuration. The uniqueness is proved in Theorem 1.

– If a is not odd, the bot-mid configuration belongs to $\mathcal{SP}4$.
– If $a = b-2$, the two configurations reachable from the initial configuration are isomorphic but the proposed moves are different (*i.e.* mid-left and top-mid).

3.2   Phase 2: Gathering

According to Fig. 3, Phase 1 of our algorithm brings the system in one of the two configurations of Fig. 4. From these configurations, we first gather the two symmetrical robots separated by the odd block (see Fig. 5), and then the two remaining robots (not on Fig. 5 but trivial as soon as a single tower exists). Figure 5 represents the case $b = 7$ (hence $b-6 = 1$). It can easily be generalized for general case, as done in the proof (Sect. 3.3).

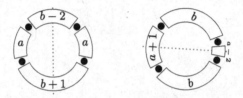

Fig. 4. Two final configurations reached after phase 1.

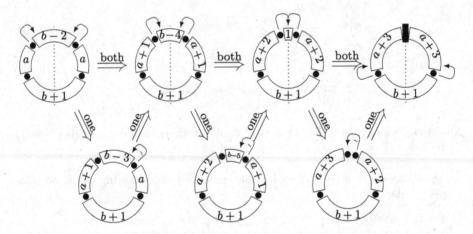

Fig. 5. Algorithm – phase 2: Gathering.

3.3   Proof of Correctness

First, we prove that the configurations in Phase 1 (algorithm described in Fig. 3) are all different, hence the execution has no loop and no conflict (incoherent moves computed for the same configuration). Then we prove that the configurations in Phase 2 (algorithm described in Fig. 5) are unique and all different from the ones of the Phase 1. That is, there is no loop and no conflict between configurations of Phase 1 and Phase 2 and also no loop and no conflict inside the set of configurations used by the Phase 2 algorithm. It follows that the system always converges to a final configuration. Finally we prove that from the two final configurations, it is always possible to gather the robots to a single node.

Theorem 1. *The configurations of $\mathcal{L}_1$ (the ones depicted in Fig. 3) are unique.*

Proof. There are two classes of configurations: asymmetrical and symmetrical.

- **Asymmetrical configurations.** There are only two asymmetrical configurations, $(b-2, a+1, b, a)$ and $(b-1, a-1, b, a+1)$.[3] Since a and b are both odd; the first configuration contains three blocks of odd size and one block of even size, while the second configuration contains one block of odd size and three blocks of even size. These configurations cannot therefore be the same.
- **Symmetrical configurations.** There are five symmetrical configurations. Contrarily to asymmetrical configurations, it is now possible to order the sequence of blocks; starting, for example, from the odd block crossed by the symmetrical axis: $(b, a, b-1, a)$, $(b-2, a+1, b-1, a+1)$, $(b-2, a, b+1, a)$, $(a, b-1, a+1, b-1)$, $(a-2, b, a+1, b)$.
  - The first configuration is the only one belonging to $\mathcal{SP}4$.
  - The third and fifth configurations contain both three odd blocks and a single even block; they correspond to the very same configuration when $a = b$. As mentioned earlier, it simply means that the two final configurations are in fact the same configuration.
  - The second and fourth configurations contain both a single odd block and three even blocks; they correspond to the very same configuration when $a = b - 2$. From the assumption, $a \neq b - 2$ so these configurations are different.   □

In the following we examine the configurations generated by the execution of the second phase of the algorithm described in Fig. 5. Phase 2 uses the following configurations:

- When Phase 2 starts from the configuration $(b-2, a, b+1, a)$:
  - Symmetrical configurations without tower $(b-4, a+1, b+1, a+1)$, $(b-6, a+2, b+1, a+2)$, ..., $(1, a + \frac{b-3}{2}, b+1, a + \frac{b-3}{2})$. More generally it corresponds to configurations:

$$\forall i \in \left\{ 1, \ldots, \frac{b-3}{2} \right\} \qquad (b-2i-2, a+i, b+1, a+i)$$

---

[3] The order of blocks is arbitrarily chosen.

- Asymmetrical configurations without tower $(b - 3, a + 1, b + 1, a)$, $(b - 5, a + 2, b + 1, a + 1)$, ..., $(0, a + \frac{b-3}{2} + 1, b + 1, a + \frac{b-3}{2})$. More generally it corresponds to configurations:

$$\forall i \in \left\{ 0, \ldots, \frac{b-3}{2} \right\} \qquad (b - 2i - 3, a + i + 1, b + 1, a + i)$$

- When Phase 2 starts from the configuration $(a - 2, b, a + 1, b)$:
  - Symmetrical configurations without tower $(a - 4, b + 1, a + 1, b + 1)$, $(a - 6, b + 2, a + 1, b + 2)$, ..., $(1, b + \frac{a-3}{2}, a + 1, b + \frac{a-3}{2})$. More generally it corresponds to configurations:

$$\forall i \in \left\{ 1, \ldots, \frac{a-3}{2} \right\} \qquad (a - 2i - 2, b + i, a + 1, b + i)$$

  - Asymmetrical configurations without tower $(a - 3, b + 1, a + 1, b)$, $(a - 5, b + 2, a + 1, b + 1)$, ..., $(0, b + \frac{a-3}{2} + 1, a + 1, b + \frac{a-3}{2})$. More generally it corresponds to configurations:

$$\forall i \in \left\{ 0, \ldots, \frac{a-3}{2} \right\} \qquad (a - 2i - 3, b + i + 1, a + 1, b + i)$$

Notation. Similarly to $\mathcal{L}_1$ for phase 1, let us denote $\mathcal{L}_2$ the list of configurations used during this phase 2. $\mathcal{L}_2$ is the union of the four cases described above.

Theorem 2. *The configurations of $\mathcal{L}_2$ are unique and there is no common configuration in $\mathcal{L}_1$ and $\mathcal{L}_2$.*

Proof. Uniqueness can be checked similarly as done for Theorem 1. It remains to check that all configurations of $\mathcal{L}_2$ are not already in $\mathcal{L}_1$. The two intermediate symmetrical configurations of Fig. 3 are $(b - 2, a + 1, b - 1, a + 1)$ and $(a, b - 1, a + 1, b - 1)$. For both configurations, the difference between the lengths of the two empty blocks crossed by the symmetrical axis equals one. Conversely, this difference equals $(b + 1) - (b - 2i - 2) = 2i + 3$ for the symmetrical configurations defined above. Since $2i + 3 > 1$, there is no overlap between symmetrical configurations of $\mathcal{L}_1$ and $\mathcal{L}_2$.

Similarly, for asymmetrical configurations one can observe that the two configurations used in the algorithm of Fig. 3 contain two opposite empty blocks whose difference of lengths equals 2; $b - (b - 2)$ and $(a + 1) - (a - 1)$. Conversely, these differences equal either $(a + i + 1) - (a + i) = 1$ or $(b + 1) - (b - 2i - 3) = 2i + 4$ for the asymmetrical configurations defined above. Since $1 \neq 2$ and $2i + 4 \neq 2$, there is no overlap between asymmetrical configurations of $\mathcal{L}_1$ and $\mathcal{L}_2$.   □

The following theorem states the correctness of the algorithm. The proof is a direct consequence of Theorems 1 and 2 and Figs. 3 and 5.

Theorem 3. *The algorithm described in Figs. 3 and 5 started in a configuration of $SP4_g$ brings the system in a gathered configuration in the ASYNC model of execution.*

4   Gathering from Initial Configurations of $\mathcal{A} \cup \mathcal{SP}4_g$

In the previous section, we assumed that the initial configuration is known and belongs to $\mathcal{SP}4_g$. In the following we assume that this assumption is not satisfied anymore. Recall that $\mathcal{A}$ denotes the set of admissible configurations (*i.e.* configurations already known to be gatherable).

Let C be a configuration of $\mathcal{SP}4_g$. If we restrict the system to start in an initial configuration belonging to $\mathcal{A} \cup \{C\}$, it is possible to solve the gathering problem by combining our algorithm with algorithms proposed in [2].[4]

Let C_1 and C_2 be two configurations of $\mathcal{SP}4_g$. In the following we investigate the situations when it is possible to compose our algorithm for C_1 with the algorithm for C_2 to solve the gathering problem if the initial configuration belongs to the set $\{C_1, C_2\}$. The following theorem exhibits the cases when the composition is not possible.

Theorem 4. *Let n be an odd value, let C_1 and C_2 be two (distinct) configurations of $\mathcal{SP}4_g$. Let $C_1 = (a_1, b_1, a_1, b_1 - 1)$ and $C_2 = (a_2, b_2, a_2, b_2 - 1)$ such that $2a_1 + 2b_1 - 1 = n - 4$ and $2a_2 + 2b_2 - 1 = n - 4$. The algorithm for C_1 cannot be composed with the algorithm for C_2 if and only if $a_1 + a_2 = \frac{n-7}{2}$. In all other cases, it is possible to compose these algorithms.*

Proof. In Sect. 3.3 we listed the configurations that are used in our algorithm:

$$
\begin{cases}
(b - 1, a - 1, b, a + 1) \\
(b - 2, a + 1, b, a)
\end{cases} \Bigg\} \text{Asymmetrical}
$$

$$
\begin{cases}
(b, a, b - 1, a) \\
(b - 2, a + 1, b - 1, a + 1) \\
(b - 2, a, b + 1, a) \\
(a, b - 1, a + 1, b - 1) \\
(a - 2, b, a + 1, b)
\end{cases} \Bigg\} \text{Symmetrical}
$$

To check if the algorithm for C_1 and the algorithm for C_2 can be composed, we need to check if there is an overlap between the above configurations.

- For asymmetrical configurations, as in Sect. 3.3, a simple parity argument proves that there is an intersection only if $a_1 = a_2$ or $b_1 = b_2$ which implies $C_1 = C_2$.
- For symmetrical configurations, a pairwise analysis shows an overlap in the following cases:
  - $(b_1 - 2, a_1 + 1, b_1 - 1, a_1 + 1) = (a_2, b_2 - 1, a_2 + 1, b_2 - 1)$, which implies $a_2 = b_1 - 2$ and $b_2 = a_1 + 2$.
  - $(a_1, b_1 - 1, a_1 + 1, b_1 - 1) = (b_2 - 2, a_2 + 1, b_2 - 1, a_2 + 1)$, which implies $a_1 = b_2 - 2$ and $b_1 = a_2 + 2$.
  - $(b_1 - 2, a_1, b_1 + 1, a_1) = (a_2 - 2, b_2, a_2 + 1, b_2)$.
  - $(a_1 - 2, b_1, a_1 + 1, b_1) = (b_2 - 2, a_2, b_2 + 1, a_2)$.

---

[4] Informally, if a configuration belongs to $\mathcal{L}_1$ or $\mathcal{L}_2$, robots decide to move according to our rules; otherwise they follow rules of [2].

The two last cases do not raise any problem since only final configurations are involved. The two first cases raise a problem since both configurations involved are given different moves in the algorithms. It implies that it is not possible to combine our algorithm in these cases. Both cases can be summarized in the single equation $a_1 + a_2 = \frac{n-7}{2}$. Since $a_1 + b_1 = a_2 + b_2 = \frac{n-3}{2}$, it can equivalently be rewritten $b_1 + b_2 = \frac{n+1}{2}$.   □

5   Conclusion and Open Problems

We presented an algorithm for oblivious mobile robot gathering on uniform rings, which focuses on the most intricate of the remaining open cases (the configurations of $\mathcal{SP}4$) that can lead to a change of the axis of symmetry upon robot movement. While from a quantitative point of view, those initial configuration represent only a minority of the open cases in $\mathcal{SP}4$, we hope that the remaining initial configurations could be solved by marginal adjustments of our algorithm, or proven to be non-gatherable. Another path for future research would be to apply computer-assisted approaches used in [6] to help solving remaining cases.

There is an obvious analogy between condition based consensus as proposed by Mostefaoui *et al.* [14] and condition based gathering as we perform in this paper. Integrating a complete set of input configurations (that is, all configurations that are gatherable) into a unified algorithm is an interesting open problem.

Acknowledgements. This work was supported in part by LINCS and by JSPS KAKENHI Grant Number 26870228. The authors would like to thank the anonymous reviewers for their constructive comments.

A   Appendix: Case Studies

In this section we investigate the gathering for some specific values of ring size n. For each value we identify the configurations for which our algorithm solves the problem and the configurations that are still open. We consider here only rings of size $n = 4x + 3$. For rings of size $n = 4x + 1$, the set $\mathcal{SP}4_g$ is empty and therefore we do not solve any $\mathcal{SP}4$ configurations in that case.

A.1   Ring of Size 7

There are 4 towerless configurations ($|\mathcal{I}| = 4$). There is no *obviously-non-gatherable* configuration ($|\mathcal{NG}| = 0$). The set of admissible configurations contains 1 asymmetrical and 1 symmetrical ($|\mathcal{A}| = 1 + 1 = 2$). There are therefore 2 configurations of $\mathcal{SP}4$:

- C_1 defined by $(a_1, b_1, a_1, c_1) = (0, 3, 0, 0)$. $C_1 \notin \mathcal{SP}4_g$ since $c_1 \neq b_1 - 1$.
- C_2 defined by $(a_2, b_2, a_2, c_2) = (1, 1, 1, 0)$. $C_2 \in \mathcal{SP}4_g$.

There is an algorithm solving the gathering problem if the initial configuration belongs to $\mathcal{A} \cup \{C_2\}$. The problem is open if the initial configuration is C_1.

A.2   Ring of Size 11

$|\mathcal{I}| = 20$, $|\mathcal{NG}| = 0$, $|\mathcal{A}| = 10 + 4 = 14$, and $|\mathcal{SP}4| = 6$ partitioned as follows:

- C_1 defined by $(a_1, b_1, a_1, c_1) = (0, 7, 0, 0)$. $C_1 \notin \mathcal{SP}4_g$ since $c_1 \neq b_1 - 1$.
- C_2 defined by $(a_2, b_2, a_2, c_2) = (1, 5, 1, 0)$. $C_2 \notin \mathcal{SP}4_g$ since $c_2 \neq b_2 - 1$.
- C_3 defined by $(a_3, b_3, a_3, c_3) = (0, 5, 0, 2)$. $C_3 \notin \mathcal{SP}4_g$ since $c_3 \neq b_3 - 1$.
- C_4 defined by $(a_4, b_4, a_4, c_4) = (2, 3, 2, 0)$. $C_4 \notin \mathcal{SP}4_g$ since $c_4 \neq b_4 - 1$.
- C_5 defined by $(a_5, b_5, a_5, c_5) = (3, 1, 3, 0)$. $C_5 \in \mathcal{SP}4_g$.
- C_6 defined by $(a_6, b_6, a_6, c_6) = (1, 3, 1, 2)$. $C_6 \notin \mathcal{SP}4_g$ since $a_6 = b_6 - 2$.

There is an algorithm solving the gathering problem if the initial configuration belongs to $\mathcal{A} \cup \{C_5\}$. The problem is open if the initial configuration is C_1, C_2, C_3, C_4, or C_6.

A.3   Ring of Size $n = 4x + 3$ for $x \in \{1, 2, 3, \dots\}$

$|\mathcal{I}| = \frac{(n^2 - 1)(n - 3)}{48}$, $|\mathcal{NG}| = 0$, $|\mathcal{A}| = \frac{(n-1)(n-3)(n-5)}{48} + \frac{(n-3)^2}{16} = \frac{(n-4)(n-3)(n+5)}{48}$, and $|\mathcal{SP}4| = \frac{(n-3)(n+1)}{16}$ are partitioned as follows:

- $C_{b,c}$ defined by $(\frac{n-4-b-c}{2}, b, \frac{n-4-b-c}{2}, c)$ for $b \in \{1, 3, 5, 7, \dots, n-4\}$ and $c \in \{0, 2, 4, \dots, b-1\}$.
  All $C_{b,c}$ for $c \neq b - 1$ do not belong to $\mathcal{SP}4_g$.
- Depending on the parity of x, there exists or not a configuration $C_{b,c}$ such that $c = b - 1$ and $\frac{n-4-b-c}{2} = a = b - 2$:
  - If x is odd; there is no such configuration. There are x configurations in $\mathcal{SP}4_g$:
    $C_b = C_{b,b-1}$ defined by $(\frac{n-3}{2} - b, b, \frac{n-3}{2} - b, b-1)$ for $b \in \{1, 3, \dots, 2x-1\}$.
  - If x is even; there is a unique such configuration $C_{b,c}$ for $b = x + 1$. There are $x - 1$ configurations in $\mathcal{SP}4_g$:
    $C_b = C_{b,b-1}$ defined by $(\frac{n-3}{2} - b, b, \frac{n-3}{2} - b, b-1)$ for $b \in \{1, 3, 5, 7, \dots, 2x - 1\} \setminus \{x + 1\}$.

Note that two of our algorithms for C_b and $C_{b'}$ are incompatible if $b + b' = \frac{n+1}{2} = 2x + 2$. In conclusion there is an algorithm solving the gathering problem if the initial configuration belongs to the set:

- $\mathcal{A} \cup \{C_1\} \cup \{C_3 \text{ or } C_{2x-1}\} \cup \{C_5 \text{ or } C_{2x-3}\} \cup \{C_7 \text{ or } C_{2x-5}\} \cup \dots \cup \{C_x \text{ or } C_{x+2}\}$, if x is odd.
- $\mathcal{A} \cup \{C_1\} \cup \{C_3 \text{ or } C_{2x-1}\} \cup \{C_5 \text{ or } C_{2x-3}\} \cup \{C_7 \text{ or } C_{2x-5}\} \cup \dots \cup \{C_{x-1} \text{ or } C_{x+3}\}$, if x is even.

The problem is open if the initial configurations belong to $\{C_{b,c}\}$ for $c \neq b - 1$, or if two incompatible configurations are included in the set of initial configurations (such as the set $\{C_3, C_{2x-1}\}$).

References

1. D'Angelo, G., Di Stefano, G., Navarra, A.: How to gather asynchronous oblivious robots on anonymous rings. In: Aguilera, M.K. (ed.) DISC 2012. LNCS, vol. 7611, pp. 326–340. Springer, Heidelberg (2012)
2. D'Angelo, G., Di Stefano, G., Navarra, A.: Gathering on rings under the look-compute-move model. Distrib. Comput. **27**(4), 255–285 (2014)
3. D'Angelo, G., Di Stefano, G., Navarra, A.: Gathering six oblivious robots on anonymous symmetric rings. J. Discrete Algorithms **26**, 16–27 (2014)
4. D'Angelo, G., Di Stefano, G., Navarra, A., Nisse, N., Suchan, K.: Computing on rings by oblivious robots: a unified approach for different tasks. Algorithmica **72**(4), 1055–1096 (2015)
5. D'Angelo, G., Navarra, A., Nisse, N.: Gathering and exclusive searching on rings under minimal assumptions. In: Chatterjee, M., Cao, J., Kothapalli, K., Rajsbaum, S. (eds.) ICDCN 2014. LNCS, vol. 8314, pp. 149–164. Springer, Heidelberg (2014)
6. Di Stefano, G., Montanari, P., Navarra, A.: About ungatherability of oblivious and asynchronous robots on anonymous rings. In: Lipták, Z., Smyth, W.F. (eds.) IWOCA 2015. LNCS, vol. 9538, pp. 136–147. Springer, Heidelberg (2016). doi:10.1007/978-3-319-29516-9_12
7. Di Stefano, G., Navarra, A.: Optimal gathering of oblivious robots in anonymous graphs. In: Moscibroda, T., Rescigno, A.A. (eds.) SIROCCO 2013. LNCS, vol. 8179, pp. 213–224. Springer, Heidelberg (2013)
8. Flocchini, P., Prencipe, G., Santoro, N.: Distributed Computing by Oblivious Mobile Robots. Morgan & Claypool Publishers, San Rafael (2012)
9. Izumi, T., Kamei, S., Ooshita, F.: Time-optimal gathering algorithm of mobile robots with local weak multiplicity detection in rings. IEICE Trans. Fundam. Electron. Commun. Comput. Sci. **E96–A**(6), 1072–1080 (2013)
10. Kamei, S., Lamani, A., Ooshita, F., Tixeuil, S.: Asynchronous mobile robot gathering from symmetric configurations without global multiplicity detection. In: Kosowski, A., Yamashita, M. (eds.) SIROCCO 2011. LNCS, vol. 6796, pp. 150–161. Springer, Heidelberg (2011)
11. Kamei, S., Lamani, A., Ooshita, F., Tixeuil, S.: Gathering an even number of robots in an odd ring without global multiplicity detection. In: Rovan, B., Sassone, V., Widmayer, P. (eds.) MFCS 2012. LNCS, vol. 7464, pp. 542–553. Springer, Heidelberg (2012)
12. Klasing, R., Kosowski, A., Navarra, A.: Taking advantage of symmetries: gathering of many asynchronous oblivious robots on a ring. Theor. Comput. Sci. **411**(34–36), 3235–3246 (2010)
13. Klasing, R., Markou, E., Pelc, A.: Gathering asynchronous oblivious mobile robots in a ring. Theor. Comput. Sci. **390**(1), 27–39 (2008)
14. Mostéfaoui, A., Rajsbaum, S., Raynal, M., Roy, M.: Condition-based consensus solvability: a hierarchy of conditions and efficient protocols. Distrib. Comput. **17**(1), 1–20 (2004)
15. Ooshita, F., Tixeuil, S.: On the self-stabilization of mobile oblivious robots in uniform rings. Theor. Comput. Sci. **568**, 84–96 (2015)
16. Suzuki, I., Yamashita, M.: Distributed anonymous mobile robots: formation of geometric patterns. SIAM J. Comput. **28**(4), 1347–1363 (1999)

To Mesh or not to Mesh: Flexible Wireless Indoor Communication Among Mobile Robots in Industrial Environments

Elnaz Alizadeh Jarchlo, Jetmir Haxhibeqiri, Ingrid Moerman,
and Jeroen Hoebeke$^{(\boxtimes)}$

Department of Information Technology (INTEC), Ghent University – iMinds,
Technologiepark-Zwijnaarde 15, 9052 Ghent, Belgium
jeroen.hoebeke@intec.UGent.be

Abstract. Mobile robots such as automated guided vehicles become increasingly important in industry as they can greatly increase efficiency. For their operation such robots must rely on wireless communication, typically realized by connecting them to an existing enterprise network. In this paper we motivate that such an approach is not always economically viable or might result in performance issues. Therefore we propose a flexible and configurable mixed architecture that leverages on mesh capabilities whenever appropriate. Through experiments on a wireless testbed for a variety of scenarios, we analyse the impact of roaming, mobility and traffic separation and demonstrate the potential of our approach.

1 Introduction

Industry is continuously looking for ways to further automate processes, improve efficiency, reduce energy consumption, increase economic benefits etc. This ongoing evolution is often referred to as Industry 4.0 [1], where everything becomes connected to a network (e.g., the Internet or a private factory network) by means of communication infrastructure. This not only involves field devices or machines, but also involves mobile robots such as Automated Guided Vehicles (AGVs) [2]. Automated Guided Vehicles (AGVs) facilitate transporting various types of goods automatically and handling materials in automated manufacturing systems. The earliest AGVs were essentially line following mobile robots, but more recent solutions consist of autonomously guided robots that act based on information about where they are and which destinations to reach.

A key technology enabling such autonomously operating robot systems is wireless communication. Wireless communication between robots or between robots and a controller system is crucial for their operation, but challenging at the same time. As robots may move around quite fast through the network area, communication paths may change frequently. On top of this, some of the communication pertains to the real-time coordination of robots and requires sufficiently low latency. Further, radio wave propagation in industrial environments is generally vulnerable and may result in

N. Mitton et al. (Eds.): ADHOC-NOW 2016, LNCS 9724, pp. 325–338, 2016.
DOI: 10.1007/978-3-319-40509-4_23

communication in industrial environments is challenging and may result in network coverage problems or packet loss.

Within this challenging context, robust and reliable wireless communication must be realized. In practice, such robots are very often foreseen to become part of the enterprise wireless network, a network consisting of multiple access points that aims to provide coverage on the entire production or warehouse floor. In this paper, we discuss the potential problems that might arise in such a wireless setting, taking the requirements from a real-life use case. We motivate the potential benefit of adding mesh capabilities to the mobile robots. The resulting mixed architecture aims to provide maximal flexibility and configurability in order to be able to meet the performance and quality requirements for a wide range of scenarios.

The outline of the paper is as follows. Section 2 further details the potential problems that might arise when solely relying on the presence of a wireless infrastructure network and motivates our decision of adding meshing capabilities. Section 3 discusses related work in this domain, whereas Sect. 4 presents our resulting node and network architecture. In Sect. 5, we illustrate through experiments using a wireless testbed the potential performance issues in infrastructure networks and show how our combined solution can deliver improved performance and flexibility. Finally in Sect. 6 conclusions are formulated together with potential improvements and future work.

2 Problem Statement

It is no surprise that the communication solution used by mobile robots such as AGVs to communicate with each other and with other actors in their environment must be a wireless one. In today's deployments, IEEE 802.11 is typically used as the underlying communication technology as it is widely adopted, is able to offer sufficient throughput and allows connecting to an enterprise infrastructure already present. However, a number of particular challenges arise when relying solely on already available wireless infrastructure, i.e. a network consisting of multiple access points providing coverage on the entire floor.

First of all it is very reasonable to assume that in some situations no wireless infrastructure is present at the factory floor or in a warehouse. This implies that the solution provider that delivers the mobile robots must enforce its customers to make significant investments in order to rollout a wireless network that provides decent coverage across the entire floor. Even if wireless infrastructure is in place, it might not be allowed to make use of it in order not to interfere with ongoing processes that already make use of this infrastructure, in particular when the mobile robot communication heavily relies on broadcast traffic. If wireless infrastructure is present and can be used, another problem may arise. In many situations coverage will not be perfect because of the challenging wireless environment with a lot of metal, reflections, etc. These coverage holes may lead to malfunctioning of the system, e.g. in case mobile robots require permanent connectivity to the wireless network and, in lack of connectivity, stop moving as safety cannot be guaranteed.

Thirdly, mobile robots can drive at reasonably fast speeds (0-2 m/s). Considering a challenging wireless environment that requires a multitude of access points to provide decent coverage, this will result in very frequent handovers. Such handovers significantly contribute to the communication latency. For the particular real-life use case we consider here, frequent time-critical broadcast exchanges between mobile robots are required for their distributed coordination, next to less time-critical, but reliable unicast traffic to and from controllers. More specifically, broadcast packets have a strict upper bound to the latency of 20 ms in order to arrive in time at nearby mobile robots. Every handover involves a series of packet exchanges, which consumes valuable time. Hence, frequent handovers may have a detrimental impact on the required performance, as we will show in Sect. 5. Finally, as requirements to the mobile robot system may change over time, e.g. when scaling up the network, it must be possible to dynamically adapt the communication behavior.

The above observations and performance requirements, lead to a challenging set of functional requirements for our mobile robot system, which we have summarized in Table 1. Based on the above requirements, it is clear that we need to target a design that is capable of connecting either to existing enterprise networks (RQ2), to create its own mesh network (RQ1) or to do both (RQ3). This requires the incorporation of two wireless network interfaces in every mobile robot. Next to this, also the other requirements have to be fulfilled, requiring sufficient intelligence and flexibility in order for the system to be deployed in a variety of scenarios, with minimal configuration, having sufficient performance and with the possibility of future extensions.

Table 1. Functional requirements for our mobile robot system

RQ1. Function in the absence of fixed wireless infrastructure (network of APs)
RQ2. Exploit the presence of available fixed infrastructure
RQ3. Deal with occasional/sudden coverage holes in wireless infrastructure
RQ4. Reliably deliver unicast traffic
RQ5. Timely deliver frequent broadcast traffic (< 20 ms)
RQ6. Deal with mobility (0–2 m/s)
RQ7. Adapt to future needs

These requirements have resulted in a modular and configurable communication system for mobile robots, consisting of 2 wireless interfaces that can either operate in ad hoc or infrastructure mode and offering the possibility to control in a fine-grained way how traffic is being handled. As such the system can support a variety of different networking architectures, potentially combining both infrastructure communication and mesh communication and supporting the separation or duplication of different traffic streams according to configuration settings. From an application point of view, no changes need to be made as everything is handled in a transparent way. The design of the system and the supported network architectures are discussed in more detail in Sect. 4, whereas the advantages of our architecture for our particular use case at hand are experimentally evaluated in Sect. 5.

3   Related Work

Systems consisting of multiple mobile robots form an interesting research domain that is gaining importance in manufacturing in order to improve performance and increase automation. An survey of mobile robots in manufacturing is given in [3], highlighting localization problems, coverage problems up to communication technologies and environment hardships in manufacturing environments as important open research issues. In [4] a survey is presented regarding the coordination in multi-robot systems, including here also the communication technologies. The authors highlight the importance of explicit communication, i.e. direct message exchanges between robots, to ensure accuracy of the information, opposed to implicit communication by perceiving a change in the environment through the use of sensors.

During the last years, mobile robot communication experienced an evolution in their application as well as in protocol used. Many works put forward ad-hoc or mesh communication as a promising solution for realizing inter-robot communication. For instance, [5] gives an overview of network and MAC layer protocols for ad-hoc robot wireless networks. They motivate the use of ad-hoc networking for mobile robot communication due to the fact that most of the robots most likely are equipped with wireless transceivers that do not allow them to communicate directly with the data collection point. This is true even in industrial environments, but for another reason, namely due to coverage problems from access points. For instance, [6] illustrates how an infrastructure network can be extended with multi-hop relaying functionality. We also recognize this as one of the key requirements for our communication solution, but we also consider direct ad-hoc or mesh communication between all mobile robots. So far, most research into multi robot communication has focused on ad hoc networking. For instance, in [7] four different routing protocols for ad-hoc networks are compared for realizing mobile robot teleoperation. Many other works studied how ad hoc routing protocols could be used and optimized for ad-hoc robot communication. A hybrid communication solution that is capable of combining both mesh and infrastructure communication and offering flexibility to distribute traffic has not been considered so far for such systems.

In addition, in industrial settings, it is also important to be able to meet the performance and latency requirements as we have indicated in Sect. 2. For meeting real-time requirements a routing algorithm should not provide just the next neighbor to forward the packet but has to provide also the additional QoS requirements, such as guaranteed bandwidth and end-to-end latency. In [8] a routing algorithm for mesh networks is presented for use in industrial applications. They use a QoS manager which, after a calibration phase, manages QoS flows based on the requests from stations on specific QoS flow requirements, Packet Data Unit (PDU) size and destination. The calibration phase makes the solution more difficult to be deployed in highly dynamic environments. Again, the possible use of an available infrastructure network and separation of traffic according to the requirements is not considered. Finally, [9] describes a solution for wireless mesh network infrastructure with extended mechanisms to foster QoS support for industrial applications. Like in [8] they propose a mesh network with a central admission unit to decide for the communication flows requested by different applications. They could offer with their solution streams with RTT less then 100 ms. Again, the mechanisms are only applied to a mesh case, whereas we believe that a mixed solution such as the one we propose can offer additional benefits, especially when further extended with more advanced QoS mechanisms.

4 Communication System and Network Architecture

In the following subsections we will describe the designed mobile robot communication system and potential network architectures that can be realized.

4.1 Mobile Robot Communication System Architecture

In Sect. 2 we motivated our decision to design a communication solution that makes use of 2 wireless communication interfaces. Each of these interfaces can either operate in ad hoc mode for establishing mesh communication or in infrastructure mode in order to connect to an existing enterprise network. From an application point of view it should not matter which interface is being used for transmitting packets or how this interface has been configured. Similarly, external components, such as a controller, that want to communicate with a particular mobile robot, should also not be bothered with underlying communication details. To this end, we have designed an abstraction layer that transparently manages and dynamically configures the underlying network interfaces on the mobile robot. Towards the local applications running on the robot, a single virtual interface with one IP is being offered. This way, all communication to and from the mobile robots makes use of a single IP independent of whether the resulting traffic will flow via a mesh network or an infrastructure network.

The latter also implies that additional logic for routing and traffic management is needed that is able to take into account the specifics of the underlying physical interfaces. Unicast and broadcast routing over a mesh network is completely different from routing over an infrastructure network. Unicast mesh routing requires a routing protocol that can establish forwarding paths over multiple hops, together with neighbor discovery and link break detection mechanisms in order to deal with mobility and trigger route recovery. Broadcasting requires appropriate mechanisms in order to stop the propagation of the broadcasts inside the network. Regarding traffic management, the node design foresees a number of traffic classification components that can be dynamically configured. According to their configuration, unicast and broadcast traffic streams can be separated and directed to different interfaces or traffic can be even duplicated for redundancy purpose.

Figure 1 gives an overview of the high-level architecture we designed for the communication system of the mobile robot. The modular Click router framework [10, 11] in addition with our own proprietary extension for event handling and dynamic interface management was used for the communication system. In terms of implementation, all components have been realized as separate modules in order to support future extensions or the replacement of existing modules with more advanced versions or different implementations (*RQ7*). Finally, the whole system can be configured

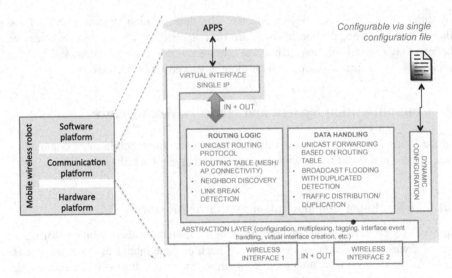

Fig. 1. Mobile robot communication architecture

dynamically, enabling administrators to define the behavior in a single configuration file (e.g. configuration of interface, how traffic must be distributed, timing values, etc.).

At this moment, a basic implementation of the DYMO routing protocol is being used for unicast mesh routing together with blind flooding for broadcast traffic. Next to this, two different neighbor discovery methods are being considered in order to recognize the occurrence of link breaks within the mesh network. The first one relies on the generation of beacons every N_{ms} seconds, the second one also takes into account the generated traffic as beacons in order to suppress real beaconing traffic. Given the fact that our particular use case heavily relies on broadcast messages, this might reduce the network load in case the same wireless interface is being used for unicast traffic as well.

4.2   Network Architecture

Depending on the particular configuration of the mobile robot communication architecture, several resulting networking architectures can be realized. This way, the solution is able to deal with the wide variety of contexts the mobile robots might have to operate in. In this subsection, we discuss a number of potential network architectures that can be easily realized by the proposed design through simple parameter reconfigurations and which are shown in Fig. 2.

Figure 2a shows a first architecture that can be realized in case no fixed wireless infrastructure is present or when fixed wireless infrastructure cannot be used (*RQ1*). Both wireless interfaces can then operate in ad-hoc mode, forming a mesh network with parallel links that operate on different frequencies. In case wireless infrastructure is present and can be used, a mixed network can be established as shown in Fig. 2b (*RQ2*). One of the interfaces is used to connect to the existing network, whereas the

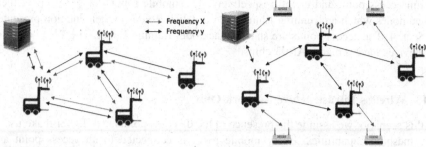

a) *Multi-interface mesh network in the absence of fixed infrastructure or in case fixed infrastructure cannot be used*

b) *Mixed mesh (single-interface) and infrastructure network*

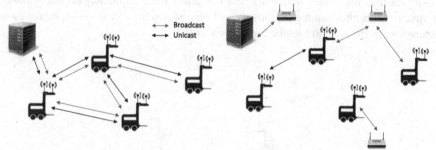

c) *Multi-interface mesh with traffic separation*

d) *Infrastructure network with mesh support for handling coverage problems*

Fig. 2. Potential network architectures that can be realized by reconfiguring the designed mobile robot communication system (Color figure online)

other interface is used to form a mesh network. Depending on additional configuration settings, it can be further decided how traffic is distributed over the different interfaces. This is shown in Fig. 2c for the case of a multi-mesh configuration, where one of the interfaces is used for unicast traffic and the other interface is used for broadcast traffic. Finally, Fig. 2d shows how the communication can be configured in order to tackle coverage problems by making use of mesh functionality in the specific area that experiences these coverage problems (*RQ3*).

5 Performance Analysis

It is clear that the proposed communication system enables several networking topologies. Combined with the flexibility on how to distribute the traffic it is interesting to investigate how this flexibility can be exploited in order to deal with the other requirements that are specific for our targeted use case (*RQ4–6*). For this, we conducted a set of experiments on the w.iLab.t wireless testbed [12], which are now discussed in the following subsections. Hostapd and wpa-supplicant are used as user space daemon to run access point and client, respectively. The mobile robots have on top of them Zotac nodes which are running Linux and our Click Router implementation presented in Sect. 4. The access points are static Zotac nodes running Linux. The Wi-Fi cards of all devices have Atheros AR93 chips.

5.1   Wireless Infrastructure Network Only

In this scenario, we assume the presence of fixed access points and do not make use of any meshing capabilities. Every mobile robot is connected to an access point and selection of the most suitable access point is based on signal strength. Mobile robots move around the environment covered by access points and get attached and detached to/from access points. As mobile robots can drive at relatively high speeds, such handovers may take place frequently and will affect the communication performance. To quantify this effect on the performance of unicast and broadcast traffic, we set up an experiment in the w.iLab.t testbed as shown in Fig. 3.

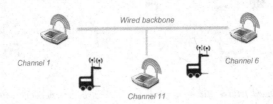

Fig. 3. Experimental setup to assess the impact of handovers on the communication performance. Three APs and two mobile robots are used.

Three non-overlapping channels (1, 6 and 11) in the frequency of 2.4 GHz have been used. To enforce handovers from one access point to another access point, we remotely control the transmit powers of the access points. The mobile robots are limited to scan only over the mentioned channels to prevent time and energy consuming procedure for scanning all available channels. During the experiment, both mobile robots are communicating with each other via the infrastructure wireless network.

Figure 4 shows the latency distribution of 10000 unicast packets during a run of 200 s. Unicast packets are exchanged every 20 ms and the frequency of roaming among access points is configured to be 10, 20 and 30 s. As can be seen, in most cases the latency is lower than 4 ms, which is close to the median amount. However, it can become as high as 78 ms during the roaming procedure. Further, the more frequent roaming happens among the access points, the higher the packet latency can become. The reason behind this is that every time a client performs a handover between access points, it gets dissociated, has to look for stronger signal strength and needs to associate to a new access point. Table 2 shows the latency statistics plot, presenting the first and third quartile of the results shown in Fig. 4.

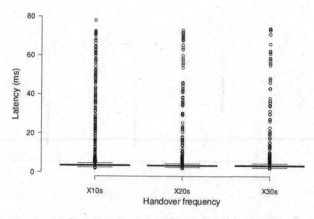

Fig. 4. Latency of unicast traffic for different roaming frequencies

Table 2. Unicast latency statistics plot considering min, max and median (in ms)

Statistics	10 s roaming freq.	20 s roaming freq.	30 s roaming freq.
Median	3.45	3.42	3.39
1st quartile	3.21	3.18	3.13
Min	2.13	2.12	2.09
Max	78	73	73.8
3rd quartile	3.75	3.66	3.66

Figure 5 presents the latency of 10000 broadcast packet transmission within the same 200 s time period. Again, the roaming procedure happens every 10, 20 and 30 s. As it is shown in Table 3, in contrast to the unicast latency, the broadcast latency is not mostly around the median number but around the third quartile number. The results also show a much more profound negative impact of handovers on the broadcast latencies, due to the way broadcasts are disseminated through the network. Every broadcast from a mobile robot needs to be rebroadcast to other devices connected to the same access point as well as to all other devices connected to the other access points. This is visible in Fig. 6 where every time the mobile robots were connected to the same access point the latency was around 5 ms while when roaming took place the latency increased up to 100 ms. It is clear that even in this simple setup our mobile robot solution will never be able to meet the envisioned latency requirements (< 20 ms) of broadcast traffic.

5.2 Mesh Network Only

In this scenario, only a mesh network is being used as shown in Fig. 2a. As mentioned, unicast traffic uses a simple reactive routing protocol, whereas broadcast traffic uses

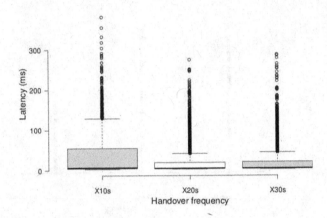

Fig. 5. Latency of broadcast traffic for different roaming frequencies

Table 3. Broadcast latency statistics plot considering min, max and median (in ms)

Statistics	10 s roaming freq.	20 s roaming freq.	30 s roaming freq.
Median	6	4.89	4.97
1st quartile	4.44	4.18	4.46
Min	3.06	3.17	3.26
Max	381	275	288
3rd quartile	54.5	19.2	20.6

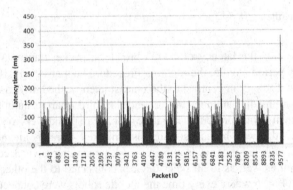

Fig. 6. Latency of broadcast traffic for a roaming frequency of 10 s (Color figure online)

blind flooding with duplicate detection. Using this setup we again measure the impact of mobility of mobile robots on the latency of packet transmissions. In order to be able to mimic a variety of speeds and thus link breaks, we used a forced mobility approach, where MAC filtering is being used to artificially change the mesh topology as showing in Fig. 7. Nodes c1 and c5 are communicating. While communicating, c1 establishes a new link with node c2, c3, c4 and c5 respectively, breaking the old link and gradually changing the number of hops over which the packets need to travel.

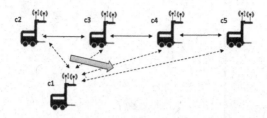

Fig. 7. Fully mesh network among mobile robots.

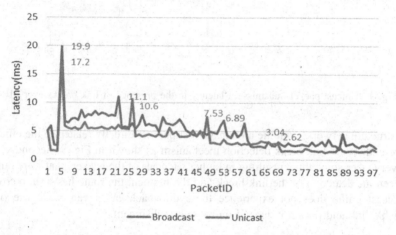

Fig. 8. Latency of broadcast and unicast traffic with link breaks occurring every 20 s. (Color figure online)

Figure 8 presents the impact of link breaks and the resulting change in topology and hop count on unicast and broadcast packet transmissions with transmissions being generated every second. In this experiment, latency for unicast and broadcast traffic varies between 17.2 ms/19.9 ms and 2.62 ms/3.04 ms and is directly related on the number of hops between the sender and receiver, which decreases from 4 to 1.

The performance of unicast traffic however, is also strongly affected by the link break detection and routing mechanism. In the scenario shown in Fig. 8, the beacon interval was set to a very small value (20 ms), making it possible to very quickly react to link breaks in this small topology. In addition, with traffic only being generated every second, no significant unicast packet losses occurred, illustrating only the impact of hop count on latency in a mesh setting.

In reality the protocol might react slower, traffic generation can happen more frequently or the topology is more complex. These first two aspects are shown in Fig. 9, where unicast traffic is being generated every 120 ms. Keep-alive beacons are sent less frequently, every 500 ms, with the detection of a link break in the absence of beacons after 2500 ms. Further, upon the detection of a link break, all traffic for a destination that has become unreachable is being buffered until the route has been established. This has two consequences. First of all, unicast traffic in the presence of link breaks in the mesh network exhibits much higher packet losses than in an

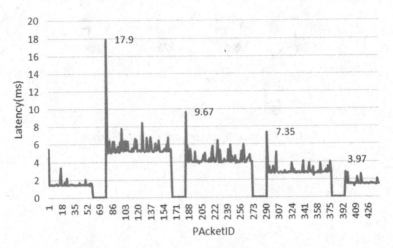

Fig. 9. Unicast packet transmission latency in the presence of link breaks every 10 s

infrastructure network, with the amount of lost packets directly related to the efficiency of the underlying link break detection mechanism as shown in Fig. 9. Secondly, route recovery takes some time, resulting in higher latencies of the packets that were buffered between the detection of the link break and the moment the route has been recovered. Broadcast traffic does not experience these drawbacks as it can make use of any available link and does not depend on route establishment.

5.3   Combined Network

The third scenario being considered is a hybrid setup, where every mobile robot uses 1 interface to connect to the infrastructure network and one interface to set up a mesh network as shown in Fig. 2b. In order not to overload the wired network with broadcast traffic, the communication system is configured to send broadcast traffic over the mesh interfaces. To avoid frequent rerouting inside the mesh network, unicast traffic is configured to run over the other wireless interface. Again, we measure the latency of unicast and broadcast traffic in order to investigate the advantages and feasibility of a hybrid configuration with traffic separation. In this scenario we use three interconnected access points (as in Fig. 3) and four mobile robots. Two of them are communicating using unicast traffic via access points while two others are generating broadcast traffic. All of them are connected to one of the access points. One mobile robot is configured to reply to the broadcast packets. Channel 6 is used for communication within the mesh network. The handover and link break frequency in this case are both 10 s.

Figure 10a shows the latency of 10000 unicast transmissions during 200 s, whereas Fig. 10b shows the latency of 10000 simultaneous broadcast transmissions. As it is shown in Table 4, the mixed scenario that exploits the possibility to separate different traffic streams, combines the best of both worlds. Broadcast traffic can meet the strict latency requirements by using the mesh network, whereas unicast traffic achieves low latency by avoiding the complexity of ad hoc routing.

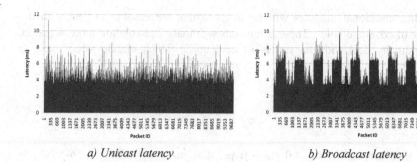

a) Unicast latency   b) Broadcast latency

Fig. 10. Unicast and broadcast latency in a mixed scenario with traffic separation

Table 4. Latency statistics of unicast and broadcast packet transmission (in ms)

Statistics	Unicast traffic	Broadcast traffic
Median	3.33	4.9
1st quartile	3.2	3.4
Min	2.26	2.73
Max	7.98	10.7
3rd quartile	3.49	5.3

6   Conclusions

Many existing solutions in industrial settings that make use of mobile robots make use of an available enterprise network. In this paper we discussed the potential drawbacks of such an approach. For our particular use case at hand, a key requirement was the ability to delivery broadcast traffic with very low latencies, a requirement that could not be fulfilled in an enterprise network where handovers take place frequently as shown on our testbed. Other requirements, such as the ability to function in the absence of infrastructure of to tackle coverage holes, made it necessary to design a flexible and modular networking architecture that is able to exploit both the advantages of the presence of an enterprise network and the advantages of a mesh network. In this paper we showed the feasibility and a proof-of-concept implementation of this architecture. The architecture supports a variety of setups and we evaluated three of them, thereby measuring the impact of mobility on unicast and broadcast traffic. The design and evaluation clearly shows the advantages of being able to exploit a mixed architecture.

This paper presented the foundations and feasibility of such an architecture, but at the same time reveals some open issues and possible improvements. More research is needed to see how unicast routing and blind flooding can be improved. One path that will be investigated is the incorporation of position information, distributed using the frequent broadcasts, in order to improve unicast routing performance and to reduce overhead. For this, connectivity will be analyzed in a realistic industrial environment.

Next to this, additional modules will be foreseen that are capable to deal with the occurrence of coverage holes. These extensions will make our solution more versatile, able to optimally deal with the variety of contexts in which mobile robots have to operate.

References

1. Drath, R., Horch, A.: Industrie 4.0: Hit or Hype? IEEE Ind. Electron. Mag. **8**, 56–58 (2014)
2. Arumugam, S., Kalle, R., Prasad, A.: Wireless robotics: opportunities and challenges. Wireless Pers. Commun. **70**, 1033–1058 (2013). doi:10.1007/s11277-013-1102-3
3. Schneier, M., Bostelman, R.: Literature Review of Mobile Robots for Manufacturing. National Institut of Standards and Technology, NewYork (2015)
4. Yan, Z., Jouandeau, N., Cherif, A.A.: A survey and analysis of multi-robot coordination. Int. J. Adv. Robot. Syst. **10**(399), 1–18 (2013)
5. Wang, Z., Liu, L., Zhou, M.: Protocols and applications of ad-hoc robot wireless communication networks: an overview. Int. J. Intell. Control Syst. **10**(4), 296–303 (2005)
6. Saghezchi, F.B., Radwan, A., Rodriguez, J.: Energy efficiency performance of WiFi/WiMedia relaying in hybrid Ad-hoc networks. In: 3rd International Conference on Communications and Information Technology, pp. 285–289 (2013)
7. Zeiger, F., Kraemer, N., Schilling, K.: Commanding mobile robots via wireless Ad-Hoc networks - a comparison of four Ad-hoc routing protocol implementations. In: Proceedings of the 2008 IEEE International Conference on Robotics and Automation (2008)
8. Herms, A., Nett, E., Schemmer, S.: Real-time mesh networks for industrial applications. In: Proceedings of 17th International Federation of Automatic Control World Congress (IFAC 2008), Seoul (2008)
9. Lindhors, T., Lukas, G., Nett, E.: Wireless mesh network infrastructure for industrial applications - a case study of tele-operated mobile Robots. In: IEEE 18th Conference on Emerging Technologies & Factory Automation (ETFA), pp. 1–8 (2013)
10. Morris, R., et al.: The click modular router. In: ACM SIGOPS Operating Systems Review, vol. 33, no. 5. ACM (1999)
11. http://read.cs.ucla.edu/click/click
12. w-iLab.t Zwijnaarde generic wireless testbed http://wilab2.ilabt.iminds.be/

Improving QoE Using Link Stability for Video Streaming in MANETs

Lokesh Sharma[✉], Chhagan Lal[✉], and D.P. Sharma[✉]

Department of Computer Engineering, Manipal University, Jaipur, India
{lokesh.sharma,chhagan.lal,deviparsad.sharma}@jaipur.manipal.edu

Abstract. Achieving expected level of Quality-of-experience (QoE) at receiving end during real-time video streaming is a challenging task in inherently error-prone multihop mobile ad hoc network (MANET). Continuously fluctuating link conditions, dynamicity in bit rates of transmitted video and topology change due to node mobility are some major issues in providing required QoE to multimedia traffic in MANETs. In this paper, we propose a reactive Link Stability Based Routing (LSBR) protocol. LSBR discover route consisting of links having high lifetime. The stability of a link is estimated dynamically during the route discovery process using its signal strength and interference along with the relative mobility of nodes forming the link. A multi-metric mathematical model has been developed using the aforementioned parameters to select a link from the candidate links at each intermediate node in such a way that leads to selection of high lifetime route. To prove effectiveness of LSBR protocol for real-life video streaming, a hybrid MANET scenario consisted of simulated nodes and physical machines is designed using an industry standard emulation tool called EXata-Cyber. Correctness of LSBR protocol for video streaming is established by simulation, hybrid testbed as well as users' experience.

Keywords: Link stability · Mobile networks · Route stability · Quality-of-Experience · Video streaming · Emulation

1 Introduction

Due to its infrastructure–less, inexpensive and easy–to–deploy nature, mobile ad-hoc networks (MANETs) are being employed in many diverse fields. On–the–fly deployment of MANETs is very useful in emergency and rescue operations required during disasters such as floods, earthquakes and war. For example, in flooded areas, the rescue team can use live video streaming to relay impact of disaster to domain experts and decision makers. Based on these video feeds, the rescue and support teams can assess the extent of damage, decide and monitor requisite relief operations. MANETs are well suited in such scenarios as these can be deployed quickly with no infrastructure requirement. These may be only available option for quick measures where existing infrastructure is either non-operational or too damaged to be restored quickly.

© Springer International Publishing Switzerland 2016
N. Mitton et al. (Eds.): ADHOC-NOW 2016, LNCS 9724, pp. 339–353, 2016.
DOI: 10.1007/978-3-319-40509-4_24

Unlike best effort services, real-time video streaming applications require quality-of-experience (QoE) [1] which needs to be maintained upto a satisfactory level. Managing an acceptable level of QoE support in MANETs for real-time video streaming is a challenging task due to the highly varying characteristics of wireless channel [2] and transmitting video data [3]. The most widely used measure to evaluate the quality of multimedia traffic is quality-of-service (QoS) [4]. In QoS evaluation only system factors related to network layer such as throughput, bandwidth, delay, jitter, error rate etc. are used to evaluate the received multimedia traffic quality. However, it has been widely accepted that the evaluation of quality of multimedia traffic transmission should be based on the quality as perceived by the end user i.e., QoE. This is because using network layer parameters alone for assessing the quality of received multimedia traffic ignores the subjective aspects of video/audio content related to expectation, sensation and perception of end user. Due to this, QoE which considers human factors (e.g., physical and mental constitution, emotional state) as well as wide range of system factors (e.g., content related, media related, network related and device related) is considered a metric for overall evaluation of a multimedia service as perceived by end users subjectively [5,6].

In this paper, we have proposed and evaluated a link-stability based routing (LSBR) protocol for enhancing the QoE of received video traffic in MANETs. LSBR uses following constraints during its improved route discovery phase to estimate link-stability:

- Signal to noise and interference ratio (SINR): SINR over a link provides the information about the condition of the link in terms of interference, noise and received signal strength.
- Relative node mobility (ζ): When combined with the SINR, ζ will give an estimation of the lifetime of the underlying wireless link.
- Traversed route stability $(\Re)$: Enforces high route stability and low route length during route selection phase.

A mathematical model that consider the above three parameters to estimate the link stability as well as route stability is developed in order to select a stable link at intermediate or destination node from the set of candidate links. Performance of LSBR has been evaluated using simulation as well as emulation tools and compared with the existing similar methods proposed in [7,8].

The rest of the paper is organized as follows. In Sect. 2, the related work done on improving the QoS as well as QoE for multimedia services in MANETs is presented. The methodology of our LSBR protocol is presented in detail in Sect. 3 along with the design and implementation of link stability metric. The emulation setup, design procedure of hybrid emulated network and result analysis for performance evaluation of LSBR protocol during transmission of real-life video streaming is presented in Sect. 4. Finally, Sect. 5 contains the conclusion and tentative direction for future work.

2   Related Work

In this section, we present the existing routing solutions proposed for managing or enhancing the QoE of video streaming services over MANETs. Subjective evaluation required for QoE measurement is done with the help of widely used metric called Mean Opinion Score (MOS) [1]. MOS is calculated by summarizing the responses for perceived video quality collected from various viewers in laboratory environment. The MOS is represented using an integer value between one and five, where one is worst and five is excellent.

In [9], authors present an analytical model for the evaluation of QoE of HTTP video streaming in TCP-based single hop MANETs. Two metrics namely underflow times (buffer empty times during the playout) and fluency of playout (ratio of idle watching time to experienced watching time) is used as the QoE evaluation metrics. The drawbacks are that only simulation results are used to verify the proposed models effectiveness and its usefulness is limited for single hop MANETs. In [10], a QoE-aware architecture that uses QoE mapping and QoE video quality estimator for smooth handover in heterogeneous wireless networks (consists of 802.11e and 802.16e specifications) is proposed. It has been shown through simulation results using the subjective as well as objective metrics that the proposed architecture provides the best connection and considers the QoE needs of mobile users and available wireless resources service differentiation classes of 802.11e and 802.16e.

In order to enhance QoE of perceived video, various QoS metrics needs to be evaluated. To reduce bandwidth and energy consumption in wireless networks, the control overhead caused by these QoS metric evaluation processes should be low. To address this issue, authors in [11] presented a joint utility function to efficiently characterize the fairness of user's QoE and power consumption. Simulation results provided in terms of QoE per power consumption in a multicell network shows the effectiveness of the proposed scheme. In [12], authors proposed evaluation of the coverage performance of heterogeneous wireless networks for multimedia traffic using SINR and QoE coverage metrics. Based on the simulation results, authors established that QoE coverage outperforms SINR coverage by adjusting target QoE thresholds, modulation and coding schemes with different videos.

A realistic reception model to calculate the threshold SINR ($SINR_{th}$) based on bit-error rate (BER) and frame error rate (FER) over a link is proposed in [8]. The $SINR_{th}$ is used during the reactive route discovery process to selectively forward RREQ messages having SINR greater than the $SINR_{th}$. Each intermediate node buffers the received RREQ messages for a specified time period. When the timer expires, the RREQ message with the highest SINR is selected for forwarding. However, the effect of mobility is not included in the results and how the end-to-end delay is affected due to increased length of the route (caused by selecting the highest SINR link) is also not considered. In [7], authors present a metric called 'Path Encounter Rate' (PER) to discover routes consisting of less congested links having high lifetime. PER employs velocity and direction information about neighbors' movement patterns. The proposed PER based routing is effective on networks with moderate mobility but fares poorly in static and high mobility MANETs. Effects of interference, shadowing and varying multimedia traffic characteristics are not considered in PER.

Several solutions for improving QoE using QoS provisioning have been proposed in recent years for efficient transmission of multimedia traffic over wireless networks [13,14]. Most of these solutions deploy simulated synthetic video traffic to represent real video data [15], due to which the ability of these methods to handle the dynamics of video traffic remains partially tested. QoS metrics are used instead of QoE to measure the service level received by the end users which questions their effectiveness for video streaming applications. In our hybrid network, the source and receiver nodes are real-life video traffic generators and receivers running on physical devices. To accurately estimate the performance of our proposed link-stability based routing (LSBR) protocol for video streaming in MANETs, we have used subjective scores to measure quality of experience (QoE) using emulation metrics (such as MOS and snapshots from received video) as well as objective scores to measure QoS using simulation metrics (such as packet delivery ratio, routing overhead and average route lifetime).

3   Design and Implementation of LSBR Protocol

This section describes the design of our multi-constraint link stability metric and implementation of proposed reactive link-stability based routing (LSBR) protocol. The goal of LSBR protocol is to choose a route from the set of possible routes which has highest lifetime.

3.1   Network Model

A MANET scenario can be represented by an undirected weighted graph $G = \{N, L\}$ where $N = \{N_1, N_2, N_3 \dots N_n\}$ is the set of mobile nodes in the network and $L = \{l_1, l_2, l_3 \dots l_m\}$ is the set of bi-directional radio links. The weight on the links can represent the link bandwidth, delay, stability etc. Each node has a unique identifier (node id or IP address) so that it can be identified by routing protocols during the route discovery and data transmission process. The links connecting the nodes are subjected to establish and broke during the network lifetime due to network mobility and congestion. Each node creates and dynamically update (using periodic control messages) a neighbor link table (NL_t) which consists of links that it creates with its neighbor nodes at any instance of time during the communication process. It is also assumed that each node is configured with same set of software (i.e., protocols at various layers) and hardware (i.e., receiving/transmitting circuit, antenna) components.

Let us assume that a route between two random nodes s and d is given by $R(s,d) = \{l(s, v_1), l(v_1, v_2), l(v_2, v_3) \dots l(v_j, d)\}$, where $v_1, v_2 \dots v_j$ are intermediate nodes and j is the length of route in terms of number of hops. As MANET is an undirected graph, a multihop path selected for routing between any two nodes is a subset of all the possible routes existing between them. Formally,

$R(s, d) = R_i$, where R_i is the route which is selected between nodes s and d during the route discovery process from the possible k candidate routes $\{R_1,$ $R_2, R_3 \ldots R_k\}$ existing between s and d at current instance of time.

In MANETs, traditional routing protocols uses minimum hop count metric [16,17] to select a candidate route from the available route set. In LSBR protocol, a link $x \rightarrow y \in L$ during a route discovery process is characterized using following metrics: (a) signal-to-interference and noise ratio ($SINR_{x \rightarrow y}$), (b) Relative node mobility ($\zeta_{x \rightarrow y}$), (c) route length (i.e., number of hops on route between node s and y), and d) traversed route stability ($\Re_{s \rightarrow y}$) here $s \rightarrow y$ is the current partial route traversed between source node s and intermediate node y. The method used for the estimation of link stability ($LS_{x \rightarrow y}$) metric over a link will be discussed in following sections. For a route $R_i \in R(s, d)$ between nodes s and d, its route stability (RS_i) can be calculated as follows:

$$RS_i = \prod_{n=1}^{j} LS_{l_n} \tag{1}$$

where l_n is the nth link on some candidate route and j is the route length in number of hops.

3.2   Link-Stability Metric Design

Unlike most of the existing link stability aware routing protocols, our proposed LSBR protocol selects a stable route between a source-destination pair by considering link stability as well as the route stability. Using only hop-to-hop stability to select a stable route may increase the route length. On the other hand, using route stability alone can not guarantee that all the links on the selected route are of high stability.

Our goal is to design a link stability metric (ℓ) that can address the aforementioned problems while at the same time also considers the dynamic nature of MANETs caused by node mobility. The predicated remaining lifetime of a link can be estimated based on our proposed ℓ. Let's assume that during the route discovery process any node (y) have to select a forwarding link based on their ℓ values from a candidate set of links i.e., $C_{set}^{link} = \{l_1, l_2, l_3 \ldots l_i\}$ where $C_{set}^{link} \subseteq NL_t (y)$.

To estimate the ℓ of a link (say $x \rightarrow y$, where $x \rightarrow y \in C_{set}^{link} (y)$) at any node y, first y calculate the link stability value called link stability indicator ($\Im$) for link $x \rightarrow y$. To estimate the $\Im$ for $x \rightarrow y$ at y, following three parameters are used:

- Signal to noise and interference ratio (SINR): At physical (PHY) layer, when a node receives a packet (either control or data) from its neighbors, it calculate the SINR of the received packet. We have calculated the SINR of a received packet (or signal) at the PHY layer as follows:

$$SINR_{x \rightarrow y} = \frac{RSS}{\sum_{i=1}^{n} I_{pow} + Noise} \tag{2}$$

where n is the number of interfering signals during the reception of current signal, RSS is the received signal strength and I_{pow} is the interference power of an interfering signal.

- Route length (R^{ℓ}): It is defined as the length of the selected route (partial or complete) in terms of number of hops.
- Traversed route stability ($\Re$): It is defined as the route stability of the partial route upto the current node from the source node. $\Re$ at node y is calculated as follows:

$$\Re_{s \to y} = \frac{(\Re_{s \to x} \cdot (R^{\ell}_{s \to y} - 1) + SINR_{curr})}{R^{\ell}_{s \to y}} \tag{3}$$

Here $R^{\ell}_{s \to y}$ is the length of the route from source node s to current node y and $SINR_{curr}$ is the current SINR of link $x \to y$. The value of $\Re_{s \to x}$ is provided to node y by node x by piggyback it with the recently received packet.

Finally, $\Im$ of link $x \to y$ is calculated using $SINR_{x \to y}$ (should be high), $\Re_{s \to x}$ (should be high) and $R^{\ell}_{s \to y}$ (should be low) metrics. To incorporate this, all the above metrics are scaled in range $[\alpha \cdots \beta], \alpha > 0, \beta > 0$ and $\beta > \alpha$ and named as $\mathscr{U}_{x \to y}, \mathscr{V}_{x \to y}$ and $\mathscr{W}_{x \to y}$. Computations are as follows:

$$\mathscr{U}_{x \to y} = \alpha + (\beta - \alpha) \times \frac{SINR_{x \to y} - SINR_{min}}{SINR_{max} - SINR_{min}} \tag{4}$$

$$\mathscr{V}_{x \to y} = \alpha + (\beta - \alpha) \times \frac{\Re_{s \to x} - \Re_{min}}{\Re_{max} - \Re_{min}} \tag{5}$$

$$\mathscr{W}_{x \to y} = \beta - (\beta - \alpha) \times \frac{R^{\ell}_{s \to x} - R^{\ell}_{min}}{R^{\ell}_{max} - R^{\ell}_{min}} \tag{6}$$

$$\Im_{x \to y} = (w_1 \cdot \mathscr{U}_{x \to y}) + (w_2 \cdot \mathscr{V}_{x \to y}) + (w_3 \cdot \mathscr{W}_{x \to y}) \tag{7}$$

Here, w_1, w_2, w_3 are the weights (such that $w_1, w_2, w_3 \in \{0, 1\}; w_1 + w_2 + w_3 = 1$) assigned to the metrics with values 0.25, 0.25 and 0.5 respectively. Higher weight is given to the number of hops because increase in route length decreases the route lifetime and increase the end-to-end delay. Before assigning the weights to all three parameters (i.e., $SINR$, $\Re$ and R_{ℓ}), we have mapped their values on the equal scale (i.e., between 0 to 100 where 0 is least desirable outcome for each parameter and 100 is most desirable outcome).

Once node y have the $\Im$ values for all the links in its C^{link}_{set}, it calculate relative node mobility (ζ) of all links in C^{link}_{set} to identify whether the nodes forming the links are moving away or towards each other. To calculate the ζ for a link $x \to y \in C^{link}_{set}(y)$, we use the previous known SINR ($SINR_{prv}$) value of link $x \to y$ which is calculated and stored at node y in the resent past with the help of periodic control messages and use it as follows:

$$\zeta_{x \to y} = SINR_{curr} - SINR_{prv} \tag{8}$$

Here $SINR_{curr}$ is the current SINR of the received packet at node y. The value of $\zeta_{x \to y}$ can be positive or negative. $\zeta_{x \to y} > 0$ indicates that both the nodes are moving towards and $\zeta_{x \to y} < 0$ means both nodes are moving away from each other.

Node y will remove all the links with negative ζ values from its C_{set}^{link}. This is because even with the highest $\Im$, if the nodes on the selected link are moving away from each other it indicates that link has low lifetime. From the remaining positive ζ valued links, a link $x_j \to y$ with the highest link stability metric (ℓ) is included in rest of the route discovery process. If all the links in C_{set}^{link} of node y has negative ζ values, then a link $x_i \to y$ with highest ℓ is selected for further processing. Once a link is selected, C_{set}^{link} is removed from the memory of node y. To estimate the ℓ of a link (say $x \to y$) that belongs to C_{set}^{link} of node y, following equation is used:

$$\ell_{x \to y} = (\beta_1 \cdot \Im_{s \to y}) + (\beta_2 \cdot \zeta_{x \to y}) \tag{9}$$

Here β_1 and β_2 are equal weights. Also, like the values of parameters in Eq., the values of $\Im$ and ζ used in Eq. are the values obtained after they are mapped on similar scale. During the mapping, we have mapped lowest positive and highest negative ζ values with the higher values on scale. This is because if the distance between two nodes forming a link is very low as compared to its previous recent know distance, the probability that they will cross paths and move away from each other in the near future is high. On the other hand, if the small change in distance (or SINR) depicts that the both ends of a link are not moving exact vertically towards each other which increases the probability that the link have high future lifetime.

3.3 Implementation

The reactive routing protocol which we have used to implement our proposed ℓ measurements is known as ad-hoc on-demand distance vector (AODV) protocol. Three types of control messages known as route request (RREQ), route reply (RREP) and route error (RERR) are used in AODV protocol for route discovery and rerouting processes. Following two custom data structures are used at all the nodes for the implementation of ℓ:

- *RREQ message buffer table* ($RREQ_{table}$): This table buffers a received RREQ message received from a neighbor node. Entries in $RREQ_{table}$ of a node are created and erased during a route discovery process.
- Neighbor link table (NL_t): This table is used to store the dynamically updated $SINR_{prv}$ values for the links that a node forms with its neighbors. An entry is created or updated in NL_t of a node every time it receives a periodic HELLO message from one of its neighbor node.

Algorithm 1 shows how our proposed LSBR protocol discovers a stable route between a given source-destination pair by using ℓ measurements during its route discovery process. When a source node (s) receives a data packet to send to destination node d and it does not have an active route for node d in its routing table, node s initiates a route discovery process by broadcasting a RREQ message into the network. when a node receives a RREQ message it calculates the $SINR_{curr}$ (using Eq. 2) of the link from which RREQ is received. If the $SINR_{curr}$ is greater than the $SINR_{min}$, node will buffer the RREQ message in its $RREQ_{table}$ else the RREQ message is discarded. $SINR_{min}$ is estimated using the same procedure as given in [8]. If this is the first RREQ message received for a route discovery process (each route discovery process is identified using a unique identifier in the network) the receiving node also set a timer (Δ) of τ milliseconds with it. All the RREQ messages belongs to the same route discovery phase are buffered until the Δ expires.

The parameter ℓ for all the RREQ messages are calculated upon the expiration of Δ using Eqs. 4, 5 and 6. Based on the values of the ℓ, a RREQ message which is received from a link with highest ℓ is processed and re-broadcasted. All the other RREQ messages in the C_{set}^{link} of the node are discarded. The traversed route stability ($\Re$) parameter calculated from source to current node using Eq. 3 is also appended with the selected RREQ message before it is re-broadcasted.

Algorithm 1. LSBR protocol route discovery process

INPUT at any node u: Neighbor table (NL_t), RREQ Buffer Table $(RREQ_{table})$
OUTPUT: A high remaining lifetime link is selected

1: **if** Node u receives a RREQ message **then**
2:      Calculate $SINR_{curr}$ using Eq. 2 at PHY layer
3:      **if** $SINR_{curr} < SINR_{min}$ **then**
4:          Discard the RREQ
5:      **end if**
6:      Create or update an entry in its $RREQ_{table}$
7:      Set Δ, if RREQ received is first for a route discovery phase
8:      Buffer the RREQ in $RREQ_{table}$
9:      Update $SINR_{curr}$ in its NL_t with the SINR of the received RREQ message
10:     **while** Δ **do**
11:         **if** u receives duplicate RREQ message **then**
12:             Repeat steps in line 2 to 6 and 8 to 9
13:         **end if**
14:     **end while**
15:     **for** Each $u \rightarrow v_i \in RREQ_{table}$ **do**
16:         Extract $\Re_{u \rightarrow v_i}$ and calculate $\Im_{u \rightarrow v_i}$ using Eq. 4
17:         Calculate $\zeta_{u \rightarrow v_i}$ and $\ell_{u \rightarrow v_i}$ using Eq. 5 and Eq. 6
18:     **end for**
19:     Process the link with highest $\Im$
20:     Calculate $\Re$ using Eq. 3 and piggyback it with the forwarding RREQ message
21: **end if**

For better understanding, the working methodology of route discovery process of our proposed LSBR protocol is explained with the help of Fig. 1. We have shown the ℓ calculations at node H for the links in its C_{set}^{link} (i.e., $A \rightarrow H$, $G \rightarrow H$ and

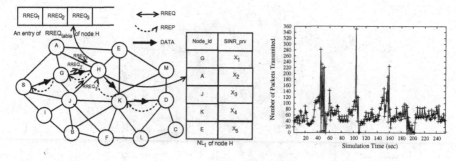

Fig. 1. RREQ message processing in LSBR protocol

Fig. 2. Varying per second data rates of transmitting video file

$J \rightarrow H$). Node S has data to send to node D and it does not have an active route for node D. Node S initiates a route discovery process by broadcasting the RREQ message. Figure 1 shows how an intermediate node H handles the received RREQ messages. When node H receives the first RREQ message (i.e., $RREQ_1$), it creates an entry in its $RREQ_{table}$ and buffer the $RREQ_1$. After storing $RREQ_1$, node H associate and initiate Δ. Until the Δ expires H receives $RREQ_2$ and $RREQ_3$ from its neighbor nodes G and J. When Δ is expired, node H traverse the RREQ packets in its $RREQ_table$ entry and calculates the $\Im$ using Eq. 4 for links $A \rightarrow H$, $G \rightarrow H$ and $J \rightarrow H$. Then with the help of its NL_t node H will calculate ζ using Eq. 5 for all three links. Finally, H will calculate ℓ for the links using Eq. 6 and select a RREQ message for processing. In this case, node H will select the RREQ message received from link $G \rightarrow H$ because the ℓ of $G \rightarrow H$ link is highest. In the same way, node K select $H \rightarrow K$ and node D select $K \rightarrow D$ which results in the selection of route $S \rightarrow G \rightarrow H \rightarrow K \rightarrow D$ for data communication between nodes S and D.

4   Emulation and Performance Evaluation

In this section, results obtained from extensive emulations are presented and discussed in order to evaluate the performance of our proposed link-stability based routing (LSBR) protocol. Emulations are performed using an industry standard emulator called EXata-Cyber v2.0 [18]. To show the effectiveness of proposed approach, we compare our LSBR protocol with the previous work known as link quality aware AODV ($LA - AODV$) proposed in [8] and PER [7] routing protocol. To increase the accuracy of emulation results and perform statistical analysis, we averaged the results of ten random emulation processes generated by changing the seed value. Furthermore, to avoid any non-representative singularities, all the presented results are an average of 25 simulation runs (each run initialized with a random seed) for each evaluated scenario. For each evaluation, confidence interval level (i.e., 98 %) as well as variance (± 0.92) is determined but

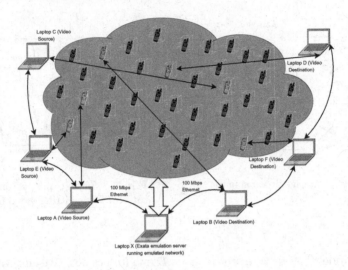

Fig. 3. Target hybrid network scenario setup in laboratory

not included in this paper due to their lower significance values. All emulations are performed on the network that is modeled using parameters listed in Table 1.

4.1 Emulation Network Design

The emulated hybrid network testbed used for performance analysis of traditional LA-AODV, PER and LSBR protocols for real-time video streaming in MANETs is shown in Fig. 3. As seen, our hybrid network testbed scenario consists of two

Table 1. Simulation parameters

Parameters	Values
Simulator	EXata-Cyber v2.0
Simulation time	5 min
Scenario dimension	1000x1000 sq.meter
Number of nodes	65
Routing protocols	LA-AODV [8], PER [7], LSBR
Mobility model	Random way-point
Node mobility speed	5 - 25 m/s
Node pause time	5 s
Radio type	802.11a/g
Data rate	24 Mbps
Timer (Δ) Duration	5 ms
Simulation results averaged over	25 runs

Table 2. Hardware configuration of machines

Parameters	Values
Number of machines	7
Hardware configuration	Intel-i5, 620@3.2 GHz
Local area network	100 Mbps ethernet
Third party softwares installed	Wireshark, VLC

Table 3. Transmitted video configuration

Parameters	Values
Application software	VLC-2.1.0 [19]
Transcoding (Active)	H.264 + ACC (MP4)
Encapsulation	MPEG-PS
Video codec	MPEG-4
Streaming bitrate	2028 kbps
Streaming frame rate	15 fps
Transmitted file type	MP4
Video run time	4.14 min
Video resolution	600 x 312
Receiver VLC Buffer	1 s

types of networks: the real world network (consisting of various desktop/laptops devices) and the emulated network (consisting of several emulated nodes). The lines that are connecting the laptops to the virtual nodes in Fig. 3 show the IP address mapping between these two types of nodes. Once a virtual node is mapped with some physical device, it can send and receive real world traffic with that physical machine. The parameters configured at each simulated node in emulated network are same as given in Table 1. The hardware configuration of physical machines and transmitted video file along with the details of streaming software are specified in Tables 2 and 3.

Figure 2 shows the varying characteristics of real-time video traffic that we have used at source nodes for transmission. As it can be seen from Fig. 2, the number of bits transmitted at each consecutive second varies largely without following any pattern. Due to this, provisioning QoS while transmitting such traffic is a difficult task. The QoS guarantees provided at the time of admission are violated due to these highly fluctuating transmission rates.

4.2   Result Analysis

In this section, results obtained from various emulations that are performed using LA-AODV, PER and LSBR protocols against increasing network mobility are discussed. For performance evaluation and comparison of proposed LSBR protocol, results are collected using various metrics such as packet delivery ratio (PDR), routing overhead (RO), route transmission efficiency ($\Re$) and mean opinion score (MOS). In the MANET scenario the number of video source-destination pairs are kept constant (i.e., 3) and network mobility is increased by 5 m/s.

Figure 4 shows the effect on average PDR for LA-AODV, PER and LSBR protocols with increase in network mobility. As it can be depicted from Fig. 4 that our proposed LSBR protocol is highly unaffected with the increase in network mobility when compared with the LA-AODV and PER protocols. This is because the route discovery process of LSBR uses relative mobility information before selecting a link. Due to this, LSBR ignore the links that are formed by the nodes which will move away from each other in near future, thus ensuring high route

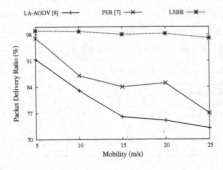

Fig. 4. PDR with increasing network mobility

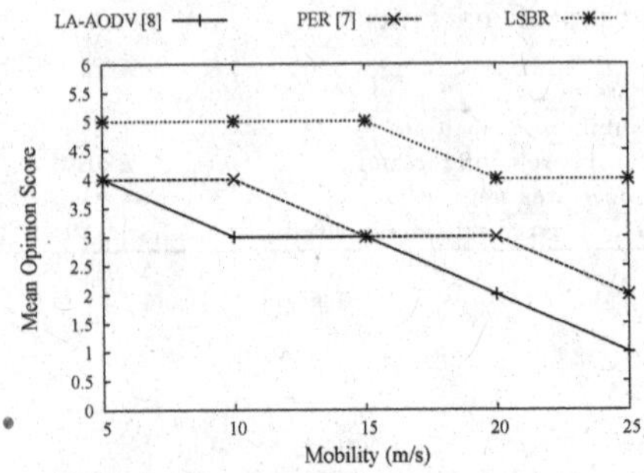

Fig. 5. MOS with increasing network mobility

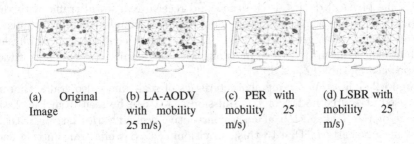

(a) Original Image

(b) LA-AODV with mobility 25 m/s)

(c) PER with mobility 25 m/s)

(d) LSBR with mobility 25 m/s)

Fig. 6. Received video quality snapshots

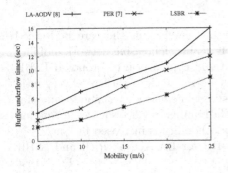

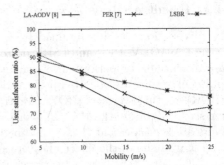

Fig. 7. Buffer underflow times (U_{bt}) with increasing network mobility

Fig. 8. User satisfaction ratio (U_{sr}) with increasing network mobility

lifetime. The lower PDR values for LA-AODV and PER protocols are clearly identified by the end users in the perceived video and same is reflected in the MOS values given by them (referring to Fig. 5). Following are the effects reported by the end users that are caused by decreased PDR in the received video file: (a) jitter in received video/audio, (b) gaps in the live received video feed and, (c) grainy images.

Figure 6 shows snapshots from the received video files for LA-AODV, PER and LSBR protocols with 25 m/s network mobility. The distortion in the received image can be seen when it is compared with the original image which is taken from the source video file. As it can be seen from Fig. 6, the degradation in received image quality of LSBR protocol is very low with respect to other comparing protocols.

Figure 7 shows the average time U_{bt} during the emulation process in which the VLC buffer remains underflow with increasing network mobility. As the underflow time increases for a communication session, the number of pauses and distortion in frames of the received video at the end-user increases. As we can see from Fig. 7 that the U_{bt} for LSBR is lower as compared to the LA-AODV and PER protocols, the reason behind this behavior is the use of node mobility information along with the minimum number of hops during the route discovery process of LSBR. The high lifetime and low delay of selected routes decreases the U_{bt} in LSBR.

The effect of mobility on user satisfaction ratio (U_{sr}) for all the comparing protocols is shown in Fig. 8. The U_{sr} of a video session is the received video percentage for which user's MOS is more than three (i.e., good). This metric provides indication of user satisfaction for the perceived video stream. Better the value of this metric, it indicates happiness of end user regarding the quality of received video. As it can be seen from Fig. 8 that LSBR has higher U_{sr} when compared to the LA-AODV and PER protocols. This is because the route stability metric used for route selection in LSBR keeps the number of fluctuations lower, thus higher MOS.

Effect of increase in network mobility for LA-AODV, PER and LSBR protocols on routing overhead (RO) is shown in Fig. 9. As it can be observed from Fig. 9, the RO of LSBR protocol is lower than LA-AODV and PER protocols because LSBR has lower number of route breaks during the transmission process. This is due to the fact that LSBR uses multi-parameter based link stability metric (ℓ) during its route discovery process. To prefer the selection of routes that includes less intermediate hops and high lifetime links in LSBR protocol decreases re-routing processes. Finally, the effect on average lifetime of the routes selected for data transmission in all three comparing routing protocols with increasing network mobility is shown in Fig. 10. We have measured the lifetime of a route in terms of the amount of traffic it can transmit before it is broken so, we are able to measure the lifetime of a selected route in terms of both, time as well as the amount of traffic carried to the destination.

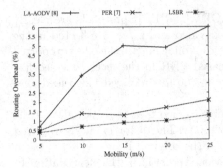
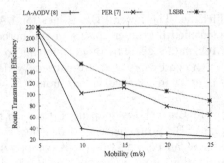

Fig. 9. Routing overhead with increasing network mobility

Fig. 10. Route transmission efficiency with increasing network mobility

5   Conclusion and Future Work

In this paper, we have presented a multi-parameter based link stability estimation method for enhancing the QoE for perceived video quality in MANETs. The emulation results show that the proposed link stability based routing (LSBR) protocol dynamically adapts the changing network topology and varying transmitted video data rates. To show the effectiveness of our proposed protocol, it is evaluated on real time video streaming which is done on a hybrid MANET scenario that is designed using an emulation process. It has been observed during the reception of live video at destination that the small delay and delay variations are flawlessly handled by the VLC buffer as long as the packet losses are very low. But, the high data packet drops are clearly visible to the end users due to missing images and lack of synchronization (that causes fluctuations) in the received video. These fluctuations are more visible, if audio is also considered along with the video in the received video traffic.

References

1. TD 109rev2 (PLEN/12), ITU - International Telecommunication Union, Definition of Quality of Experience (QoE) (2008)
2. Lindeberg, M., Kristiansen, S., Plagemann, T., Goebel, V.: Challenges and techniques for video streaming over mobile ad hoc networks. Multimedia Syst. **17**, 51–82 (2011)
3. Anand, P.S., Dutta, V., Arya, J., Kurose, M., Chetlur, S., Kalyanaraman, S.: On managing quality of experience of multiple video streams in wireless networks. Mobile Comput. **14**, 619–631 (2015)
4. ITU-T Recommandation, p.10/g.100. Vocabulary for Performance and Quality of Service (QoE) (2008)
5. Hewage, C.T.E.R., Martini, M.G.: Quality of experience for 3D video streaming. IEEE Commun. Mag. **51**(5), 101–107 (2013)

6. Fiedler, M., Hossfeld, T., Tran-Gia, P.: A generic quantitative relationship between quality of experience and quality of service. IEEE Netw. **24**(2), 36–41 (2010)
7. Son, T.T., Le Minh, H., Sexton, G., Aslam, N.: A novel encounter-based metric for mobile ad-hoc networks routing. Ad Hoc Netw. **14**, 2–14 (2014)
8. Moh, S.: Link quality aware route discovery for robust routing and high performance in mobile ad hoc networks. In: 11th IEEE International Conference on High Performance Computing and Communications, HPCC 2009, pp. 281–288 (2009)
9. Zhou, H., Li, B., Qu, Q., Yan, Z.: An analytical model for quality of experience of http video streaming over wireless ad hoc networks. In: 2013 IEEE International Conference on Signal Processing, Communication and Computing (ICSPCC), pp. 1–5, August 2013
10. Jailton, J., Carvalho, T., Valente, W., Natalino, C., Frances, R., Dias, K.: A quality of experience handover architecture for heterogeneous mobile wireless multimedia networks. IEEE Commun. Mag. **51**(6), 152–159 (2013)
11. Shao, H., Jing, W., Wen, X., Lu, Z., Zhang, H., Chen, Y., Ling, D.: Joint optimization of quality of experience and power consumption in OFDMA multicell networks. IEEE Commun. Lett. **20**(2), 380–383 (2016)
12. Shao, H., Lu, Z., Wen, X., Zhang, H., Chen, Y., Hong, Y.: Content-aware video qoe coverage analysis in heterogeneous wireless networks. Wirel. Pers. Commun., 1–16 (2015)
13. Macit, M., Gungor, V.C., Tuna, G.: Comparison of QoS-aware single-path vs. multi-path routing protocols for image transmission in wireless multimedia sensor networks. Ad Hoc Netw. **19**, 132–141 (2014)
14. Sanchez-Iborra, R., Cano, M.-D., Garcia-Haro, J.: performance evaluation of batman routing protocol for voip services: a QoE perspective. IEEE Trans. Wirel. Commun. **13**(9), 4947–4958 (2014)
15. Van der Auwera, G., David, P.T., Reisslein, M.: Traffic, quality characterization of single-layer video streams encoded with the h.264, mpeg-4 advanced video coding standard, scalable video coding extension. IEEE Trans. Broadcast. **54**(3), 698–718 (2008)
16. http://tools.ietf.org/html/draft-ietf-manet-olsrv2-11
17. Perkins, C.E., Royer, E.M.: Ad-hoc on-demand distance vector routing. In: Proceedings of Second IEEE Workshop on Mobile Computing Systems and Applications, WMCSA 1999, pp. 90–100, February 1999
18. SCALABLE NETWORK TECHNOLOGIES. http://www.scalable-networks.com
19. VLC MEDIA PLAYER. http://www.videolan.org/vlc/index.html

Author Index

Printed in the United States
By Bookmasters